# Ionization in High-Temperature Gases

# Progress in ASTRONAUTICS and ROCKETRY

*A series of volumes sponsored by*

**American Rocket Society**

*500 Fifth Avenue, New York, New York, 10036*

---

*Progress Series Editor*

Martin Summerfield

*Princeton University, Princeton, New Jersey*

*Titles in the Series*

Volume 1. SOLID PROPELLANT ROCKET RESEARCH. 1960
*Editor:* MARTIN SUMMERFIELD, *Princeton University, Princeton, New Jersey*

Volume 2. LIQUID ROCKETS AND PROPELLANTS. 1960
*Editors:* LOREN E. BOLLINGER, *The Ohio State University, Columbus, Ohio;* MARTIN GOLDSMITH, *The RAND Corporation, Santa Monica, California;* AND ALEXIS W. LEMMON JR., *Battelle Memorial Institute, Columbus, Ohio*

Volume 3. ENERGY CONVERSION FOR SPACE POWER. 1961
*Editor:* NATHAN W. SNYDER, *Institute for Defense Analyses, Washington, D. C.*

Volume 4. SPACE POWER SYSTEMS. 1961
*Editor:* NATHAN W. SNYDER, *Institute for Defense Analyses, Washington, D. C.*

Volume 5. ELECTROSTATIC PROPULSION. 1961
*Editors:* DAVID B. LANGMUIR, *Space Technology Laboratories, Inc., Canoga Park, California;* ERNST STUHLINGER, *NASA George C. Marshall Space Flight Center, Huntsville, Alabama;* AND J. M. SELLEN JR., *Space Technology Laboratories, Inc., Canoga Park, California*

Volume 6. DETONATION AND TWO-PHASE FLOW. 1962
*Editors:* S. S. PENNER, *California Institute of Technology, Pasadena, California;* AND F. A. WILLIAMS, *Harvard University, Cambridge, Massachusetts*

Volume 7. HYPERSONIC FLOW RESEARCH. 1962
*Editor:* FREDERICK R. RIDDELL, *Avco Corporation, Wilmington, Massachusetts*

Volume 8. GUIDANCE AND CONTROL. 1962
*Editors:* ROBERT E. ROBERSON, *Consultant, Fullerton, California;* AND JAMES S. FARRIOR, *Lockheed Missiles and Space Company, Sunnyvale, California*

ACADEMIC PRESS · NEW YORK AND LONDON

## Progress in ASTRONAUTICS and AERONAUTICS

*(a continuation of Progress in Astronautics and Rocketry)*

*A series of volumes sponsored by*

**American Institute of Aeronautics and Astronautics**

*1290 Avenue of the Americas, New York, New York 10019*

---

*Progress Series Editor*

Martin Summerfield

*Princeton University, Princeton, New Jersey*

*Titles in the Series*

Volume 9. ELECTRIC PROPULSION DEVELOPMENT. 1963

*Editor:* ERNST STUHLINGER, *NASA George C. Marshall Space Flight Center, Huntsville, Alabama*

Volume 10. TECHNOLOGY OF LUNAR EXPLORATION. 1963

*Editors:* CLIFFORD I. CUMMINGS AND HAROLD R. LAWRENCE, *Jet Propulsion Laboratory, California Institute of Technology, Pasadena, California*

Volume 11. POWER SYSTEMS FOR SPACE FLIGHT. 1963

*Editors:* MORRIS A. ZIPKIN AND RUSSELL N. EDWARDS, *Space Power and Propulsion Section, Missiles and Space Division, General Electric Company, Cincinnati, Ohio*

Volume 12. IONIZATION IN HIGH-TEMPERATURE GASES. 1963

*Editor:* KURT E. SHULER, *National Bureau of Standards, Washington, D.C.*
*Associate Editor:* JOHN B. FENN, *Princeton University, Princeton, New Jersey*

*(Other volumes are planned)*

ACADEMIC PRESS · NEW YORK AND LONDON

# Ionization in High-Temperature Gases

*Editor*
Kurt E. Shuler
*National Bureau of Standards, Washington, D. C.*

*Associate Editor*
John B. Fenn
*Princeton University, Princeton, New Jersey*

A Selection of Technical Papers
based mainly on
the American Rocket Society Conference
on Ions in Flames and Rocket Exhausts
held at Palm Springs, California
October 10-12, 1962

ACADEMIC PRESS · NEW YORK · LONDON · 1963

ACADEMIC PRESS INC.
111 FIFTH AVENUE
NEW YORK 3, N. Y.

*United Kingdom Edition*
Published by
ACADEMIC PRESS INC. (LONDON) LTD.
BERKELEY SQUARE HOUSE, LONDON W. 1

*Library of Congress Catalog Card Number: 63-23423*

PRINTED IN THE UNITED STATES OF AMERICA

## THE PROPELLANTS AND COMBUSTION COMMITTEE OF THE AMERICAN ROCKET SOCIETY

(Predecessor of American Institute of Aeronautics and Astronautics)

October 1962

Stanford S. Penner, Chairman
Institute for Defense Analyses, Washington, D. C.

John B. Fenn
Princeton University, Princeton, N. J.

Irvin Glassman
Princeton University, Princeton, N. J.

Leon Green Jr.
Ford Motor Company, Newport Beach, Calif.

Robert A. Gross
Columbia University, New York, N. Y.

Charles J. Marsel
New York University, University Heights, New York, N. Y.

Joseph F. Masi
Office of Scientific Research, Washington, D. C.

Ernest Mayer
National Engineering and Science Company, Pasadena, Calif.

Raffaele F. Muraca
Stanford Research Institute, Menlo Park, Calif.

W. Nachbar
Stanford University, Palo Alto, Calif.

Peter L. Nicholas Jr.
Aerojet-General Corporation, Azusa, Calif.

Antoni K. Oppenheim
University of California, Berkeley, Calif.

J. M. Patton
Office of Naval Research, Washington, D. C.

Henry M. Sheuy
Rohm and Haas Company, Huntsville, Ala.

Robert J. Thompson Jr.
Rocketdyne, Canoga Park, Calif.

Hans G. Wolfhard
Thiokol Chemical Corporation, Denville, N. J.,

# PREFACE

This volume is based to a large extent on the Conference on Ions in Flames and Rocket Exhausts sponsored by the American Rocket Society in Palm Springs, California, October 10-12, 1962. A number of the papers presented at this meeting are included in this book. In addition, the editor solicited for inclusion in this volume several review papers from a number of experts in order to present to the prospective audience a fairly comprehensive and self-contained survey of the field. Thus, this book is intended as a report on, and as an authoritative reference source of, the present status of the field of ionization in high-temperature gaseous systems.

With the advent of space flight and space exploration and with the current research and development activities in plasma physics and magnetohydrodynamics, the subject of ionization in high-temperature gases has assumed an important place in the forefront of research. A perusal of the table of contents will indicate the wide scope of problems in which ionization, i. e., the production of free electrons and ions, plays an important role. To set the stage and to orient the reader, who will not necessarily be an expert in this field, we have provided in Section I two authoritative review articles which are concerned with the thermodynamics and the basic elementary processes of gaseous ionization. The information contained therein is basic to the more complex problems discussed in the subsequent sections.

The study of the mechanism and energetics of the elementary chemical reactions in flames, as detailed in the papers in Section II, has contributed greatly to our understanding of ionization in combustion systems. Among the many topics discussed in these papers are: chemi-ionization, i. e., the production of ions and electrons in nonequilibrium concentration through specific chemical reactions, the kinetics and thermodynamics of electron attachment and detachment reactions, electron-ion recombination reactions, and conductivity measurements in flames. The fundamental information obtained from these flame studies has provided us with a better understanding and control of the many complex phenomena in aerospace applications.

The study of shock waves and detonation waves also has furnished valuable information on the elementary processes of ionization in high-temperature systems. Of particular importance in the detonation wave research have been the interesting effects due to the interaction of the shock wave with the chemical reactions. In addition to these important basic research aspects, the study of ionization in shock and detonation waves is also of great practical importance in connection with supersonic flight, re-entry phenomena, and other aerospace developments. The recent research activities and the status of this field are reviewed in Section III.

The chemistry of ionization in rocket exhausts is considered in Section IV. From the basic point of view, one meets here again the complicated interplay between chemical reactions and hydrodynamics which gives rise to a number of interesting phenomena in the ionization in high-temperature reactive gas streams. From a more applied point of view it is of great importance to have this basic knowledge in order to understand and subsequently control the resulting electromagnetic effects which can cause serious interference with communication and guidance.

The papers in Section V are devoted to the problems of the production, control, and utilization of the free electrons produced in high-temperature ionization processes. For purposes of magnetohydrodynamic power production it is essential to have an electrically conducting working fluid such as, for instance, a weakly ionized plasma. Whereas in some of the earlier papers in this volume the emphasis might have been on the suppression of ionization, the emphasis in these studies is instead on the enhancement of ionization, with its concomitant increase in the concentration of free electrons, through the seeding of gases or through the use of specialized combustion techniques. The final two papers in this section are devoted to a most novel application of ionization, namely, the scattering and reflections of radio waves from high-altitude chemical plasma clouds. An important aspect of this research effort is again the generation of free electrons and the enhancement of their concentration.

The material presented in this volume in essence, then, summarizes much of our present knowledge in the field of ionization in high-temperature gaseous systems. As such, it stands as an impressive testimonial to the ingenuity and industriousness of the many active workers in this field, and to our increased understanding and knowledge over the past decade.

The editor would like to express at this point his appreciation to the many contributors for their wholehearted cooperation in the preparation of this volume. Particular thanks are due a number of authors who at the request of the editor, and on short notice, prepared contributions for inclusion in this book on subjects not covered during the conference. The editor also gratefully acknowledges the assistance of Dr. John B. Fenn, Princeton University, in the preparation of this volume. Finally he wishes to express his appreciation to Dr. Martin Summerfield, Series Editor, and Miss Ruth F. Bryans, Managing Editor,

Scientific Publications Department, of the AIAA, and members of her staff, for their help and guidance in bridging the gap from good intentions to actual completion.

Kurt E. Shuler
Assistant Director and Senior
Research Fellow
National Bureau of Standards
Washington, D. C.

August 5, 1963

# CONTENTS

# I

# ELEMENTARY PROCESSES

# THERMODYNAMICS AND ELEMENTARY PROCESSES OF GASEOUS IONS

R. Stephen Berry*

Yale University, New Haven, Connecticut

Abstract

The thermodynamics and microscopic processes associated with ions in hot gases are surveyed. Processes are classified and discussed according to whether they are charge production, charge conserving or recombination processes. Emphasis is placed on specification of the important physical variables and magnitudes governing the various processes. An attempt is made to describe the present theoretical and experimental situation in each area, and to point out recently-solved and outstanding unsolved problems.

## 1. Introduction

This section is intended as an elementary survey and review of the conditions associated with the presence of ions in gases, and of the processes involved in their production and destruction. With as sweeping a topic as this, the discussion must necessarily be exceedingly selective and brief in treating any single subject. The material divides itself into three parts: the equilibrium conditions associated with gaseous ions, the microscopic processes associated with the production, destruction and reactions of ions, and the macroscopic transport properties of ionized gases.

This chapter is devoted only to the thermodynamics and microscopic phenomena, and omits discussion of the transport properties. Throughout, the principal aim will be the presentation of the physical basis of each phenomenon, its magnitude, and the main physical factors involved. An attempt will be made to discuss, in each case, the present state of experimental and theoretical work, but without any presumption of

---

Submitted to AIAA May 1, 1963.

*Assistant Professor, Department of Chemistry; Alfred P. Sloan Fellow.

reviewing the literature in detail. Pertinent books and review articles are cited as often as possible. Current literature references are sparse, and the author must apologize at the outset to the many people working in this broad field whose contributions are not mentioned explicitly.

## 2. Equilibrium Conditions

The presence of gaseous ions and electrons under conditions of equilibrium is a phenomenon easily understood and one which lends itself to very straightforward and detailed computation. Our knowledge of atomic and molecular energy levels is very detailed; and, in the limiting case in which potential energies of ionic interaction are small compared with the kinetic energy of individual ions, we know from the theory of electrolytes the laws governing the equilibrium and transport properties. The condition of equilibrium is simply the requirement that the mass-action laws (or Saha equation, in the case of ions and electrons) be satisfied. If there are n independent reactions possible, then there are n equations stating the conditions that the chemical potential be stationary. Each of these has the form of an expression for an equilibrium constant:

$$K = \frac{\prod_{i=1}^{\ell} (Z_i/V)}{\prod_{j=1}^{m} (Z_j/V)} = \frac{\prod_{i=1}^{\ell} \nu_i \gamma_i}{\prod_{j=1}^{m} \nu_j \gamma_j} \qquad (2.1.)$$

where V is the volume, i refers to products ($\ell$ in number), j refers to each of the m reactants, $Z_k$ is the partition function of the $k^{th}$ species, $\nu_k$ is the number density of the $k^{th}$ species and $\gamma_k$ is its activity coefficient. The units of $\nu_i$ are those of $V^{-1}$.

When the partition $Z_k$ are evaluated from atomic and molecular data, one can supply the necessary conditions of composition, such as total amount of material or amounts of specific species, and solve the equilibrium equations for the remaining concentrations. These concentrations naturally appear in the activity coefficients for a dilute plasma of equally charged ions,

$$\gamma_k = \exp \frac{-Z_k^2 e^2 \kappa}{2\epsilon kT} \tag{2.2}$$

where $Z_k e$ is the ionic charge, $\epsilon$ is the dielectric constant of the surrounding medium and $\kappa$ is the inverse of the Debye length:

$$\kappa = \left[\frac{4\Pi e^2}{\epsilon kT} \sum_{\substack{\text{all} \\ \text{ionic species}}} \nu_k Z_k^2\right]^{1/2} \tag{2.3}$$

$$= 0.1491 \left[\frac{1}{\epsilon T} \sum \nu_k Z_k^2\right]^{1/2}$$

For gases in which 1.2 is valid, $\epsilon$ can generally be taken as unity. An error of about a part per thousand is introduced by this assumption when the density of uncharged particles is about $10^{20}$ - $10^{21}$ per $cm^3$.

The concentrations can be determined either by iteration or by direct solution of the simultaneous equations 1.1, provided the interionic interactions are neglected. Iteration of the mass-action equations and the expressions 1.2 does not lead to a self-consistent set of concentrations. Rather, because 1.2 is based on a linearized Poisson-Boltzmann equation, iteration leads to a divergence in which all particles appear to become ionized. Approximate values of 1.2 can be obtained from the concentrations derived from 1.1, to give approximate corrections to these concentrations, if ion densities are relatively low ($\leq 10^{17}$ $cm^{-3}$). Considerable attention is still devoted to trying to develop the thermodynamics of the partly-ionized gas.[16,44,129]

At very high temperatures, the only species of interest are positive ions and electrons. This is a consequence of the large entropy contribution to the free energy, in the translational part of the product of partition functions. Neglecting the activity coefficients,

$$\frac{\nu(M^+)\,\nu(e^-)}{\nu(m^\circ)} \approx \left(\frac{2\Pi m_e kT}{h^2}\right)^{3/2} e^{-\frac{I(M)}{kT}}$$

or

$$2.416 \times 10^{15}\ T^{3/2} \times 10^{-[5.040 \times 10^3\ I(ev)/T(^\circ K)]} \qquad (2.4)$$

(the Saha equation),

where I(M) is the ionization potential of M. The exponential factor is small at temperatures up to $10^4$ °K because ionization potentials are all about 4eV or more. Nevertheless, the pre-exponential factor, which arises from the entropic contribution of the free electron, can be quite large, large enough to give quite sizeable electron concentrations at temperatures of a few thousand degrees. For example, $\nu(e^-)$ is about $2 \times 10^{16}$ $cm^{-3}$ in sodium vapor at 5000°K, if $\nu(Na^\circ)$ is about $10^{17}$ $cm^{-3}$. More extensive calculations are given in the section following, by Smith and Gatz.

If temperatures are much below 5000°K, then molecular dissociation processes, either directly to ions or to favorable ion sources, may predominate over electron production. This results from the small electron mass in the pre-exponential factor of 1.4, which keeps the entropic contribution of translational motion low for electron production. The much larger atomic masses which replace $m_e$ in the corresponding expressions for dissociation make these processes relatively easy at temperatures of the order of 1000°K in many cases. The entropy contributions of vibrations and rotations tend to stabilize the bound molecules and to hinder dissociation, but at high temperatures these factors are less important than the mass effect in the translational factor.

In most molecular systems, equilibrium dissociation of molecules to smaller molecules, atoms, or radicals predominates over dissociation to ions, simply because of the large differences between most ionization potentials and most electron affinities. The alkali halides provide the most extreme and simplest species that do dissociate to ions in significant proportions. The very low ionization potentials of the alkalis and the high electron affinities of the halogen atoms make the energetic contribution very favorable, and the large masses make dissociation easy. Figure 1 shows the ion and electron concentrations as functions of concentration and temperature, for a typical alkali halide. As the temperature rises, the negative ion concentration rises to a maximum at a temperature $T_{max}$ which is a rather insensitive function of salt concentration. Then it falls steadily. Meanwhile, the positive ion and electron concentrations rise monotomically with T. In effect, at high temperatures, electron boil off of the negative ions and a normal plasma is attained.

Among the simple species which can contribute to the formation of atomic and molecular negative ions, the most important after the halogen atoms are probably O atoms, OH radicals,[112] S atoms, SH radicals, and perhaps C atoms and halogen molecules. Polyatomic species with sizeable electron affinities appear to be $NO_2$ and probably $N_3$, $SO_3$, $SO_2$, and CNO. The polyatomic species and even the non-hydridic diatomics are unlikely to contribute to negative ion formation in hot gases at equilibrium, because of their proclivity for dissociation. On the other hand, halogens, O, S, C, OH and SH all can be important in the supression of free electron production. Their affinities are all well above 1 eV, and the tendency for OH and SH to dissociate is relatively small.

## 3. Ionic Processes in Gases

The processes involving production, transformation and destruction of ions span a tremendous range of molecular physics and chemistry. Even in the short time since the publication of the volume Atomic and Molecular Processes,[B2] a number of phenomena have been found or elucidated. Only the basic phenomena can be summarized here, in order to provide a little background and perspective for the following chapters. The number of topics prohibits even the amount of detail of the foregoing discussion of equilibrium.

The discussion is divided into three main sections, which deal in turn with the production of charged gaseous species, with transformations in which charge is conserved, and with recombination processes. These, in turn, are divided into processes involving radiation, nonradiative processes involving electron collisions, and collision processes of heavy particles. It will sometimes be convenient to use the notation introduced by Hasted. Integers denote charges on atomic or molecular particles; barred charges are negative, unbarred are positive. Electrons are denoted by e, excited states by primes, and photons by $\nu$. Incoming particles of a collision are written to the left of the bar and outgoing particles (in the same order as the incoming particles when any ambiguity would be possible) to the right. Hence, a single positive charge transfer becomes 10/01, the radiative capture of an electron is $e0/\bar{1}\nu$ and one sort of excitation with charge transfer is 10/0'1.

## 3.1. Production of Charged Particles

For the purposes of this review, the discussion of charged-particle production can be confined to photoprocesses, including photoionization and photodissociation of molecules to ionized states, electron impact processes, and impact processes involving heavy particles, which must include chemionization.

### 3.1.1. Photoproduction of Charged Particles

Photoproduction and especially photoionization of atoms has been studied extensively. The photoionization problem is one of finding the shape of the continuum curve of the cross-section for the process υ0/el as a function of the light frequency $\upsilon$. The threshold frequency is the first ionization potential. The cross-section $\sigma$ can be defined, for an optically-thin sample, by the Beer's Law expression

$$\sigma(\upsilon) = (1/n\ell)\ \ln\ (I_o/I) \qquad (3.1)$$

where n is the concentration of species under study, $\ell$ is the path length in the sample which a light beam of intensity $I_o$ enters, and from which a beam of intensity I emerges. The total cross-section for frequency $\upsilon$ can be expressed in terms of the initial state $\Psi_i$ (1,..., n) of the n-electron atom, its final state $\Psi_f$ (1,..., n) and the n-electron coordinate operator, $\sum_{j=1}^{n} r_i$:

$$\sigma(\upsilon) = \frac{8\pi^3\upsilon}{3c}\ \frac{1}{\omega_i}\ \times$$

$$\sum_{fi}\left|\int \Psi_i^*(1,\ldots,n)\ (\sum_{j=1}^{n} er_j)\ \Psi_f(1,\ldots,n)\ d\tau_1\ldots d\tau_n\right|^2 \qquad (3.2)$$

where $\omega_i$ is the degeneracy of the initial state. The final state in photoionization involves one electron in a continuum state, so that the normalization of $\Psi_f$ must be defined. The generally-used energy normalization based on the limit of a normalized wave packet $\Phi$ of energy eigenfunctions $\Psi_f$ is taken

$$\Phi = \int a(E)\ \Psi_f(E)dE \tag{3.3}$$

so that

$$\int_E a(E)^2 dE = 1 \tag{3.4}$$

or $a(E)^2dE$ is the probability that the packet function have energy in the range dE about E. Insofar as it can be interpreted in terms of one-electron states, then $\Psi_f$ is a continuum function for only one electron; all the others are in negative-energy (bound) states very similar to those appearing in $\Psi_i$.

The cross-section has been evaluated exactly for hydrogenic atoms, and is given in terms of the principal quantum number n of the bound state as

$$\sigma(\upsilon, n) = 4e^6 RZ^4 g/3^{3/2}\hbar^3\upsilon^3 n^5 \pi \tag{3.5}$$

The Gaunt factor g is a tabulated function with values within 10 to 20% of 1 except near the threshold energy, and which approaches unity as n goes to infinity. R is the Rydberg constant ($e^4m/2\hbar^3c$).

Many-electron atoms are not so readily solved. The Hartree-Fock method provides a basis for calculations of cross sections in these systems. The initial state can be determined as a conventional n-electron atom problem, and the final state, as an (n-1)-electron ion interacting with a positive-energy electron. The expressions for $\sigma(\upsilon)$ apply equally for photoionization of molecules, provided that $\Psi_i$ and $\Psi_f$ also include the vibrational part of the wave function describing the nuclear motion, and that the integration is carried out over the nuclear coordinates. The vibrational part is particularly important for molecules because the initial and final states may have rather different geometries. This in turn means that an optical transition at frequency $\upsilon_1$ takes the molecule from a geometry at or near the equilibrium configuration in the initial state i to a geometry in an excited final state f, a state of the positive ion which may or may not be the lowest vibrational state. The positive ion state may even be a dissociating state, in which case the process could be υ0/e10, for example. The three processes are shown schematically in terms of one-dimensional potential surfaces in Fig. 2. The time of passage of a photon is at least as short as the passage time for ordinary absorption processes, so that the Franck-Condon principle will be obeyed reasonably well. That is, the nuclei will not move large distances or change their momenta greatly during the time of passage of the photon. The transition can be represented in first approximation as a "vertical line" on a diagram such as Fig. 2, starting from the position of greatest probability in the electronic-vibrational ground state and terminating only at points in the upper state corresponding to probability maxima -- the center of a Gaussian vibrational ground state function or points near classical turning points for excited vibrational or continuum states. If the transition does link a lowest vibrational state with the initial state, then the observed photoionization energy threshold will be the true thermodynamic or adiabatic ionization energy. This condition seems to be obeyed in the 1-alkenes, simple aldehydes and simple ketones, for example.[125] If the vertical transition takes the molecule-ion to an excited vibrational state, then the lowest observed photoionization energy will just be an upper bound for the thermodynamic ionization potential. This appears to be the situation in the photoionization of alkanes and alkyl radicals like ethyl and propyl, $C_2H_5$ and $C_3H_7$.[42] The two sorts of behavior sometimes can be distinguished experimentally by their behavior at the photoionization threshold, since the former gives a very sharp onset.[92]

The theory of photoionization of atoms, particularly hydrogenic atoms, has been developed rather extensively, but calculations for many-electron atoms generally have fallen short of impressive success, largely because of the unavailability, until recently, of reasonably accurate atomic wave functions. The recent calculations of Cooper,[28] which make use of Hartree-Fock orbitals and their corresponding potentials, are successful enough to predict the existence of maxima for several cross sections, and to match the shapes and magnitudes of a number of such curves, within 20 to 50%. Calculations for molecular cases are very rare; they have been carried out for $H_2^+$ and $CH_4$. These require either a knowledge of electronic wave functions throughout the relevant part of the vibrational potential surface, or the assumption of perfectly vertical transitions. The subject has been recently reviewed by Ditchburn and Öpik.[B2]

A collection of typical ionization potentials is given in Table 1. These include both photo- and electron-impact ionization values, several of which have been reported since the review by Ditchburn and Öpik.

The cross-sections for photoionization of atoms are finite at the threshold, and generally have negative initial slope which in some cases changes to positive.[28] Their magnitudes at threshold fall between 0.01 x $10^{-18}$ $cm^2$ for Rb and K, through ~ 6 x $10^{-18}$ for H and He to 35 x $10^{-18}$ $cm^2$ for Ar. Molecular photoionization cross-sections are zero at threshold if the transition is at all "nonvertical" and may rise to a maximum of 8 x $10^{-18}$ $cm^2$ (for $H_2$) to 20-50 x $10^{-18}$ $cm^2$ (for $O_2$ and $CH_4$, respectively).

Photoionization occurs as an important mechanism for ion production in two kinds of systems: equilibrium systems whose temperatures are so high that the internal radiation field is strong at wavelengths as short as those of ionization energies, and nonequilibrium systems in which intense radiation is generated in one part of the system, which then is absorbed and produces ionization in another part of the system. The equilibrium systems must satisfy the condition that the rate of photoionization is comparable to or larger than the rate of impact ionization; this occurs in very hot and rather rarefied gases. As an illustration, one can estimate $\omega_{10}$ the number of radiative transitions which an atom makes in one second in a typical system. Choose a wavelength of 5000 Å, and let the transition be an electric-dipole allowed transition between bound states, so that the Einstein B-coefficient, the number of transitions per second, per unit of energy density $U(\nu)$:

$$B\,(0,1) = \omega_{10}/U(\nu) = \frac{2\pi}{3\hbar^2}\left|(\Psi_1 \mid \sum_j er_j \mid \Psi_o)\right|^2 \approx 5 \times 10^{19}\ \text{sec}^{-1}\ (\text{erg sec}^{-1}\ \text{cm}^{-3})^{-1} \qquad (3.6)$$

The energy density

$$U(\nu) = \frac{8\pi\nu^3}{c^3}\ \frac{1}{e^{h\nu/kT} - 1} \qquad (3.7)$$

which is about 5 x $10^{-17}$ at 3000°K and about 7 x $10^{-12}$ at 30,000°K, in the 5000 Å region. Consequently, the atom may make about $2.5 \times 10^3$ induced transitions per second at 3000°K and about $3.6 \times 10^8$ transitions per second at 30,000°K. (These figures show clearly how the increased frequency of induced transitions populates the upper state at high temperatures, since the temperature-independent spontaneous transition frequency is about $10^8$ $sec^{-1}$.) The rate of collisional excitation cannot exceed the rate of collisions having sufficient energy; and, as later discussion shows, it is often considerably lower. The collision frequency of atoms is about $2 \times 10^9$ $sec^{-1}$ at 1 atm and 3000°, and drops to about $6 \times 10^8$ at 30,000 at the same pressure. Of these collisions, about one in four at 30,000° has sufficient energy for a 5000 Å excitation, but only one in 10,000 has enough energy at 3000°K. At 1 atm, collisional processes clearly can be important even at 30,000°. If the pressure is lower by a factor of 100 or more, then, clearly, radiative transfer may become the dominent mechanism at high temperatures.

The photoionization process itself has been studied in the laboratory in recent years by a mass spectrometric technique in which the conventional ion impact source is replaced by the emergent beam from a monochromator, as well as by measurement of the optical absorption coefficient and of the number of ions produced by a beam of light. These means all have been discussed by Ditchburn and Opik. Some more recent measurements of polyatomic molecules are those of Steiner, Giese and Inghram[125] and of Schoen;[111] Inghram and his coworkers[42] have also studied relative ionization cross section curves of some free aliphatic radicals, e.g. $CH_3$ and $C_2H_5$. The method of ion current measurement which represents one of the oldest and most widely explored techniques, has been used recently in a simple and especially elegant way by Turner and Al Jobory.[131] The conventional Lozier tube, a metal cylinder, contains the gas being studied. The cylindrical wall of the tube acts as an electron collector and is screened by a grid which can be placed at a variable potential to permit velocity analysis of the electrons. Turner and Al Jobory simply used a beam of ionizing light (the helium 21.3 ev resonance line) which entered the tube axially, so that the electric vector of the radiation, perpendicular to the tube axis, forced the electrons produced by absorption to travel to the tube wall. With this device, the velocity analysis of the photoelectrons could be done with enough resolution that the $^2P_{3/2}$ and $^2P_{1/2}$ peaks of $Kr^+$ could be distinguished and the corresponding $Xe^+$ peaks could be resolved.

Nonequilibrium photoionization is exemplified by precursor radiation in electromagnetically driven shock tubes, T-tubes. Precursor radiation is a phenomenon responsible for ionization ahead of very strong shock waves. The subject has been reviewed by Kolb and Griem.[B2] They estimate that in a typical experiment with electron temperature about 4 ev and electron density of $10^{17}$ per $cm^3$, the total photon flux can rise well above $10^{22}$ $cm^{-2}$ $sec^{-1}$. This is enough to cause sufficient atomic or molecular excitation ahead of the shock front that ionization by the shock is much enhanced and the temperature of the gas behind the shock front is considerably higher than it would be without the assistance of the radiation.

## 3.1.2. Ionization by Electron Impact

The means of ionization used longest and studied most extensively is ionization by electron impact. Ionization by electron impact is an effective process, in the sense that cross-sections for the process are large and are not subject to the stringent selection rules of photoionization. Crudely, one electron can excite another efficiently because the identity of their masses makes momentum transfer as effective as it can be. The actual cross sections for single ionization of light atoms reach broad maxima of about $2 \times 10^{-16}$ $cm^2$ at energies of about 100 ev.

Relatively few absolute cross-sections for ionization of atoms by electron impact have been measured. The rare gases, mercury and oxygen were studied in the 1930's; oxygen and hydrogen have been reexamined by Fite and Brackmann[48,49] and by Rothe, Marino, Neynaber and Trujillo,[106] and nitrogen has been studied by Smith, Caplinger, Neynaber, Rothe and Trujillo.[123] The problem was until relatively recently one of choosing to measure a total cross section for production of all charged particles over a wide energy range, or the relative cross section of a single charged species over an energy range narrow enough to preclude production of more than one species. Moreover, the problem of producing electron beams with narrow energy distributions traditionally plagued the investigators in this field. Retarding potentials and electrostatic analyzers both have been used; these are discussed by Fite in his review of excitation and ionization cross sections.[B2] The other problem, particularly important for the cross section measurements made in mass spectrometers, was the uncertainty of geometry and of collecting efficiency, which led to the relative nature of the cross sections collected by this method.

The most powerful technique now available is, of course, the crossed beams method, which was used by Fite and Brackmann and by Rothe et. al., and which traditionally suffered only the problem of low intensity. Modulation and ultrasensitive detectors have proven to be the final technological advances which have made this an entirely practical method.

The theory of atomic ionization by electron impact can hardly be called a finished subject. Thomson applied this classical argument; whenever the colliding electron has initial energy and impact parameter such that the energy it can transfer to a bound electron is greater than the ionization-potential I, then ionization will occur; otherwise it will not. We present the argument in a form given by Seaton.[B2] Let W be the initial kinetic energy of the projectile electron, and R be its impact parameter, so that the energy transferred in collision is

$$\epsilon = W \frac{1}{1 + (RW/e^2)^2} \tag{3.8}$$

and if $\epsilon$ is just equal to I, we let $R = R_o$. Then the cross section is

$$Q_{cl} = \pi R_o^2 n \tag{3.9}$$

if there are n equivalent ionizable electrons (i.e., n electrons in the outer shell). If one uses units of $I_H = e^4 m/2\hbar^2$, and expresses $R_o$ in Bohr radii $a_o = h^2/me^2$, then for a species with ionization potential I,

$$Q_{cl} = 4n\left(\frac{I_H}{I}\right)^2 \left(\frac{I}{W}\right) (1 - \frac{I}{W}) \quad \pi a_o^2 \tag{3.10}$$

A useful empirical relation which Equation 3.10 suggests is the universal functional dependence

$$\frac{1}{4n}\left(\frac{I}{I_H}\right)^2 \quad Q = f\ (I/W) \tag{3.11}$$

This has been suggested by Elwert[43] and is useful for order-of-magnitude estimates but is generally low.

Another relation similar to 3.11 derived by Drawin[40] is

$$Q = 2.66\,n \left(\frac{I_H}{I}\right)^2 \left(\frac{1}{W}\right)\left(1 - \frac{1}{W}\right) f_1 \ln (1.25\, f_2 W)\ \pi a_0^2\ \text{cm}^2 \tag{3.12}$$

where $f_1$ and $f_2$ are functions whose values are close to unity.

Impact ionization may be treated as a radiative process, in which the projectile electron makes a transition from energy $W_i$ to $W_f$, giving a photon of energy $\epsilon = W_i - W_f$ to the electron being ionized. The process so construed is a sort of resonant bremsstrahlung, in which the photon remains a virtual particle, present only in the intermediate state. The outgoing electron carries kinetic energy $I - \epsilon$. This formulation was used by Seaton[120] to express the electron impact cross section for ionization $Q_i$, in terms of the photoionization cross-section $\sigma(\epsilon)$:

$$Q_i = \frac{I_H}{a\pi W_i} \int_0^{W_i - I} \frac{\sigma(\epsilon)}{I + \epsilon} \ln\left(\frac{4W_i\tau}{I + \epsilon}\right) d\epsilon \tag{3.13}$$

The constant $\tau$ is of the order of unity; $\alpha$ is the fine structure constant. From this expression the more general functional relationship

$$Q_i = (A/\epsilon_i) \ln (B\epsilon_i) \tag{3.14}$$

has been inferred and seems to be rather successful at high energies in accord with the Bethe-Born approximation[B2] used to introduce the logarithmic term in the integrand; at lower energies, 75 to 200 ev above I, a scaling law has been inferred from 3.13 by Seaton; he suggests that

$$\frac{I_A Q_{iA}(W_i)}{\sigma(0)} \tag{3.15}$$

is approximately a universal function, so that measured photodetachment cross sections, together with the ionization cross section for one substance allow one to derive approximate cross sections for a number of others. When neon was used to provide the oxygen impact cross section, the theoretical curve (Seaton) of Fig. 3 was obtained. Clearly the relationship is accurate enough to be quite useful.

The quantum mechanical treatment of impact ionization as a collision process has been carried out only for hydrogen. Peterkop[97] and more recently Geltman, Rudge and Seaton[57] have explored refinements of the Born approximation to include effects of exchange and of shielding. The results of the Geltman, Rudge and Seaton are summarized in Fig. 4, where comparison is made with the experimental cross sections of Rothe et al.,[106] Fite and Brackmann[48] and Boyd and Boksenberg.[15] One sees readily that the simple Born approximation is almost as good as the Born-Exchange approximation at high energies. (The Born-Exchange (BE) method uses the Born scattering amplitude with electron momenta permated to represent the exchange scattering amplitude.) It can also be seen in the figure that only the SA curve gives the correct linear threshold behavior, albeit with incorrect slope. (The threshold problem is discussed in more detail later.) The SA approximation includes a spherically-averaged shielded Coulomb potential for the potential seen by the faster electron, and therefore uses Coulomb waves instead of plane waves for the final state functions.

The Born approximation is expected to be less applicable as a high-energy limit for heavy atoms than for hydrogen atoms. This difficulty arises because of the increasing number of processes which become available as the impact energy is increased. Its simplicity limits the number of final channels, so that the hydrogen atom becomes unusually suitable for a Born approximation treatment. If it were possible to observe the energy dependence of single-channel processes such as valence shell ionization, then the Born approximation would be useful for heavy atoms. It is a bit surprising that impact ionization of the methane molecule is rather well represented by the Born approximation (see below).

A rather successful treatment of ionization by impact, as well as of many other inelastic collision processes of electrons and ions with atoms is the classical formulation of Gryzinski.[59] The formulation is based on the direct interaction of the projectile with the electron with which it is interesting, and so is closely related to the Thomson derivation. The trajectories are handled explicitly, however, so that the target electron is not simply standing still. The results calculated by the method seem to be rather better than more elaborate computations based on the quantum mechanical interaction of the free electron wave with the stationary state wave function of the bound electron. This point is discussed again at the close of this section.

Although quantum theory cannot yet give absolute cross sections for electron impact ionization near threshold, it has seen definite successes in the prediction of the shapes of cross sections near threshold. Geltman[55] developed a general relationship between the energy dependence of the ionization cross section near threshold and the number n of new electrons produced in the collision; specifically,

$$Q_i \sim (E - I)^n + \text{higher terms.} \qquad (3.16)$$

This relationship was derived on the assumption that the new outgoing electron has no angular momentum (i.e. goes as an s-wave), and can be described by the wave function for a simple Coulomb field; it was shown, however, that the higher angular momentum states do not affect the functional dependence of $Q_i$ near $E = I$, although they may charge the coefficient of the leading term. After several years of uncertainty, the experimental evidence is now strong enough to establish the threshold law quite firmly. Fox,[53] studying single ionization of He, Ne, Ar, Hg, CO and $N_2$, Foner and Nall[51] studying the rare gases, Kiser[78] examining impact production of $Kr^{+2}$, $Kr^{+4}$, $Xe^{+2}$ - $Xe^{+5}$ and others have amply verified the relation. Dorman and Morrison[39] have discussed the coefficient of $(E - I)^n$ in terms of an integral over a structural factor of the form S-exp ((E - I)/I), where S is a function proportional to n, and an energy distribution function for the electron beam. This formulation eventually may be particularly useful for molecules, where the structural term can be related in part to a Franck-Condon factor.

A crossed-beam method has been applied recently by Dolder, Harrison and Thoneman[36] to study the ionization of $He^+$ to $He^{+2}$ by electron impact. The measured cross-section for this process is not reproduced by any single theory, but fits reasonably well with the classical cross-section computed by the method of Gryzinski at low energies, and with the Coulomb-Born approximation at high energies.

The electron impact ionization of molecules is a vast topic, the subject of many treatises, because of its role in mass spectrometry. A proportionate amount of space cannot be devoted to the subject here. Several recent books have presented the basic problems and summarized the current interests very thoroughly,[B1, B2, B4, B5, B6, B10] and some of the pertinent aspects are discussed in the chapter by Franklin, Munson and Field.

The Geltman threshold law seems to be obeyed by molecules, according to the results of Fox quoted previously,[53] and of Dorman and Morrison,[38] who examined the double- and triple-ionization of a number of molecules including CO, $N_2$ and NO.

Ionization of molecules has the challenging complication that the ion produced and the parent molecule each have their own effective potential energy surfaces that govern the motion of the nuclei. Ionization by electron impact is a fast process, so that the nuclei of a molecule effectively do not move during the process. The new ion-molecule is created in a geometry which in isolated cases (e.g. $CH_3^+$) is its equilibrium geometry, but more generally (e.g. $C_2H_5^+$) is produced in some distorted and therefore vibrationally-excited state. Sometimes the excited state has no potential minimum or only a shallow minimum, so that ionization can lead to dissociation of all or most of the ion molecules formed initially. An example of this sort of behavior is neopentane. The ionized molecule need not decompose with certainty; the probability for dissociation depends, in general, on electron energy.

The dissociation process may occur rapidly or slowly; if the lifetime of the parent ion is in the range of about $10^{-5}$ - $10^{-7}$ sec, then the lifetimes can be measured by mass spectrometric methods; the fragments appear as peaks of intermediate mass because they are produced during the acceleration period in the ion source. Metastable ions have been examined recently by Coggeshall, for example,[26] who finds evidence for as many as three decay times associated with the production of a given pair of dissociation products. The ion $C_4H_{10}^+$ of normal butane dissociates to $C_3H_7^+$ and $CH_3$ with time constants $8.9 \times 10^{-8}$, $6.1 \times 10^{-7}$ and $2.7 \times 10^{-6}$ sec; the three observed decay times to $C_3H_6^+$ and $CH_4$ are very similar.

The behavior of vibrationally excited polyatomic ion-molecules has been treated rather successfully as a quasi-equilibrium phenomenon. The ion-molecule is presumed to distribute its excess energy among its vibrational degrees of freedom; the fraction of fragment ions of any given type, and particularly isotope ratios, can then be estimated from known and estimated vibrational frequencies and bond energies. Recent treatments by Vestal, Warhaftig and Johnston[134] and by Rosenstock[104] have overcome the main objectionable aspects of the theory, namely the apparently arbitrary number of effective or active vibrational degrees of freedom. The inactivity of certain modes now seems to result from the fact that the vibrational excitation energy available is not large enough to distribute itself equally among high- and low-frequency modes. Purely statistical considerations lead directly to the conclusion that most of the important states are those in which high-frequency modes are unexcited.

Energy transfer from electrons to molecules seems to be most effective when the impact energy is 25 to 100 ev. High-energy electron collisions (up to 10 kev), studied by Melton[89] and by Kebarle and Godbole,[73] lead to fragmentation patterns which are almost identical to those from impact by 70 ev electrons, except that the quantities of rearrangement products are much lower from high-energy impact processes. The rearrangements presumably require excess internal energy, if intramolecular potential barriers are to be surmounted. Just as with atoms, as the impact energy increases into the Born approximation region, the cross section for excitation of weakly-bound electrons decreases.

Those rearrangements that do occur after ionization by low-energy electrons seem to be unaffected by the electron energy;--there does not seem to be a variety of paths having very different energy dependencies. This was the conclusion of the study of isotopic distributions of products from deuterated xylenes and toluenes, carried out by Meyerson.[90] It is as though all available rearrangement paths (leading to a given species) open at energies very near the appearance potential for that species (i.e. within 1 v or less). Furthermore, no new channels open at higher energies, up into the range where energy transfer from the electron to the molecular skeleton becomes inefficient.

Electron impact cross sections for ionization and excitation of molecules have not been widely calculated. One example of a recent quantum mechanical treatment is that of Inokuti,[70] who was able to carry out a calculation in the Born approximation limit for ionization of methane. This case was made tractable by the use of a one-center expansion of the molecular wave function. The result was a cross section in fairly good agreement with the observed high-energy behavior, in the range from 100 ev and above. Aside from the quasi-equilibrium calculations of fragmentation patterns and Born approximation calculations, the theory of electron impact of ionization of molecules has remained relatively unpopular.

An interesting result of electron impact ionization of diatomic molecules arose from recent work of Dorman and Morrison[39] on double and triple ionization. They were able to demonstrate that the potential curves of $CO^{+2}$, $N_2^{+2}$ and $NO^{+2}$ ground states are actually repulsive curves with metastable minima 5.9, 4.66 and 5.17 ev above the energies of the separated singly charged ions. The existence of the minima (with depths as yet unknown), superimposed on Coulombic repulsive curves, can be interpreted in terms of short range valence forces between the +1-ions. Alternatively, one can develop an interpretation in terms of contributions from higher, bonding configurations in which one atom is neutral and the other, doubly-charged. For example, $N_2^{+2}$ can be formed in a $^3\Pi$ state analogous to the first excited state of $C_2$, by combinition of two $N^+$ ions. At the internuclear distance equal to the equilibrium distance for neutral $N_2$, the two $N^+$ ions repel each other with an energy of approximately 13.1 ev. The dissociation limit corresponding to $N^{+2}$ + N lies 15.07 ev above the $N^+ + N^+$ limit. These two species, isoelectronic with BN, presumably lead to a bound state analogous to the $^3\Pi$ ground state of BN, which has a dissociation energy of about 3.44 ev, and an equilibrium internuclear distance of 1.281 Å. The corresponding $N_2^{+2}$ state presumably has a shorter equilibrium distance; if this distance is less than 1.22 Å, and if the dissociation energy in the $N^\circ N^{+2}$ state is comparable to that of BN, then the noncrossing rule will force the existence of a local minimum in the $^3\Pi$ state of $N^+ + N^+$. In a similar way, one can expect doubly- and perhaps multiply-charged molecules to exhibit bound metastable states. A discussion similar to this by Bates and Carson[2] suggested the existence of metastable $O_2^{+2}$.

It is pertinent here to examine the difference between photoionization and electron impact ionization in terms of the time scales associated with the two processes. Photoionization can be treated successfully in conventional first-order radiation theory, which implies that photon absorption is associated with a time long compared to a single period of oscillation of the classical electromagnetic radiation (but short compared with the time required to ionize a large number of molecules). Since ionization energies of most molecules are about 10 to 15 ev, this means that the time for ten oscillations of the radiation field is about $3 \times 10^{-15}$ sec, which is in accord with the usual Franck-Condon assumption, that nuclei oscillating at $10^{12}$ to $10^{13}$ $sec^{-1}$ move slowly compared with times for electronic transitions. The electrons interacting with incident photons, on the other hand, certainly have time to behave as though they are moving in the average field of the other electrons, so that Hartree-Fock orbitals, having the symmetry of the molecule, are appropriate for describing the bound electron states.

Electron impact presents a different situation. A 50-volt electron has a velocity of $4.2 \times 10^{8}$ cm/sec; classically it passes an atom in a time of about $3 \times 10^{-17}$ sec; if it is given a finite wavelength (of 1.7 Å), this time can be stretched a bit. To stretch it still further, a packet representation can be chosen; if the energy uncertainty of an electron is 0.1 to 0.01 ev, then the wave packet of the electron will be about 2.7 - 8.6 Å long, so that the transit time can be as long as $2 \times 10^{-16}$ sec. A 10 ev electron could take as long as about $5 \times 10^{-16}$ sec to pass an atom. These times are obviously very short compared with vibrational periods, so that the Franck-Condon principle can be invoked with confidence for molecular ionization.

Let us carry this line of thought one step further and examine the time scales of electronic motions. A typical time for an atom is the circuit time for an electron in a Bohr orbit of helium, say. This is $3.8 \times 10^{-17}$ sec, fast compared with the electron transit time, so that one can expect electron impact ionization of atoms to be well-represented by ionization of electrons from their Hartree-Fock orbitals. In a molecule, the Hartree-Fock orbitals are much more spread out than in an atom, and the circuit or delocalization times are correspondingly longer. In particular, the Hartree-Fock skeletal orbitals, built from linear combinations of local bonding orbitals, can show rather long characteristic times over which the motion of an electron must be averaged, if it is to be described properly by a Hartree-Fock orbital. The time for exchange of two electrons between localized bond orbitals $\phi$ and $\phi$ is given approximately by $\hbar/J$, where

$$J = \int \phi_1^*(1)\phi_2^*(2) \frac{e^2}{r_{12}} \phi_1(2)\phi_2(1) \, d\tau_1 d\tau_2 \qquad (3.17)$$

This integral is also the measure of the energy separation between the localized, average-energy, nonstationary-state orbitals and the Hartree-Fock orbitals. If J is about 0.1 ev, the size of such integrals between orbitals on nonadjacent bonds, then the associated $\Delta t$ is $3.3 \times 10^{-15}$ sec. Any process much faster than this time will interact with localized orbitals, and will exhibit the properties of the localized orbitals. These properties are the average properties of the Hartree-Fock orbitals of which they are composed; one such property is the orbital energy or ionization potential. One is led to the conclusion that electron impact ionization is inherently a rapid enough process that it cannot distinguish one-electron orbitals separated by tenths of volts, and that it must give ionization energies equal to a sort of average energy of the closely-spaced Hartree-Fock orbitals. It is concluded further that this diabatic phenomenon is not a characteristic of photon impact since it is so much slower. One infers, finally, that the diabatic character of electron impact is at least one of the factors responsible for the fact that ionization potentials from electron impact are almost invariably a few tenths of an electron volt higher than the values obtained by photon impact.

### 3.1.3. Charge Production by Impact of Heavy Particles

Charge production resulting from collisions of atoms, molecules or ions with each other can be divided into two rather different areas. First, there is the primitive process of direct ionization or charge transfer in a simple collision, typical cases of which can be represented as 10/1le and 00/1$\bar{1}$. It will be convenient to include some mention of direct charge transfer reactions like 20/11 in this context, although they are discussed in a little more detail in the next section. Second, there are many chemical reactions, particularly in flames and detonations, in which ions are produced in some step of the reaction, as a result of the relative positions ionic and molecular potential surfaces. This is the chemionization process.

At low energies, below 100 ev, ionization by simple collisions of heavy particles is considerably less efficient than by electron-atom collisions. This is a direct consequence of the lower velocity of the heavy particle, in a process that is almost a pure single-particle excitation. If ionization could be considered a pure direct interaction of projectile and electron, then ionization in ion-atom or atom-atom collisions would occur only in the kilovolt range, and in ion-atom collisions, would be much like electron-atom collisions at the same velocity.[63] The classical Thomson theory would require the projectile velocity to be of the same magnitude as electron orbit velocities, about $2.2 \times 10^8$ cm/sec. Cross sections for ionization by heavy particles do reach their maximum values, ca. $2 \times 10^{-16}$ to $15 \times 10^{-16}$ cm$^2$, in the kilovolt range; the maximum for $H^+ + H_2 \rightarrow H^+ + H_2^+ + e$ occurs at about 80 ev. In fact, ionization also occurs at lower energies, whenever the relative kinetic energy T of projectile (mass $m_1$) and target (mass $m_2$, ionization potential $I_2$) satisfied

$$T > \left[(m_1 + m_2)/m_2\right] I_2 \qquad (3.18)$$

The high-energy problem has been treated by Born approximation for proton collisions with H, He and Li, and for hydrogen atoms with H and $He^+$. The Born approximation leads to a quantitative relationship between the cross section $Q_e$ and $Q_i$ for inelastic collisions of electrons and bare nuclei with the same target:[B2]

$$Q_e(v) = \frac{1}{1 + \frac{\Delta E}{2mv_1{}^2}} \left[ Q_i(v_1) - \left(\frac{\Delta E}{2mv_1{}^2}\right)^2 Q_i\left(\frac{\Delta E}{2mv_1}\right) \right] \qquad (3.19)$$

Here $\Delta E$ is the threshold energy for the process, v is the electron velocity, $v_1$ and $\Delta E/2mv_1$ are ion velocities and m is the electron mass.

The impact parameter form of the Born approximation treats the incoming projectile as a particle moving with velocity v, and the target as a quantum-mechanical system undergoing a time-dependent perturbation. Direct application of time-dependent perturbation theory and substitution of the distance variable $Z = tv$ for the time t in the integration leads to a very general and useful characterization of impact collisions. The projectile velocity v, threshold energy $\Delta E = E_{final} - E_{initial}$

for the inelastic process, and the range a of the interaction appear to be related by the Massey criterion

$$\Delta E/h \approx v/a$$

in the region of the maximum cross section. In other words, the excitation frequency must be approximately the inverse of the time interval during which interaction occurs. This relation is valid for processes with single inelastic channel, and for which $\Delta E$ is relatively large. Resonant or near-resonant collision processes like symmetric charge exchange show cross sections that sometimes oscillate with increasing energy, as in the experiments of Everhart and co-workers[88,136] or which may show maxima from several final channels, as in the charge-transfer reactions of atomic hydrogen with various gases carried out by Donahue and Hushfar.[37]

Theoretical treatments aimed at handling many-electron systems have been developed by Firsov,[47] Gryzinsky,[59] Russek and co-workers,[18,108,109] and by Lee and Hasted.[84] These are all based on the direct interaction of individual charged particles within the target and the projectile. Russek's approach, which is somewhat similar to Firsov's, is called an evaporation model. Energy is presumed to be transferred by direct interactions; calculation of the amount of energy transferred constitutes one part of the problem. This energy is then permitted to distribute itself statistically among the outer electrons, and from this energy distribution, the relative numbers of ions of various positive charges is computed in the second part of the problem. The results of this semi-empirical treatment are rather successful and are in much better agreement with the observations of Everhard et al.[88,136] than are results of a single-step ionization model. The Gryzinsky method was described previously, in connection with electron impact. Lee and Hasted's calculations of the ionization of atoms by protons attempts to set the limits of impact energy E between which the ionization cross section maximum falls. They conclude that

$$1.46\ I \leq (m_e/m_p)\ E \leq 2.25\ I \qquad (3.20)$$

It was shown by Tuan and Gerjuoy[130] that one cannot treat processes like the collision of H with $H_2$ simply as the collision of H with H, so that some detailed structure of the electron orbit remains even when the projectile can be treated by Born approximation.

Experimental results on ion, atom and molecule charge-changing-collisions were reviewed recently by Fedorenko[45] and Fite.[B2] Some of the more recent observations are, for example, the collision studies of Utterback[132] and Utterback and Miller.[133] These represent some, perhaps the first experiments carried out on ionization of neutral molecules by beams of low-energy neutral atoms. Utterback has measured the ionization cross sections of $N_2$ and $O_2$ by beams of $N_2$ and $O_2$, at energies from 30 to 1000 ev. The cross sections for $N_2 - O_2$, $O_2 - N_2$ and $O_2 - O_2$ are smooth curves, rising from as low as $2 \times 10^{-20}$ at an energy of about 4.1 ev (center of mass) above threshold for ionization of $O_2$ by a beam of $N_2$. The ionization of $N_2$ by $N_2$, uniquely, showed a clear inflection at an energy about 10.5 ev above threshold.[132] There is no explanation at present for this structure.

In the absence of chemionization, the kinetics leading to equilibrium concentrations of electrons and ions in hot gases are governed by the cross sections for ionization by electron and ion impact. In a heating process (but not in an electrical discharge), atomic and molecular collisions are the only ones, initially; the rate of production of electrons is very small so long as the only mechanisms available are these collisions, at any chemically meaningful temperature. The rate will be determined by cross sections of the order of $10^{-20}$ $cm^2$, and the rate will be given by

$$\frac{d(e^-)}{dt} \approx 10^{-20}\, v \cdot n(m)n(m^*) \qquad (3.21)$$

We let v be the velocity of the excited particle $m^*$; n(m) and $n(m^*)$ are the densities of molecules and of molecules with sufficient extra energy to produce ionization. In a typical system, with $T \sim 3000°$ or $\sim 0.3$ ev, and $n(m) \sim 10^{18}$ $cm^3$, if we assume an ionization potential of 10 ev and require $m^*$ to have all this excess energy in translation, then we have $n(m^*) \sim 10^6$ $cm^{-3}$ and v, about $10^6$ cm/sec. Hence

$$\frac{d(e^-)}{dt} \approx 10^{10}\ cm^{-3}\ sec^{-1} \qquad (3.22)$$

which would be quite a low rate if it were the only one to lead to ionization equilibrium. Fortunately, the cross sections for electron impact ionization are $10^4$-$10^5$ larger than for atom or ion impact ionization, so that at high temperatures, ionization by electron impact can rapidly overtake the slower heavy particle process. Such phenomena are known to occur in the ionization of gases behind shock waves; argon is a well-studied example.[B3,60,98] They also occur in discharges: Ben Daniel and Bishop [5] have shown that in a Cs-seeded discharge in helium, the degree of ionization corresponds to the electron temperature (2000-4000°C), and not to the gas temperature (250°C).

Many common gaseous systems containing molecules do not reach charge equilibrium by the process just described. Either or both of two very important pathways may completely dominate the kinetics of charge formation. One of these is thermal dissociation of molecules into ions; the other is chemionization. Thermal dissociation into ions is important whenever the ground state of a molecule is rather ionic in character, or when there is some ionic dissociation limit at an energy rather close to that of the ground state dissociation products. The alkali halides present a rather extreme example of this situation, as we noted in the first section of this chapter. Their dissociation might occur adiabatically, and lead to a distribution of separate atoms which rapidly come to translational equilibrium and, as soon as they have gained enough kinetic energy, can experience electron-transferring collisions and become dissociated ion pairs. This is not the most likely mechanism in cases as extreme as the alkali halides. It is much more probable that in hot gases, dissociation occurs too fast to be adiabatic, going through quantum jumps in which the vibrational quantum number v'' increases by two or more units. Multiple jumps are probably most important when the vibrational energy approaches the dissociation energy, because the vibrational levels are so closely spaced for high v''. The importance of multiple jumps has been shown by the recent work of Mies and Shuler,[91] Shuler and Weiss[113] and Zwanzig and Shuler.[137] When an alkali halide is excited to an unbound level of the ground electronic state, the species present are essentially alkali cation and halide anion. It has been shown that the adiabatic charge transfer (return of the electron from halide to metal) required for dissociation is very unlikely.[6] Instead, ionic molecules can be excited up to energies approaching that of the ionic dissociation limit before they will actually start to dissociate. Then, when dissociation does occur, ion-pairs and not atom-pairs are the primary products. This process is presumably the basis for the observed excess ion concentration in flames containing alkali salts, such as those discussed by King.[74,76] Molecules in which ionic character plays a smaller role will obviously be less susceptible to this sort of dissociation. Nevertheless, the extreme inefficiency of direct collisional ionization to give positive ion and electron may make diabatic dissociation of molecules to ions important, especially in halogen-containing systems.

The process whereby two atoms collide and become positive and negative ions is one example of the chemionization process discussed considerable detail throughout the later chapters of this volume. The basic process, the reaction of neutral particles to give free charged particles, can be thought of as a transition of the system in question to a continuum level of a state whose dissociation-limit species are ions. The probability of a transition to such a state, a so-called crossover process, depends on the wave functions of electrons and nuclei in the immediate region in which the potential surfaces of the initial and final states approach or cross each other. The problem was originally formulated in an approximate but rather general and useful way by Landau, Zener and London in 1932, and has been developed considerably since then by a number of authors. Limitations of the original formulation have been discussed by Bates, who has recently reviewed much of the theoretical work.[B2] The crossing probability from state 1 to state 2 is governed by the coupled equations for the wave function $\Psi_1$ and $\Psi_2$ which represent the nonstationary states that would be stationary if the nuclei moved infinitely slowly:

$$i\hbar \frac{d \cos w}{dt} = (\cos w)H_{11} + (\sin w)H_{12} \qquad (3.23a)$$

$$i\hbar \frac{d \sin w}{dt} = -(\sin w)H_{21} + (\cos w)H_{22}. \qquad (3.23b)$$

We have assumed that only the two states need be considered, and that the stationary states of the system, $\Psi_A$, $\Psi_B$, would be expressed as

$$\Psi_A = \psi_1 \cos w + \psi_2 \sin w, \tag{3.24a}$$

$$\Psi_B = -\psi_1 \sin w + \psi_2 \cos w. \tag{3.24b}$$

Then if w is small, an approximate solution can be written

$$\sin w \approx \exp(H_{22}t/i\hbar) \cdot f(t) \tag{3.25}$$

where

$$f(t) = (1/ih)\int_{-\infty}^{t} \left[\exp i(H_{22} - H_{11})\tau/h\right] H_{21}(\tau)d\tau. \tag{3.26}$$

The probability of crossing in a single traversal of the crossing region, in the Landau-Zener approximation, is given by

$$p = 1 - e^{-\frac{2\pi}{hv}\left(\frac{H_{12}^2}{H'_{11} - H'_{22}}\right)} \tag{3.27}$$

where primes indicate differentiation with respect to $\vec{R}\cdot\vec{v}$. All functions, v, $H_{12}$, $H_{11}$ and $H_{22}$ are evaluated at the crossing point. The formalism can be carried out less approximately by integrating the perturbation f(t) from $-\infty$ to $+\infty$. When time integration is transformed into integration along the reaction coordinate R (v is taken constant), the integrand has the form

$$\frac{dw}{dR}\left( \chi_1^* \frac{d\chi_2}{dR} - \chi_2^* \frac{d\chi_1}{dR} \right) \tag{3.28}$$

where $\chi_1$ and $\chi_2$ are the vibrational wave functions along the coordinate R. The derivative dw/dR is approximately Lorentzian for systems having crossings at large R, so that the magnitude of the integral itself depends on a good fit of the Lorentzian with the nuclear "momentum-overlap" contained in parentheses. The formation has been applied particularly to alkali halides.[6]

Ionization in flames, which seems to be largely chemionization, was traditionally studied by field deflection, Langmuir probe techniques and optical spectroscopy.[87,19] It has proved possible recently to insert into the reactive zones of flames sampling probes which lead directly to mass spectrometers. This method, which is discussed in more detail in later chapters, has been used by Knewstubb and Sugden,[81] DeJaegere, Deckers and van Tiggelen[35] and Calcote and Reuter.[20] It has led to the identification of the predominant ions in hydrocarbon-oxygen flames ($H_3O^+$, then $C_3H_3^+$), their relative concentrations and points of appearance in the flame. Evidence is now strong that the primary ionization step is that suggested by Hand and Kistiakowsky,[62] namely

$$CH + O \rightarrow HCO^+ + e$$

This process has been observed very recently by Fontijn and Baughman[52] in the reaction of oxygen atoms with acetylene at room temperature. The $H_3O^+$ and other ions result from secondary reactions; $C_3H_3^+$ may be either a primary or secondary product, from evidence currently available.

Most of the cases in which chemionization are known to be important have not yet been subject to theoretical treatment simply because so little is known about their potential surfaces. This is true of most of the systems occurring in flames and detonations, and in some reactions of excited atoms. Two recent examples might be cited. The acetylene-oxygen detonation was studied by Hand and Kistiakowsky.[62] They found that their results, together with the earlier results of Kistiakowsky and Richards[80] on photo-emission from the same system, allowed elucidation of the mechanism. The charge production arises from the reaction of an excited CH ($A^2\Delta$ or $B^2\Sigma$); this may emit radiation in the visible and near-ultraviolet, it may react with an oxygen atom to produce excited CO ($A^1\Pi$) which can emit short wavelength ultraviolet radiation, or it can react with an oxygen atom to give HCO and an electron. In another kind of system altogether, Dahler, Franklin, Munson and Field[34] have studied the formation of $He_2^+$, $Ar_2^+$ and $Ne_2^+$. They found that the chemionization process $A^* + A \rightarrow A_2^+ + e^-$ is quite important (although three-body reactions are very significant in these systems also). The cross-sections were estimated as $10^{-12}$ - $10^{-13}$ $cm^2$, although this does depend on the lifetime $\tau_A$ of $A^*$. If $\tau_A$ were $10^{-8}$ sec, the foregoing values would obtain; if $\tau_A$ were larger, the cross sections would decrease proportionately.

## 3.2. Charge-Conserving Processes

The charge-conserving processes include the photoprocesses of radiative capture and photodetachment, elastic and inelastic electron scattering, charge transfer and chemical reactions of ionized atoms and molecules. Experimental studies of photoprocesses and reactions of ionized molecules were rather uncommon until very recently, although theoretical work on both had been done before 1940. Electron scattering and charge transfer have been examined in many experiments and have been examined in many experiments and have been the subject of considerable theoretical labor. The photoprocesses have been reviewed recently by Branscomb;[B2] electron scattering has been discussed by several authors in Atomic and Molecular Processes,[B2] and by Burke and Smith[B4] (electron-hydrogen scattering). Charge transfer is usually treated in the same context as ionization by ion and atom impact. Reactions of ionic molecules we take to include processes like dissociative attachment, e.g. $e^- + O_2 \rightarrow O^- + O$, and its reverse, associative detachment. These processes are discussed in Atomic and Molecular Processes. A wide variety of ion rearrangements and ion-molecule rearrangements have been surveyed by Field and Franklin in Electron Impact Phenomena,[B6] and many are discussed in the following chapters.

We shall try to survey the basic physics of each area and to present typical examples and magnitudes. Other chapters discuss many of the topics in some detail, so that the coverage here should be taken as illustrative, and in no way exhaustive. This qualification holds even more for this section of the chapter than for the others because of the breadth of the fields.

### 3.2.1. Photodetachment and Radiative Attachment

The direct radiative attachment of an electron to an atom or its reverse, the photodetachment of an electron, is an attractive subject to study experimentally or theoretically. The potential energy of attraction, V, between electron and neutral atom varies as $-\beta^2 R^{-4}$ for large R, which is an intermediate between the long-range Coulomb attraction that dominates photoionization processes and the very short-range potentials of nuclear forces, or even of the van der Waals potentials between atoms. The rather short-range nature of V together with the stringent selection rules for electric dipole radiation lead to very simple forms for the photodetachment and radiative attachment cross sections. If the bound state of the electron in question has angular momentum quantum number $\ell$, then the continuum states into which the electron may go or from which it may come have angular momentum $\ell \pm 1$, so that a partial wave expansion of either cross section contains only two terms (one if $\ell = 0$). Moreover, the threshold behavior of the cross section with $\ell$, for photodetachment into a wave with angular momentum quantum number $\ell'$,

$$\sigma_{det}(k \to 0) \sim k^{2\ell' + 1} \qquad (3.29)$$

where $\hbar k$ is the electron momentum at large distances R from the atom.[48] Consequently, the threshold behavior is dominated by the lower angular momentum wave. For detachment of an s-electron, as in $H^-$, $\ell = 0$ and $\ell' = 1$ so that $\sigma_{det}$ increases as $k^3$ or $(E - E_o)^{3/2}$, where E is the photon energy and $E_o$ is the threshold, the electron affinity. If a p electron is detached by a photon, which occurs with $O^-$ or $Cl^-$, the threshold shape of $\sigma_{det}$ is determined by the outgoing wave for which $\ell' = 0$, the s wave. In this case $\sigma_{det}$ increases as $(E - E_o)^{1/2}$ and therefore has infinite slope at the threshold.

The threshold shape for s-wave electrons, for example, can be developed beyond the first term with the aid of the expressions derived by Spruch, Rosenberg and O'Malley[124] for the limiting phase shift in an $R^{-4}$ potential. The cross section is

$$\sigma_{det}(k) = \frac{16\pi}{3} \frac{f_o}{\lambda} k(|M_s|^2, + 2|M_d|^2) \qquad (3.30)$$

where $f_o$ is the square of an (n-1)-electron overlap integral of the bound electron

$$f_o = \left| \int \Psi_i^* (1, \ldots, n-1) \Psi_f (1, \ldots, n-1) d\tau_1 \ldots d\tau_{n-1} \right|^2, \qquad (3.31)$$

and $\lambda$ is the optical wavelength, which can be expressed in terms of the electron affinity $E_o$ as

$$\frac{1}{\lambda} = \frac{1}{hc} \left(E_o + \frac{\hbar^2 k^2}{2m}\right). \qquad (3.32)$$

Near k = 0, the d-wave contribution is negligible and the phase shift $\eta$ of the outgoing s-wave electron can be expressed in terms of the scattering length a and the polarizability parameter $\beta^2$;[121]

$$\cot \eta = \frac{1}{k}\left[ -\frac{1}{a} + \frac{\pi\beta^2 k}{3a^2} + \ldots \right] \tag{3.33}$$

Then the detachment cross section is given by

$$\sigma_{det} \approx \frac{16\pi}{3} \frac{f_0}{\lambda} k\left[(aI_0 - I_1)^2 + \frac{2}{3}\pi\beta^2 I_0(aI_0 - I_1) + \ldots \right] \tag{3.34}$$

near threshold. The integrals $I_0$ and $I_1$ are integrals of the bound-state orbital, the r operator and successive terms in the k-expansion of the outgoing wave.

Calculations of photodetachment cross sections for p electrons going into s waves have not been extensive. Cooper and Martin recently computed $\sigma_{det}$ for $C^-$, $O^-$, $F^-$ and Cl,[29] and obtained rather good agreement with the experimental cross section obtained by Branscomb and coworkers, particularly at energies less than 2 ev above threshold. The observed and calculated values rise smoothly to a knee at about 2 ev, at which point $\sigma_{det}$ is just over 6 x $10^{-18}$ $cm^2$. This seems to be the magnitude of $\sigma_{det}$ for the first-row atoms. The agreement between calculated and experimental[121] curves for $C^-$ is less impressive; the experimental value is roughly constant at about 14 x $10^{-18}$ $cm^2$ between 1.5 and 3.0 ev above threshold while the calculated curve has a definite slope, rising from 8 to 15 x $10^{-18}$ $cm^2$ in the same range. Close comparisons with experiment for $F^-$ and $Cl^-$ cannot yet be made, but the calculations do agree with the sixfold increase of $\sigma_{det}$ $(Cl^-)$ over $\sigma_{det}$ $(F^-)$,[8,9,10] about 0.1 ev above threshold, the former is about 15 x $10^{-18}$, but the latter, only about 2.5 x $10^{-18}$ $cm^2$. By contrast with the detachment into an s state, detachment of an s electron into a p state has been the subject of detailed computation, since it offers the simplest system -- $H^-$. The most recent of these calculations were performed by John[71,72] and by Geltman and Krauss.[56] The cross section was measured by Smith and Burch.[122] The three are in quite good agreement and also are consistent with the sum rules

$$\frac{mc}{\pi e^2}\int_{\nu_0}^{\infty}\sigma(\nu)\,d\nu = n \tag{3.35}$$

and

$$\frac{mc}{\pi h}\int_{\nu_0}^{\infty}\frac{\sigma(\nu)}{\nu}\,d\nu = \frac{2}{3a_0}\ \langle\,(r_1^{\,2} + r_2^{\,2})\,\rangle \tag{3.36}$$

where n = 2, the number of electrons, and the expectation value is taken over the ground state of $H^-$. The cross section itself is a smooth, slightly asymmetric function of $\lambda$, with a maximum of $4 \times 10^{-17}$ cm$^2$ at approximately 8000 Å. Because of the low cross section near threshold, the expected zero initial slope of $\sigma_{det}$ and the difficulty of working in the 17,000 Å region, photodetachment does not offer a powerful tool for the experimental determination of the electron affinity of the hydrogen atom.

The experimental study of p electron photodetachment has become the most powerful method for precise determination of electron affinities. Branscomb and his co-workers have used crossed ion and photon beams to measure the affinities of carbon, oxygen, sulfur and iodine atoms,[126] and of the OH radical.[B2] The intensities of ion current and electron current as functions of the wavelength of the incident radiation are used to determine $\sigma_{det}$. Berry, Reimann and Spokes[9,10] and Berry and Reimann[8] have observed the continuous absorption spectra of the ions $F^-$, $Cl^-$, $Br^-$ and $I^-$ directly, by taking flash spectra of dissociated alkali salt vapors behind shock waves. A typical curve of the relative cross section for photodetachment is that of $Cl^-$, shown in Fig. 5. The two methods are in good agreement for the affinity of iodine, the one example done by both methods.

Photodetachment from molecular ions shows the influence of the Franck-Condon principle (or its violation) just as photoionization does. The $OH^-$ ion seems to be a case in which detachment produces very little or no vibrational excitation, according to the results discussed by Branscomb.[B2] The $O_2^-$ ion, by contrast, might photodetach to give $O_2$ with some vibrational excitation. Since the extra electron is weakly antibonding, it is quite probable that $O_2^-$ has a larger equilibrium internuclear separation than $O_2$.

The photodetachment and radiative attachment cross sections are simply related by the condition of microscopic reversibility; the total probability for reaction in one direction, per unit density of final states in a single encounter must be equal to the corresponding probability for the reverse reaction, per unit density of its final states. Thus

$$\sigma_{att} = \sigma_{det}\left(\frac{h\nu}{mcv_e}\right)^2 \tag{3.37}$$

where $v_e$ is the electron velocity. This relation requires, for example, that the radiative attachment cross sections of Cl, Br and I be about $10^{-21}$ $cm^2$ for capture of an electron having about 0.1 ev of kinetic energy.

The rate of attachment is proportional to the frequency of collisions for which reaction occurs, and therefore to the product of the cross section and the electron velocity. One commonly defines

$$\alpha = \sigma_{att} \cdot v_e \tag{3.38}$$

the attachment coefficient. Typically, Br at 3645 Å, corresponding to an electron energy of about 0.037 ev, has an attachment coefficient of approximately $10^{-16}$ $cm^3$ $sec^{-1}$.

The measured light intensity can be expressed as a function of the photon energy E and the electron affinity $E_o$ of species x:

$$\begin{aligned} I(E)\,dE &= \sigma\cdot v_e\cdot n_x\cdot n_e(E - E_o)\cdot EdE \\ &= \left(\frac{2}{kT}\right)^{3/2}\frac{(E - E_o)e^{-\left(\frac{E - E_o}{kT}\right)}}{\pi m}\,\sigma n_x n_e\cdot EdE \end{aligned} \tag{3.39}$$

Alternatively, as a function of the wave vector k,

$$I(k)dk = \sqrt{\frac{2}{\pi}}\left(\frac{\hbar^2}{mkT}\right)^{3/2}\sigma n_e n_x\cdot\left(E_o + \frac{\hbar^2 k^2}{2m}\right) \times e^{-\frac{\hbar^2 k^2}{2mkT}}\cdot k^2 dk \tag{3.40}$$

From expressions 2.29 and 2.37, we see that the threshold dependence of $\sigma_{att}$ goes as $k^{2\ell'-1}$, so that the attachment cross section of zero-energy s-wave electrons is infinite if there is a bound p-state into which they can fall. The attachment coefficient has a finite value for this process, since it varies as $k^{2\ell'}$ near $k = 0$. The light intensity itself increases with $k^{2\ell'+2}$, because of the Boltzmann distribution of electron velocities. The function $I/(E - E_0)^{1/2}$ as a function of E should approach the limiting value of

$$\frac{2\pi}{(\pi kT)^{3/2}} \quad E_0 n_e n_x \alpha$$

for s wave capture.

Radiative capture spectra have been reported for H, N, O and Cl in stabilized arcs, by Lochte-Holtgreven,[87] Boldt[12] and Henning.[64] The continua were distinguished by subtraction of the computed contributions from bound-free and free-free transitions. It was not possible to observe thresholds in any of these cases. The fact that the observations reported by Henning on Cl include the threshold region and show no step shed doubt on his results.

Shock tubes have also been used to produce radiative capture spectra. Weber[135] has attributed part of a continuum to $H^-$, by subtracting the calculated contributions from other sources. Recently, direct observation of the sharp emission steps due to zero-energy electron capture by Cl, Br and I have been observed by Berry and David.[7] These were produced by shock dissociation of cesium and rubidium salts.

### 3.2.2. Electron Scattering

The field of electron scattering is broad, popular and frequently reviewed.[B1,B2,B4,B5,B8,B9] This discussion can concern itself with a brief examination of low-energy elastic and inelastic scattering from atoms and molecules, and touch on some recent illustrations pertinent to these aspects of the problem.

The first problem is the description of the interaction between the electron and the atom or molecule which scatters it. Lippmann, Mittelman and Watson[86] have shown how electron-atom scattering can be developed in terms of an effective one-electron potential, $\mathcal{V}$, whose expansion begins with a classical $r^{-4}$ interaction and includes electron exchange and penetration effects in higher terms. Scattering by an ion obviously is dominated at long distances by the Coulomb interaction.

If the effective potential is known, then the scattering problem can be treated by conventional methods. For example, the scattered electron wave can be expanded as a series of partial waves. Each one is a spherical harmonic having its own phase shift $\eta_\ell$ by which, at large distances from the scatterer, it differs from an "unscattered" wave originating at the same point. The entire wave, incoming and outgoing, is expressed by the function

$$\psi \sim e^{ikz} + (e^{ikr}/r)f(\theta, \phi) \qquad (3.41)$$

and in scattering by random molecules of a gas, the scattering amplitude $f = f(\theta)$ only. If the scattering amplitude is expanded as a spherical harmonic series, then

$$f(\theta) = (1/k)\sum_\ell(2\ell + 1)\ e^{i\eta_\ell}\ \sin \eta_\ell \cdot P_\ell\ (\cos \theta) \qquad (3.42)$$

The scattering cross section Q is the sum of all the partial cross sections $Q_\ell$, each of which is

$$Q_\ell = (4\pi/k^2)\ (2\ell + 1)\ \sin^2\eta_\ell \qquad (3.43)$$

The scattering amplitude and individual phase shifts can be found by direct numerical integration or by a variety of approximate methods. Among them are the Born approximation, which is useful for single channels at high energies; the effective range method, which was mentioned previously because of its utility for determining limiting forms for $\eta_\ell$ near thresholds, and the Hulthén-Kohn variation methods. These have been summarized and discussed by Moiseiwitsch[B2] in connection with elastic scattering. Inelastic collisions have been somewhat more difficult to treat than elastic collisions but for simple systems are tractable. The outgoing scattered wave is written as a superposition of waves, only one of which corresponds to the elastic wave. The coefficients of the final states for all the possible initial states comprise the S matrix or scattering matrix; at large distances

$$\Psi \sim \psi_I \phi_I - \sum_F \psi_F \phi_F, \; S(F, I) \tag{3.44}$$

where $\psi_I$ and $\psi_F$ are initial and final states of the target and $\phi_I$, $\phi_F$ are the incoming electron function and outgoing scattered waves. The sum is taken over all states. The inelastic scattering problem is one of computing the S-matrix or a function of it. For example, when the T or transmission matrix is defined by

$$T(F, I) = \delta_{FI} - S(F, I) \tag{3.45}$$

the cross section for excitation of an atomic electron from state $n\ell$ to state $n'\ell'$ is

$$Q(n\ell \rightarrow n'\ell') =$$

$$\frac{\pi}{k_n^2} \frac{1}{(2\ell + 1)} \sum_{\ell_1, \ell_1', L} (2L + 1) \, T(n\ell\ell_1 L, n'\ell'\ell_1' L)^2 \tag{3.46}$$

where $k_n$ is the initial electron momentum, $\ell_1$ and $\ell_1'$ its initial and final angular momentum and L is the total angular momentum of electron plus scatterer. As in the case of elastic scattering, the ultimate computation requires an integration of initial and final state wave functions over the effective potential.

The elastic scattering cross section for electrons colliding with hydrogen is computed (numerically) to be $36 \times 10^{-16}$ cm$^2$ at $k = 0$, and to drop smoothly, reaching about $4.7 \times 10^{-16}$ cm$^2$ when $k = 1$ Å$^{-1}$.[B2,B4] At $k = 0$, the entire cross section scatters s waves; at $k = 1$, the cross section is over 30% p-scatterer, and the d partial cross section is about 3 1/2% of the total. This scattering has been measured by several groups, using beam techniques; experimental and calculated cross sections are in rather good agreement. These results as well as those on inelastic scattering of electrons by hydrogen are reviewed by Burke and Smith, and by several writers in Atomic and Molecular Processes. Inelastic processes have been treated with some success by the impulse approximation, in which the incident electron interacts directly with the bound electron.[B4] The success of the classical method of Gryzinski depends on the same considerations as the impulse approximation: the interaction must occur quickly compared with the circuit time, or at least not very slow compared with this time. For low-energy elastic scattering, a Thomas-Fermi potential has been used by Robinson;[103] this represents the "slow" extreme. Bonham[13] has recently developed first Born approximation expressions for the differential and total cross sections for inelastic collisions of electrons with atoms in which bound states are excited. The formulation shows that exchange is not important in the Born approximation region; it also leads to approximate selection rules for the orbital electron's change of state. As one usually finds, the most important transitions involve $\Delta \ell = 1$.

Helium also has been studied, for both its elastic and inelastic cross sections. Calculations show that below 15 ev, the contribution to Q from exchange is extremely important; without exchange, Q at k = 0 is too large by a factor of about 9. The value with exchange is $7.3 \times 10^{-16}$ cm$^2$ at zero incident energy. The inelastic cross section to $2^3P$ is almost two hundredfold smaller at its peak energy of about 20.4 ev.

Recently total cross sections have been measured for atomic oxygen and atomic nitrogen, by Neynaber, Marino, Rothe and Trujillo.[95] The total cross sections are similar in magnitude to hydrogen; N has a total cross section of about $5 \times 10^{-16}$ cm$^2$ for energies of impact from 4 to 10 ev. The nitrogen cross section does appear to drop at energies below 4 ev, in accord with the semi-empirical calculation of Seaton based on an optical excitation model like that described previously for ionization cross sections. For excitation,

$$Q(I \rightarrow F) = \frac{4\pi^2}{k_I^2 \sqrt{3}} \; \frac{f(F, I)}{E_F - E_I} \; \bar{g}(k_F, k_I) \tag{3.47}$$

where f is the oscillator strength for the transition and $\bar{g}$ is an empirical effective Gaunt factor, which appears to be almost a universal function -- or at least close enough to be very useful. Relative cross sections for potassium were studied by Rubin, Perel and Bederson,[107] who were able to measure the exchange cross section also; the exchange cross section (ca. $10^{-14}$ $cm^2$ for energies of $\leq 0.5$ ev) can be measured also by optical pumping techniques.[54,96] One very recent result which is rather interesting is Schulz's observation of a resonance or maximum in the elastic scattering of electrons by helium.[118] The maximum, at 19.8 ev, is found with a fixed detector 72° from the unscattered beam. Its existence implies a bound or, more likely, a metastable level of $He^-$, probably 1s2s2p ($^4P$).

Methods other than beam scattering have been used very extensively to determine electron scattering. Total collision cross sections have been measured by direct attenuation methods since 1920; the cross section is derived from the attenuated electron beam intensity I

$$Q = (1/n\ell)\ \ln (I_o/I) \qquad (3.48)$$

where n is the number density of scatterers and $\ell$ is the path length in which scattering occurs. A related and traditional method is the measurement of drift velocities.[14] Pulsed photoproduction of electrons followed by time-of-flight measurements have been quite effective for measurement of cross sections in $SF_6$, in work done by Edelson, Griffiths, and McAfee.[41] Transport collision cross sections for electrons in $N_2$ and the rare gases were recently studied by Bowe, from drift velocities.[14] The attachment coefficient of $H_2O$ leading to dissociation was measured by an electron swarm capture method, by Hurst, O'Kelly and Bortner.[69] A method which has been used recently to determine collision cross sections is the measurement of line widths of electron cyclotron resonances. These have been carried out by Bulewicz[17] and by Bayes, Kivelson, and Wang.[4] The latter group carried out a partial wave analysis for scattering of thermal electrons by several gases having permanent moments, and obtained rather good agreement with their measurements. Observed cross sections are very large for these molecules; while $N_2$ gives only $4.96 \times 10^{-16}$ $cm^2$, $CO_2$ with its large quadrupole gives a measured $302 \times 10^{-16}$ $cm^2$, $NH_3$ has a 1560 $Å^2$ cross section and $CH_3NO_2$, 6020 $Å^2$.

When a molecule captures an electron, it effectively jumps from one potential surface to another. The negative ion-molecule may find itself in a bound state of a stable species, as is the case with $SF_6$ and electrons up to about 0.2 ev,[31] or in the case of $NO_2$ with electrons having 1.4 ev or more of kinetic energy.[30] It may find itself in a purely repulsive state, as probably occurs in $N_2$ as was previously mentioned, or in $N_2O$, which Curran and Fox[32] and Schulz[115] found to give $O^- + N_2^+$ from electrons with essentially zero energy. The ion molecule may be metastable, existing long enough to be stabilized by collisional deactivation; this was found from microwave measurements of the rates of electron capture in $O_2$ and $N_2$ by Lennan and Mulcahy[85] and earlier by Chanin, Phelps, and Biondi.[21] As an example, the effective three-body rate constant for the reaction

$$O_2 + e^- + N_2 \rightarrow O_2^- + N_2$$

is about $2.4 \times 10^{-30}$ $cm^6$ $sec^{-1}$. Attachment of electrons to molecules was reviewed by Prasad and Craggs;[B2] since the appearance of their review, several significant developments have occured, some of which are especially pertinent to a discussion of ions in hot gases.

The excitation of electronic states by inelastic collisions has been known since the 1930's, but vibrational excitation has only been observed recently, in the work of Haas on $N_2$[61] and of Schulz on $N_2$, CO and $O_2$.[114,116] It seems well-established now that an unstable $N_2^-$ ion can be formed at an impact energy of 1.5 ev or more, which rapidly goes back to $N_2 + e$; in the process, the $N_2$ molecules are deposited in their electronic ground state, but with some vibrational excitation. A similar process for $O_2$ having no energy threshold was reported by Schulz and Dowell.[119] The first 6-8 vibrational levels of $O_2$ could be detected by maxima in the trapped electron current, as could the low-lying excited states $^1\Delta g$ and $^1\Sigma g^+$. The excitation cross sections for the electronic states are of the order $10^{-20}$ $cm^2$, presumably because of the spin multiplicity change and the electric dipole forbiddeness of the transition. The lower vibrational excitation cross sections appear to be somewhat higher, but probably not a hundredfold larger; no numerical estimates were made for these because they are so intertwined with the elastic scattering cross section at very low energies. A theoretical treatment of the problem has been carried out by Chen and Magee.[23] They did computations for the $H_2$ molecule, because of its tractability, and conclude that the vibrational excitation should reach a maximum just above 7 ev, and that the cross sections for individual states rise from about $0.1 \times 10^{-16}$ for $0 \rightarrow 1$

to almost $10^{-16}$ for $0 \rightarrow 4$ and drop essentially to zero for $0 \rightarrow 7$. The principal factor in fixing the energy is the height of the $H_2^-$ curve above the ground state when $R \sim R_e$, and that the relative contributions of different vibrational states is governed by the integral of the product of the bound state vibrational function of the neutral species and the continuum nuclear wave function of the negative ion-molecule. It appears that the new interpretation of the electron "impact" excitations is correct, that negative ion-molecules are formed and that they may be very significant in contributing to vibrational relaxation. This is especially true in the case of molecules having positive electron affinities like $O_2$ or $SF_6$, which can capture zero-energy electrons. Chen[22] has developed a much more detailed and quantitative form of the theory of electron capture by molecules. He carried out computation for the energy dependence of the cross section for $H_2 + e^-$ $H + H^-$. The results are indendent on the form of the potential curve for $H_2^-$; two approximations representing $H_2^-$ as having a shallow well and large equilibrium separation give results agreeing in general shape and magnitude with the observed cross section reported by Schulz. Two maxima appear for which both H and $H^-$ are in their ground states. These correspond to two maxima in the square of the electronic transition matrix element of the nuclear kinetic energy operator, $|(\psi_I|T_{nuc}|\psi_F)|^2$. Crudely, this arises from the perturbation due to the slight inability of the electrons in state I to follow the nuclei perfectly.

Conway[27] has estimated the detailed relationship between the $O_2$ and $O_2^-$ potential curves; he places the ground vibrational state of $O_2$ between the second and third excited states of $O_2^-$, that computed collision numbers for deactivation are well above experimental lower limits (Calculated: several hundred with $O_2$, $CH_4$, $C_2H_4$ or $CO_2$, and several thousand or more with He or $N_2$), and that the probability for the reverse process $O_2 + O_2^- \rightarrow 2O_2 + e^-$ is about $8 \times 10^{-7}$ per collision.

One recent piece of work of considerable interest is Schulz's application of attachment to $O_2$, CO and $CO_2$ to measure the electron affinity of O.[117] With each of these molecules it was possible to measure the kinetic energy of the $O^-$ ion formed by dissociative attachment and to measure the kinetic energy of the electron which produced it. The ion energy was extrapolated to zero; the corresponding electron energy represents half the difference between the dissociation energy of $O_2$ and the electron affinity of oxygen. Schulz was able, therefore, to measure the affinity of O; he obtained a value of 2.0 ± 0.1 ev. The difficulty that exists at the time of this writing is the discrepancy between Schulz's value and that obtained by Branscomb et al.[B2] by photoionization, namely 1.465 ± 0.005 ev. It is possible that the photodetachment value is the result of detachment from $O^-$ to leave excited $O(^1D)$ would settle the issue; the absorption should fall at ~ 3612 Å or at about 3100 A, depending on whether the low or high figure is the true electron affinity of oxygen.

### 3.2.3. Charge Transfer and Ion-Molecule Reactions

Charge transfer from one atom or molecule to another in a simple collision has interested theorists and experimenters for many years. This interest may be due in part to the rather large cross sections associated with the process, in an energy range that is easily accessible. For example, the resonance charge transfer $H^+ + H \rightarrow H + H^+$ has a cross section $\sigma_T$ of at least 30 x $10^{-16}$ $cm^2$ at all energies below 400 ev; the cross section for $K + K^+$ is more than 100 x $10^{-16}$ $cm^2$ in the same range. The reviews of Fedorenko,[45] Bates[B2] and Hasted[B2] summarize most of the work in this area done before 1962, but some rather important results have been achieved more recently which are especially pertinent to this context, in the realm of cross sections at very low energy.

The theory of the low-energy limit for charge transfer reactions has been the subject of considerable controversy until recently. The question of interest is the low velocity limit of $\sigma_T$ for the reaction $X + Y^+ \rightarrow X^+ + Y$, when the two dissociation limits have almost the same energy. In the case of symmetrical charge transfer, e.g. $H + H^+$ or $K + K^+$, the cross section goes logarithmically to infinity as the impact energy goes to zero. The relation can be expressed as

$$\sigma_T = -k_1 \ln v^2 + 2k_2 \tag{3.49}$$

where $k_1$ varies as $I^{-1/2}$ and $k_2$, as $I^{-1/2} \ln I$.[101] This cross section increases without limit because at infinitesimal v, the pair of particles remains in its stationary ground state regardless of the separation distance R; the true ground state is a superposition of equal parts of the nonstationary states representing $X + X^+$ and $X^+ + X$, which are degenerate for $R \sim \infty$. At no point does the system "know" its previous history, in the adiabatic limit, so that each X has a 50% probability of being found as the charged particle after the collision. In the asymmetric case, when the limits $X^+ + Y$ and $X + Y^+$ have different energies at large R, the cross section goes to zero when v goes to zero because the adiabatic limit at large R corresponds to a well-defined state $\Psi$ for which the atomic functions are exact bases; $\psi = \phi(X^+)\phi(Y)$ is an exact state (after antisymmetrization, of course). When the particle pair travels adiabatically through the collision trajectory, it enters as $X^+ + Y$, say, remains on the corresponding potential curve at all values of R and returns as $X^+ + Y$, so that there is no probability for charge transfer and $\sigma_T$ becomes zero. Now we suppose that, at large R, $E(X + Y^+) - E(X^+ + Y) = \epsilon$, a small energy. When the particles are close together, the two stationary states have an energy separation $\Delta E$ which, in most cases, is much larger than $\epsilon$, but approaches $\epsilon$ as R grows large. When $\Delta E$ is large, the stationary states are by no means pure $X^+ + Y$ or $X + Y^+$, but are mixtures of these two (and possibly other states, which are not pertinent here. The mixing coefficient for the two limiting state components is a function of R, and has a derivative that differs significantly from zero for $R < R_1$. in order that the system behave adiabatically, the particles must travel very slowly. Specifically, they must travel so slowly that the system traverses the region $R < R_1$ in a time $\tau$ which is long compared with $\hbar/\Delta E$, for $\Delta E$ everywhere in the region. Otherwise the system will find itself following a diabatic path, becoming a mixture of $X^+ + Y$ and $X + Y^+$ at some small $R < R_1$, unable to maintain the equilibrium charge distribution or stationary-state wave function at all R, so that by the time R has passed $R_1$, the system is frozen into a nonequilibrium distribution in which it has a finite probability of being in either the $X^+ + Y$ or $X + Y^+$ state. The system takes on all the character of the symmetric case, for velocities greater than

$$v_1 = R_1 \Delta E/\hbar \qquad (3.50)$$

It is actually getting trapped in a nonstationary state because of the diabatic nature of the transition. In this sense the problem is closely parallel to the diabatic dissociation of ionic molecules discussed previously. The criterion for $v_1$ is formally the same as the Massey "adiabatic criterion," with a slightly different interpretation of the range parameter $R_1$. An analysis of the problem in somewhat different and quantitative terms has been given by Rapp and Francis.[101] From the observed maxima for a number of charge transfer reactions, Hasted and Lee[63] have estimated $R_1$ and found it to be about 7 Å; in all the processes they considered, the cross section maximum is found at energies over 150 ev, and for all of these, the potential curves were corrected for long-range interactions. Iovitsu-Popescu,[99] from a resonant-transfer viewpoint, has estimated a charge transfer distance of about ten times the atomic radius. If we apply the 7 Å figure to a straight-trajectory collision for which $\epsilon = 0.01$ ev (as in $O^+ + H$), then $v_1 = 1.1 \times 10^6$ cm/sec, $\tau = 6 \times 10^{-14}$ sec and the impact energy in the center-of-mass system is 0.6 ev. This should be approximately the energy of the maximum cross section $\sigma_T$. We can conclude that although the degenerate symmetric and near-degenerate asymmetric cases have different limiting cross sections, the smaller $\epsilon$ is made, the more difficult it becomes to observe the distinction. In fact, it seems that cases like $O^+ + H$ are so near to degeneracy that present experimental techniques will not distinguish them from the truly degenerate cases.

Very few cross sections for charge transfer at very low energies, below 50 ev for example, have been measured as functions of energy. Fite, Smith and Stebbings[50] recently have studied $H^+ + H$, the near-resonant $H + O^+$ and the nonresonant $H + He^{+2}$ to energies of 25 ev, in order to compare the symmetric and asymmetric near-resonant cases. Within the range covered, as v decreases, $\sigma_T$ increases for $H^+ + H$ and $H^+ + O$ in the same way. The foregoing discussion would place the maximum at far lower energies than the range studied, however. The $H + He^{+2}$ case shows a maximum, as expected.

Another low-energy charge transfer process studied recently was the electron transfer process $H^- + O_2 \rightarrow H + O_2^-$; this was accompanied by the ion-molecule reaction $H^- + O_2 \rightarrow OH^- + O$; the collision $OH^- + O_2$ was also reported at the same time, by Baker, McGuire and Muschlitz.[1] The two $H^-$ processes are nearly resonant ($\epsilon < 1$ ev) and the observed cross sections do rise as the energy falls, down to 4 ev.

One more charge transfer cross section which might be mentioned here is that for excited $N_2^+$ with O atoms, computed by Roth[105] from the nonequilibrium emission of shocked $N_2$ and $O_2$ in Xe. The total cross section, to give $N_2 + O^+$, $NO^+ + N$ and $NO + N^+$ is about $0.02 \times 10^{-16}$ for impact energies of about 1 ev.

Ion-molecule reactions are discussed in detail in the chapter by Franklin, Munson and Field; this introduction can only summarize the brief outlines and some recent developments. The problem, as it is defined here, is similar to charge transfer, except that a massive particle, an atom or molecular fragment, is transferred during a collision. The microscopic cross section for the reaction, $\sigma_R$, is governed by the attractive potential of interaction between the ion and neutral molecule, which can be expressed as $-\beta^2/R^4$ or

$$V = - e^2\alpha/2r^4 \tag{3.51}$$

This potential has a critical impact parameter $b_o$, a function of v, such that if $b > b_o$, then the projectile never comes close to the origin than $b_o/\sqrt{2}$, but always passes through the origin if $b < b_o$. This property was used by Gioumonsis and Stevenson[58] to develop a theoretical expression for the cross section based on the hypothesis of a critical reaction radius less than $b_o$. The cross section, derived originally by Eyring, Hirschfelder and Taylor, is

$$\sigma_R(v) = \frac{2\pi}{v}\left(\frac{e^2\alpha}{m}\right)^{1/2} \tag{3.52}$$

where m is the reduced mass and v is the relative initial velocity. Gioumonsis and Stevenson developed an expression for the phenomenological cross section Q measured under the conditions of the mass spectrometer, and showed the relationship between $\sigma_R$ and Q. The agreement of calculated cross sections or rate constants $v\sigma_R$ with the data, for example, of Stevenson and Schissler[127] is fairly good but not universal. More recently Boelrijk and Hamill[11] have extended the treatment by assuming that, for high-energy reactions, only head-on collisions are effective. They derive for their "two-mechanism" model a pair of expressions

$$\sigma_R = (A_1/v) + A_2 \text{ at } E < E_{crit} \tag{3.53a}$$

$$\sigma_R = (B_1/v^2) + B_2 \text{ at } E > E_{crit} \tag{3.53b}$$

This is good agreement with the data on many reactions. $D_2^+ + D_2 \rightarrow D_3^+ + D$ and $D_2^+ + O \rightarrow DO_2^+ + D$ for example, are reported by Boelrijk and Hamill; they have critical energies of 1.7 and 1.5 ev, respectively. Pottie, Lorquet and Hamill[100] have examined cyclopropane reactions, and calculated $A_1$ to be 47 Å$^2$ ev$^{1/2}$ and measured it to be 70 to 90 Å$^2$ ev$^{1/2}$, depending on the reaction; the gas-kinetic cross section $A_2$ is calculated as 51 Å$^2$ and found to be 82 ± 9 Å$^2$. Other recent extensions of this work were done by Theard and Hamill[128] and by Kubose and Hamill.[83] The isotopic reactions of $H_2^+$ with $H_2$, $D_2^+$ with $D_2$ and $HD^+$ with HD, studied by Reuben and Friedman,[102] indicated an $X_4^+$ intermediate.

The hydrogen-helium system was discussed by Hertzberg, Rapp, Ortenberger and Briglia,[65] who showed that $H_2^+$ could react with He to give $HeH^+ + H$, as well as with $H_2$ to give $H_3^+ + H$; more recently von Koch and Friedman[82] have shown that the $H_2^+$ must be in the fifth or higher vibrational state to give $HeH^+ + H$.

Formation of gaseous ion-molecules at high pressures, studied by Field, Franklin, and their collaborators,[B6,34,46,93] has been mentioned in connection with ionization. The work has led to discovery of a large number of ion-molecules, particularly of the rare gases, and of a very wide range of third- and higher-order reaction processes.

## 3.3. Recombination

Electronic and ionic recombination processes are discussed separately here, just as before, even though these cannot always be distinguished under real experimental conditions as easily as some of the processes described previously. Nevertheless, it is convenient for the viewpoint of interpreting the processes. Recombination has been reviewed by Bates and Dalgarno[B2] and by Sayers.[B2]

The electronic recombination process is, under many circumstances, very similar to the radiative attachment process discussed earlier. The two differ in two ways, principally. First, in recombination the electron is attracted by a Coulomb field, not an $r^{-4}$ field, so that the continuum wave function of the electron can be considered approximately as hydrogenic, and the hydrogenic recombination problem is soluble. Second, the atom formed by recombination has a large number of bound states available for the newly-added eledtrons; the negative ion formed by radiative attachment offers only one bound level, in general. The availability of many levels implies the possibilities of relaxation and stepwise excitation and deexcitation and therefore of a sort of pressure dependence of recombination that does not occur with attachment.

The radiative recombination rate is expressed by the effective second-order rate law

$$\frac{dn(A^+)}{dt} = -\alpha n(e)n(A^+) \tag{3.54}$$

where $\alpha$ is called the radiative recombination coefficient. When radiative processes are the only important ones, the hydrogenic expression for recombination to a level n in an ion of charge Z is very effective:

$$\alpha_n(Z, T) = 8.1234 \times 10^{-9} \frac{Z^4}{n^3 T^{3/2}} S_n \tag{3.55}$$

where

$$S_n = \int_0^{\infty} \frac{g(n, \epsilon) e^{-\frac{157,890 Z^2 \epsilon}{T}} n^2 d\epsilon}{n^2\epsilon + 1}$$

$$\approx e^{\frac{157,890 Z^2}{n^2 T}} \mathcal{E}_i \left(\frac{157,890 Z^2}{n^2 T}\right) \tag{3.56}$$

where $Z^2\epsilon$ is the free electron energy in Rydberg units. Typical radiative recombination coefficients of $H^+$, summed over all bound states, range from $10^{-13}$ at 64,000°K down to $4.8 \times 10^{-12}$ at 250°K. The high bound states are relatively more important than the deep levels at low temperatures than they are at high temperatures because the populated continuum states are the low-momentum states under these conditions.

The high-lying bound states have momenta comparable to that of the populated free states, so that the total transition probability is sizeable for these states at low temperatures.

When the density of ionized gas is high, collisional relaxation processes can play an important role, both in depopulating the lowest level and in establishing equilibrium among the excited states. This problem was treated theoretically by D'Angelo[33] and by Bates, Kingston and McWhirter.[3] If the plasma is optically thin, but collisions are frequent, then the assumption may be made that a steady-state distribution is established for the excited state population, so that the net recombination process involves the sifting of electrons from continuum to the lowest state. Thus,

$$\frac{dn_0(A)}{dt} = -\frac{dn(A^+)}{dt} = \alpha n(e)n(A^+) - Sn_0(A)n(e) \qquad (3.57)$$

The coefficient S is called the collisional-radiative ionization coefficient by Bates, Kingston and McWhirter.[3] The assumption has been made that the upper levels are in thermal equilibrium with the ions and electrons, so that their populations can be computed and from the populations and excitation probabilities from Gryzinski's model, radiative recombination coefficients are computed. The values become sensitive to temperature and electron density, so that for $H^+$ recombination, $\alpha$ is $8.8 \times 10^{-6}$ cm$^3$ sec$^{-1}$ at 600°K when $n(e) = 10^{15}$, drops to $2.7 \times 10^{-10}$ at 4000°K at the same density of electrons, and drops to $1.4 \times 10^{-12}$ if $n(E) = 10^{10}$ and $T = 4000$°K. Bates and Dalgarno[B2] quote an extensive table of $\alpha$.

Experiments by Hinnov and Hirschberg[67] are in partial accord with the above picture, insofar as there is apparently a steady state in which the high-n levels are in equilibrium with the free electrons. Low-n levels are not, it seems. The emission of high-n lines of H and He in an interrupted discharge are more intense when the discharge is off than when it is on; the reverse is true for low-n lines. The high-n levels are therefore presumed to stay in equilibrium with the continuum by three-body recombination, $He^+ + 2e \rightarrow He + e$. For $kT < 0.25$ ev, Hinnov and Hirschberg derive the expression

$$\alpha_c \approx 5.6 \times 10^{-27}(kT)^{-9/2}n(e) \text{ cm}^3 \text{ sec}^{-1} \qquad (3.58)$$

for the case in which upper levels are populated predominantly by the collision process, and

$$\alpha = \alpha_C + \alpha_R \tag{3.59}$$

with $\alpha_R = 2.7 \times 10^{-13} (kT)^{-3/4}$ cm$^3$ sec$^{-1}$ for the case that radiation is significant. Their observed and calculated values of $\alpha$ are in quite good agreement. Typically, in He at a pressure of 10μ, temperature of 2670°K, and n(e) of $3.6 \times 10^{13}$, $\alpha$ was measured to be $7.3 \times 10^{-11}$ and calculated to be $9 \times 10^{-11}$.

Electronic recombination may proceed through steps involving radiationless stabilization. When an atom has captured an electron into a weakly bound or metastable state deactivation processes are obviously competitive with steps leading to re-ionization. The deactivation may be treated as a phenomenon having a single time constant, or may be broken into separate possible steps involving direct state crossing, state crossing followed by radiation (dielectronic recombination) or by collisional deactivation. The relative importance of electronic-collisional and radiative stabilization can be estimated from the empirical relationship between their rates $k_C$ and $k_R$:

$$k_C \sim 10^{-29} \lambda^4 n(e) k_R \tag{3.60}$$

Bates and Dalgarno[B2] quote typical conditions for which the rates are probably comparable:

$$n(e) \sim 4 \times 10^{14} \text{ cm}^{-3}, \text{ for } \lambda = 4000 \text{ Å}$$

and

$$n(e) \sim 2 \times 10^{13} \text{ cm}^{-3} \quad \text{for } \lambda = 8000 \text{ Å}$$

At high pressures, 0.1 atm or more, if atoms like oxygen or the halogens, which have significant electron affinities, are present, then the free electron concentration may be small. Instead, as the discussion in the first section showed, the negative particles are negative ions. When this is the situation, recombination is predominantly accomplished by three-body collisions, as described by the classical Thomson theory, when the pressure is less than 2 atm. At higher pressures, ion association or, alternatively, many-body interactions, require a formulation in terms of mobilities -- the Langevin formulation. These two have been combined into a single unified form by Natanson[94] which is in excellent agreement with the data discussed by Sayers.[B2] Bates has suggested that dissociative recombination also can play an important role in ionic recombination, and such a mechanism is observed in nonattaching gases (c.f. Bates and Dalgarno[B2]). King's experiments,[75,76,77] on the recombination of ions in flames at 1 atm, with temperatures of 1600-2000°K give recombination coefficients of about $10^{-7}$ $cm^3$ $sec^{-1}$, consistent with the three-body or dissociative recombination mechanisms. However, the coefficients increase with T, which is consistent only with the conclusions of the high-pressure Langevin theory, which, traditionally, was not expected to be valid at pressures as low as 1 atm. The Langevin theory predicts a decrease in $\alpha$ with a decrease in mobility. A decrease in mobility is observed when chlorine is added to a flame, and the recombination coefficient does drop. King, in a later chapter, discusses the observations in more detail.

Acknowledgement

The author would like to thank Miss Mary Craig for her part in the computation of equilibrium constants and concentrations of the alkali halides. He would also like to thank Professor Willis Lamb for very helpful discussions regarding the absorption of radiation by molecules, and Professor Kenneth Greider for discussions regarding the Born approximation. The work was supported in part by the Directorate of Chemical Sciences, United States Air Force Office of Scientific Research.

Table 1 Typical Molecular Ionization Potentials (ev) Unless otherwise noted, values are taken from the compilation of Kiser.[79]

| | Photoionization | Electron Impact | Spectroscopic |
|---|---|---|---|
| $CH_2$ | | $11.82 \pm 0.05$[a] | $10.396 \pm 0.003$[b] |
| $CH_3$ | $9.82 \pm 0.04$[c] | $9.86 \pm 0.02$[a] | $9.843 \pm 0.001$[b] |
| $CH_4$ | $12.99 \pm 0.01$ | 13.1 | 12.95 |
| $C_2H_4$ | 10.510-10.57 | 10.80 | 10.507 |
| $C_2H_5$ | 8.4[c] | $8.78 \pm 0.05$ | |
| $C_2H_6$ | 11.65 | 11.8 | |
| $CH_3CH{=}CH_2$ | 9.73[d,e] | 9.81 | 9.6-9.7 |
| $C_2H_5CH{=}CH_2$ | 9.58,[e] $9.61 \pm 0.02$[d] | 9.76 | |
| $C_6H_6$ | 9.245 | 9.3-9.57 | 9.247 |
| $CH_3OH$ | 10.85 | 10.9-10.97 | |
| $CO$ | $13.9 \pm 0.1$, $14.01 \pm 0.01$ | 14.1 | 14.013 |
| $H_2O$ | 12.6 | 12.6-12.7 | 12.56-12.61 |
| $CO_2$ | 13.788 | 13.8-13.9 | 13.78 |

a See Ref. 24.

b See Ref. 66.

c See Ref. 42.

d See Ref. 125.

e See Ref. 110.

## References

Books and Reviews

B1 Allison, S. K., Rev. Mod. Phys. 30, 1137-1168 (1958).

B2 Bates, D. R., ed., Atomic and Molecular Processes (Academic Press Inc., New York, 1962).

B3 Bradley, J. N., Shock Waves in Chemistry and Physics (John Wiley and Sons Inc., New York, 1962).

B4 Burke, P. G. and Smith, K., Rev. Mod. Phys. 34, 458-502 (1962).

B5 Craggs, J. D. and Massey, H.S.W., Handbuch der Physik, Vol. 37 (S. Flügge, ed.) (Springer Verlag, Berlin, 1959).

B6 Field, F. H. and Franklin, J. L., Electron Impact Phenomena and the Properties of Gaseous Ions (Academic Press Inc., New York, 1957).

B7 Gaydon, A. G. and Wolfhard, H. G., Flames, Their Structure, Radiation and Temperature (Chapman and Hall Ltd., London, 1960), esp. Chapter XIII.

B8 Gerjuoy, E., Rev. Mod. Phys. 33, 544-552 (1961).

B9 Massey, H.S.W. and Burhop, E.H.S., Electronic and Ionic Impact Phenomena (Oxford University Press, London, 1952).

B10 Waldron, J. D., ed., Advances in Mass Spectrometry (Pergamon Press Inc., New York, 1959).

Papers

1 Baker, C. W., McGuire, J. M., and Muschlitz, E. E., J. Chem. Phys. 37, 2571-2574 (1962).

2 Bates, D. R., and Carson, T. R., Proc. Phys. Soc. (London) A68, 1199-1202 (1955).

3 Bates, D. R., Kingston, A. E., and McWhirter, R.W.P., Proc. Roy. Soc. (London) A267, 297-312 (1962).

4 Bayes, K. D., Kivelson, D., and Wong, S. C., J. Chem. Phys. 37, 1217-1225 (1962).

5 Ben Daniel, D. J. and Bishop, C. M., Phys. of Fluids 6, 300-306 (1963).

6 Berry, R. S., J. Chem. Phys. 27, 1288-1295 (1957).

7 Berry, R. S. and David, C. W., Proc. Third Intl. Conf. on Physics of Electronic and Atomic Collisions (North Holland, 1963).

8 Berry, R. S. and Reimann, C. W., J. Chem. Phys. 38, 1540-1543 (1963).

9 Berry, R. S., Reimann, C. W. and Spokes, G. N., J. Chem. Phys. 35, 2237-2238 (1961).

10 Berry, R. S., Reimann, C. W. and Spokes, J. Chem. Phys. 37, 2278-2290 (1962).

11 Boelrijk, N. and Hamill, W. H., J. Am. Chem. Soc. 84, 730-739 (1962).

12 Boldt, G., Z. Physik. 154, 319-329, 330-338 (1958).

13 Bonham, R. A., J. Chem. Phys. 36, 3260-3269 (1962).

14 Bowe, J. C., Phys. Rev. 117, 1411-1415, 1416-1420 (1960).

15 Boyd, R.L.F. and Boksenberg, A., Proc. Fourth Intl. Conference on Ionization Phenomena in Gases, Uppsala, 1959 (North Holland Publishing Company, Amsterdam, 1960) Vol. I, pp. 529-532.

16 Brunner, J., Z. Physik. 159, 288-310 (1960).

17 Bulewicz, E. M., J. Chem. Phys. 36, 385-391 (1962).

18 Bulman, J. B. and Russek, A., Phys. Rev. 122, 506-511 (1961).

19 Calcote, H. F. and I. R. King, Fifth Symposium (International) on Combustion (Reinhold Publishing Corp., New York, 1955) p. 423-434.

20 Calcote, H. F. and Reuter, J. L., J. Chem. Phys. 38, 310-317 (1963).

[21] Chanin, L. M., Phelps, A. V. and Biondi, M. A., Phys. Rev. 128, 219-230 (1962).

[22] Chen, J.C.Y., Phys. Rev. 129, 202-210 (1963).

[23] Chen, J.C.Y. and Magee, J. Chem. Phys. 36, 1407-1411 (1962).

[24] Clarke, E.W.C. and McDowell, C. A., Proc. Chem. Soc. 1960, 69 (1960).

[25] Cochran, L. W. and Forester, D. W., Phys. Rev. 126, 1785-1788 (1962).

[26] Coggeshall, N. D., J. Chem. Phys. 37, 2167-2175 (1962).

[27] Conway, D., J. Chem. Phys. 36, 2549-2557 (1962).

[28] Cooper, J. W., Phys. Rev. 128, 681-693 (1962).

[29] Cooper, J. W. and Martin, J. B., Phys. Rev. 126, 1482-1488 (1962).

[30] Curran, R. K., Phys. Rev. 125, 910-911 (1962).

[31] Curran, R. K., J. Chem. Phys. 38, 780-781 (1963).

[32] Curran, R. K. and Fox, R. E., J. Chem. Phys. 34, 1590-1594 (1961).

[33] D'Angelo, N., Phys. Rev. 121, 505-507 (1961).

[34] Dahler, J. S., Franklin, J. L., Munson, M.S.B. and Field, F. H., J. Chem. Phys. 36, 3332-3344 (1962).

[35] De Jaegere, S. Deckers, J. and van Tiggelen, A., Eighth Symposium (International) on Combustion (Williams and Wilkins Co., Baltimore, 1962) p. 155-160.

[36] Dolder, K. T., Harrison, M.F.A. and Thoneman , P. C. Proc. Roy. Soc. (London) A264, 367-378 (1961).

[37] Donahue, T. M. and Hushfar, F., Phys. Rev. 124, 138-144 (1962).

[38] Dorman, F. H. and Morrison, J. D., J. Chem. Phys. 34, 578-582 (1961).

39 Dorman, F. H. and Morrison, J. D., J. Chem. Phys. 35, 575-581 (1961).

40 Drawin, H. W., Z. Physik 164, 513-521 (1961).

41 Edelson, D., Griffiths, J. E. and McAfee, K. B., Jr., J. Chem. Phys. 37, 917-918 (1962).

42 Elder, F. A., Giese, C. Steiner, B. and Inghram, M., J. Chem. Phys. 36, 3292-3296 (1962).

43 Elwert, G., Z. Naturforsch. 7a, 432-439 (1952).

44 Engelmann, F., Z. Physik. 169, 126-140 (1962).

45 Fedorenko, N. V., Uspekhi Fiz. Nauk 68, 481-511 (1959). Translation, Soviet Physics Uspekhi 2, 526-546 (1959).

46 Field, F. H., J. Am. Chem. Soc. 83, 1523-1534 (1961).

47 Firsov, O. B., Zh. Eksper. teor. Fiz. 36, 1517-1523 (1959); Translation, Soviet Physics - JETP 9, 1076-1080 (1959).

48 Fite, W. L. and Brackmann, R. T., Phys. Rev. 112, 1141-1151 (1958).

49 Fite, W. L. and Brackmann, R. T., Phys. Rev. 113, 815-816 (1959).

50 Fite, W. L., Smith, A.C.H. and Stebbings, R. F., Proc. Roy. Soc. (London) A268, 527-536 (1962).

51 Foner, S. N. and Nall, B. H., Phys. Rev. 122, 512-524 (1961).

52 Fontijn, A. and Baughman, G. L., J. Chem. Phys. 38, 1784-1785 (1963).

53 Fox, R. E., J. Chem. Phys. 35, 1379-1382 (1961).

54 Franken, P., Sands, R. and Hobart, J., Phys. Rev. Letters 1, 52-54 (1958).

55 Geltman, S., Phys. Rev. 102, 171-179 (1956).

56 Geltman, S. and Krauss, M., Bull. Am. Phys. Soc. [27] 5, 339 (1960).

57 Geltman, S., Rudge, M.R.H. and Seaton, M. J., Proc. Phys. Soc. (London) 81, 375-378 (1963).

58 Gioumousis, G. and Stevenson, D. P., J. Chem. Phys. 29, 294-299 (1958).

59 Gryzinski, M., Phys. Rev. 115, 374-383 (1959).

60 Groening, H., Phys. Fluids 6, 142-144 (1963).

61 Haas, R., Z. Physik 148, 177-191 (1957).

62 Hand, C. W. and Kistiakowsky, G. B., J. Chem. Phys. 37, 1239-1245 (1962).

63 Hasted, J. B. and Lee, A. R., Proc. Phys. Soc. 79, 702-709 (1962).

64 Henning, H., Z. Physik 169, 467-474 (1962).

65 Hertzberg, M., Rapp, D., Ortenberger, I. B. and Briglia, D. D., J. Chem. Phys. 34, 343-344 (1961).

66 Herzberg, G., Proc. Roy. Soc. (London) 262, 291-317 (1961).

67 Hinnov, E. and Hirschberg, J. G., Phys. Rev. 125, 795-801 (1962).

68 Hooper, J. W., Harmer, D. S., Martin, D. W. and McDaniel, E. W., Phys. Rev. 125, 2000-2004 (1962).

69 Hurst, G. S., O'Kelly, L. B. and Bortner, T. E., Phys. Rev. 123, 1715-1718 (1961).

70 Inokuti, M., J. Chem. Phys. 37, 2387-2393 (1962).

71 John, T. L., Month. Not. Roy. Astron. Soc. 121 (1), 41-47 (1960).

72 John, T. L., Astrophys. J. 131, 743-744 (1960).

73 Kebarle, P. and Godbole, E. W., J. Chem. Phys. 36, 302-305 (1962).

74 King, I. R., J. Chem. Phys. 31, 855 (1959).

[75] King, I. R., J. Chem. Phys. 35, 380 (1961).

[76] King, I. R., J. Chem. Phys. 36, 553-554 (1962).

[77] King, I. R., J. Chem. Phys. 37, 74-80 (1962).

[78] Kiser, R. W., J. Chem. Phys. 36, 2964-2968 (1962).

[79] Kiser, R. W., U. S. Atomic Energy Commission Office of Technical Information TlD-6142 (June 1960).

[80] Kistiakowsky, G. B. and Richards, L. W., J. Chem. Phys. 36, 1707-1714 (1962).

[81] Knewstubb, P. F. and Sugden, T. M., Proc. Roy. Soc. (London) A255, 520-537 (1960).

[82] von Koch, H. and Friedman, L., J. Chem. Phys. 38, 1115-1122 (1963).

[83] Kubose, D. A. and Hamill, W. H., J. Am. Chem. Soc. 85, 125-130 (1963).

[84] Lee, A. R. and Hasted, J. B., Proc. Phys. Soc. 79, 1049-1057 (1962).

[85] Lennan, J. J. and Mulcahy, M. J., Proc. Phys. Soc. 78, 1543-1545 (1961).

[86] Lippmann, B. A., Mittleman, M. H. and Watson, K. M., Phys. Rev. 116, 920-925 (1959).

[87] Lochte-Holtgreven, W., Naturwiss. 38, 258-259 (1951).

[88] Lockwood, G. J. and Everhart, E., Phys. Rev. 125, 567-572 (1962).

[89] Melton, C. W., J. Chem. Phys. 37, 562-566 (1962).

[90] Meyerson, S., J. Chem. Phys. 34, 2046-2049 (1961).

[91] Mies, F. H. and Shuler, K. E., J. Chem. Phys. 37, 177-181 (1962).

[92] Morrison, J. D., Hurzeler, H., Inghram, M. G. and Stanton, H. E., J. Chem. Phys. 33, 821-824 (1960).

93 Munson, M.S.B., Field, F. H. and Franklin, J. L., J. Chem. Phys. 37, 1790-1799 (1962).

94 Natanson, G. L., Zh. Tekh. Fiz. 29, 1373 (1959); in Russian.

95 Novick, R. and Peters, H. E., Phys. Rev. Letters 1, 54-55 (1958).

96 Neynaber, R. H., Marino, L. L., Rothe, E. W. and Trujillo, S. M., Phys. Rev. 129, 2069-2071 (1963).

97 Peterkop, R. K., Zh. Eksperim. Teoret. Fiz. 41, 1938-1939 (1962); Translation, Soviet Physics - JETP 14, 1377-1378 (1962).

98 Petschek, H. E., Rose, P. H., Glick, H. S., Kane, A. and Kantrowitz, A., J. App. Phys. 26, 83-95 (1955).

99 Popescu, I.V., Proc. Phys. Soc. 78, 584-587 (1961).

100 Pottie, R. F., Lorquet, A. J. and Hamill, W. H., J. Am. Soc. 84, 529-531 (1962).

101 Rapp, D. and Francis, W. E., J. Chem. Phys. 37, 2631-2645 (1962).

102 Reuben, B. G. and Friedman, L., J. Chem. Phys. 37, 1636-1642 (1962).

103 Robinson, L. B., Phys. Rev. 117, 1281-1283 (1960).

104 Rosenstock, H. M., J. Chem. Phys. 34, 2182-2183 (1961).

105 Roth, W., J. Chem. Phys. 37, 925-926 (1962).

106 Rothe, E. W., Marino, L. L., Neynaber, R. H. and Trujillo, S. M., Phys. Rev. 125, 582-583 (1962).

107 Rubin, K., Perel, J. and Bederson, B., Phys. Rev. 117, 151-158 (1960).

108 Russek, A. and Thomas, M. T., Phys. Rev. 109, 2105-2025 (1958).

109 Russek, A. and Thomas, M. T., Phys. Rev. 114, 1538-1540 (1959).

110 Samson, J.A.R., Marmo, F. F. and Watanabe, K., J. Chem. Phys. 36, 783-786 (1962).

111 Schoen, R. I., J. Chem. Phys. 37, 2032-2040 (1962).

112 Schüler, H., Reinebeck, L. and Michel, A., Z. Naturforsch. 9a, 279-285 (1954).

113 Shuler , K. E. and Weiss, G. H., J. Chem. Phys. 38, 505-509 (1963).

114 Schulz, G. J., Phys. Rev. 116, 1141-1147 (1959).

115 Schulz, G. J., J. Chem. Phys. 34, 1778-1781 (1961).

116 Schulz, G. J., Phys. Rev. 125, 229-232 (1962).

117 Schulz, G. J., Phys. Rev. 128, 178-186 (1962).

118 Schulz, G. J., Phys. Rev. Letters 10, 104-105 (1963).

119 Schulz, G. J. and Dowell, J. T., Phys. Rev. 128, 174-177 (1962).

120 Seaton, M. J., Phys. Rev. 113, 814 (1959).

121 Seman, M. L. and Branscomb, Phys. Rev. 125, 1602-1608 (1962).

122 Smith, S. J. and Burch, D. S., Phys. Rev. 116, 1125-1131 (1959).

123 Smith, A.C.H., E. Caplinger, Neynaber, R. H., Rothe, E. W. and Trujillo, S. M., Phys. Rev. 127, 1647-1649 (1962).

124 Spruch, L., O'Malley, T. F. and Rosenberg, L., Phys. Rev. Letters 5, 375-377 (1960).

125 Steiner, B., Giese, C. F. and Inghram, M. G., J. Chem. Phys. 34, 189-220 (1961).

126 Steiner, B., Seman, M. L. and Branscomb, L. M., J. Chem. Phys. 37, 1200-1202 (1962).

127 Stevenson, D. P. and Schissler, D. O., J. Chem. Phys. 29, 282-294 (1958).

[128] Theard, L. P. and Hamill, W. H., J. Am. Chem. Soc. 84, 1134-1139 (1962).

[129] Theimer, O. and Gentry, R., Phys. Rev. 116, 787-792 (1959).

[130] Tuan, T. F. and Gerjuoy, E., Phys. Rev. 117, 756-763 (1960).

[131] Turner, D. W. and AlJobory, M. I., J. Chem. Phys. 37, 3007-3008 (1962).

[132] Utterback, N. G., Phys. Rev. 129, 219-224 (1963).

[133] Utterback, N. G. and Miller, G. H., Phys. Rev. 124, 1477-1481 (1961).

[134] Vestal, M., Wahrhaftig, A. L. and Johnston, W. H., J. Chem. Phys. 37, 1276-1283 (1962).

[135] Weber, O., Z. Phys. 152, 281-293 (1958).

[136] Ziemba, F. P., Lockwood, G. J., Morgan, G. H. and Everhart, E., Phys. Rev. 118, 1552-1561 (1960).

[137] Zwanzig, R. and Schuler, K. E., J. Chem. Phys. 33, 1778-1784 (1960).

Fig. 1 Concentration of atoms, ions and electrons in typical alkali halides as functions of T at constant total dissociated salt concentration. A. RbCl, in which the dissociated salt concentration is $10^{16}$ cm-3. (Note that the scale of the chloride ion concentration is one-tenth of the others.) B. KCl, in which the dissociated salt concentration is 2 x 1017 cm-3. (Note that the scale of the $K^{\circ}$ concentration is ten times that of the others.) The higher ion density and higher dissociation energy of B lead to the maximum for the chloride concentration, in the region of 3400-3500°K.

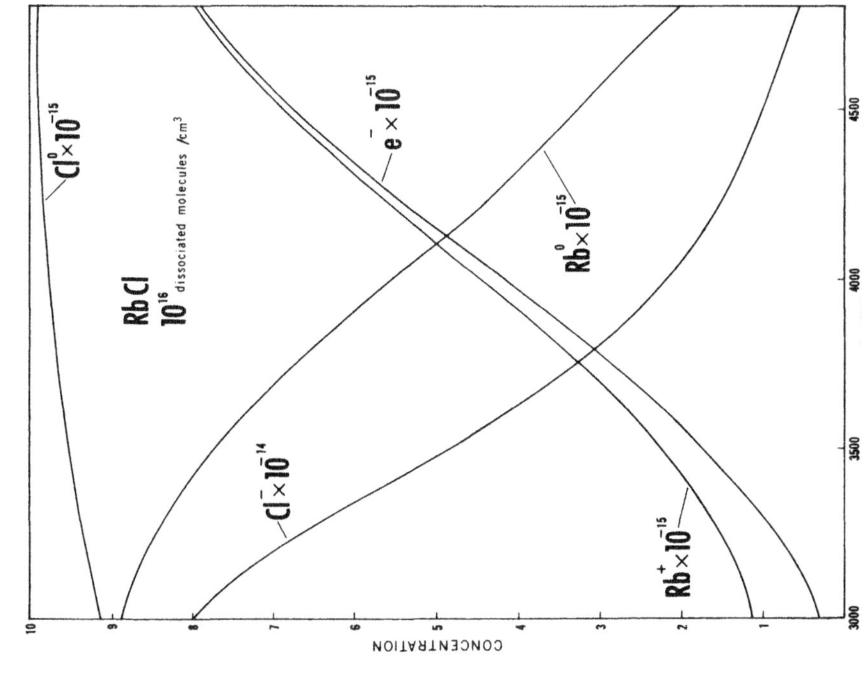
RbCl
$10^{16}$ dissociated molecules /cm$^3$
$Cl^0 \times 10^{-15}$
$e^- \times 10^{-15}$
$Rb^0 \times 10^{-15}$
$Cl^- \times 10^{-14}$
$Rb^+ \times 10^{-15}$
CONCENTRATION
T (°K)
3000
3500
4000
4500

(1a)

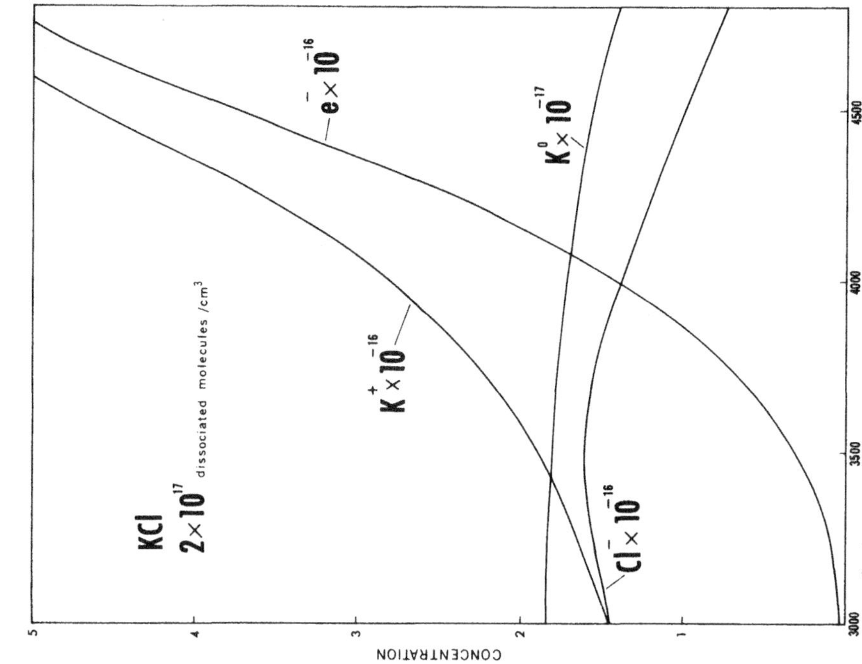
KCl
$2 \times 10^{17}$ dissociated molecules /cm$^3$
$e^- \times 10^{-16}$
$K^0 \times 10^{-17}$
$K^+ \times 10^{-16}$
$Cl^- \times 10^{-16}$
CONCENTRATION
T (°K)
3000
3500
4000
4500

(1b)

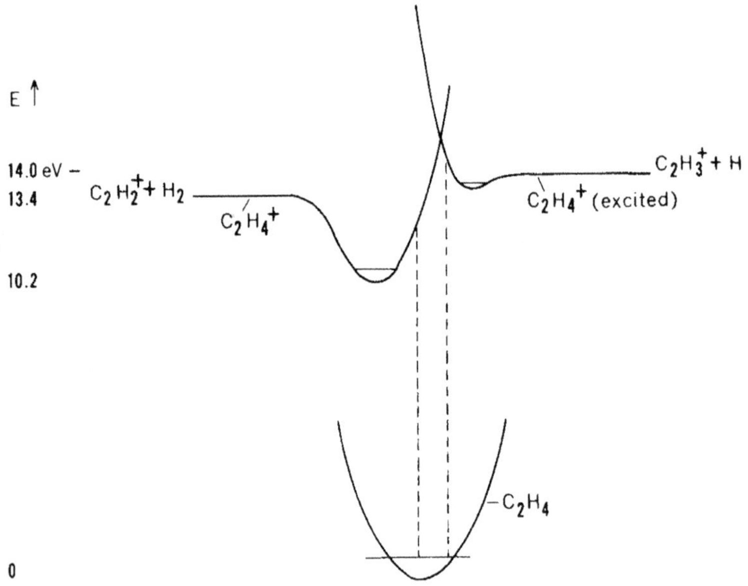


Fig. 2 Potential surfaces for a molecule and molecule ion. The illustration is hypothetical, but is drawn to be consistent with the absorption cross section, photoionization cross section and mass spectrum of ethylene as measured by Schoen.[111] The abscissa is an unspecified displacement coordinate on the vibrational potential surface.

IONIZATION IN HIGH-TEMPERATURE GASES

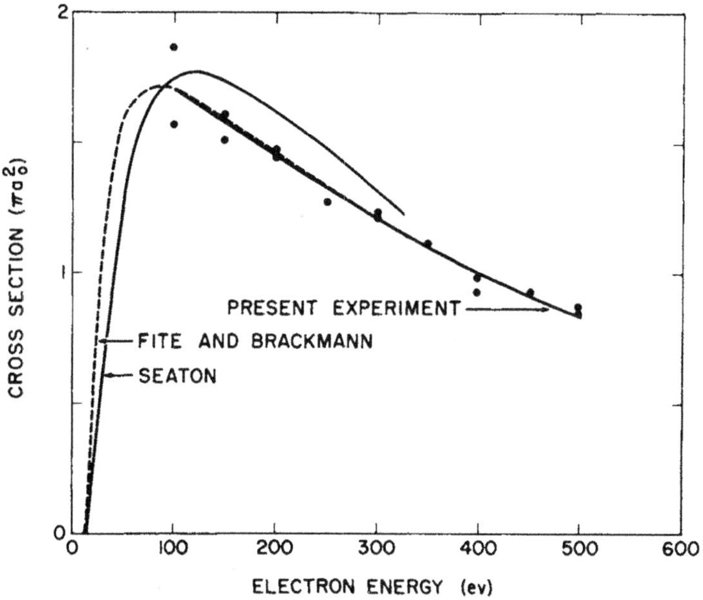


IONIZATION CROSS SECTIONS FOR ATOMIC OXYGEN

Fig. 3 Ionization cross section of oxygen by electron impact. The curve marked "Seaton" is calculated from a semi-empirical theory; the other two are experimental. "Present Experiment" refers to Ref. 106, and "Fite and Brackmann" to Ref. 49. (Reproduced by courtesy of Rothe, Marino, Neynaber and Trujillo[106] and the Physical Review.)

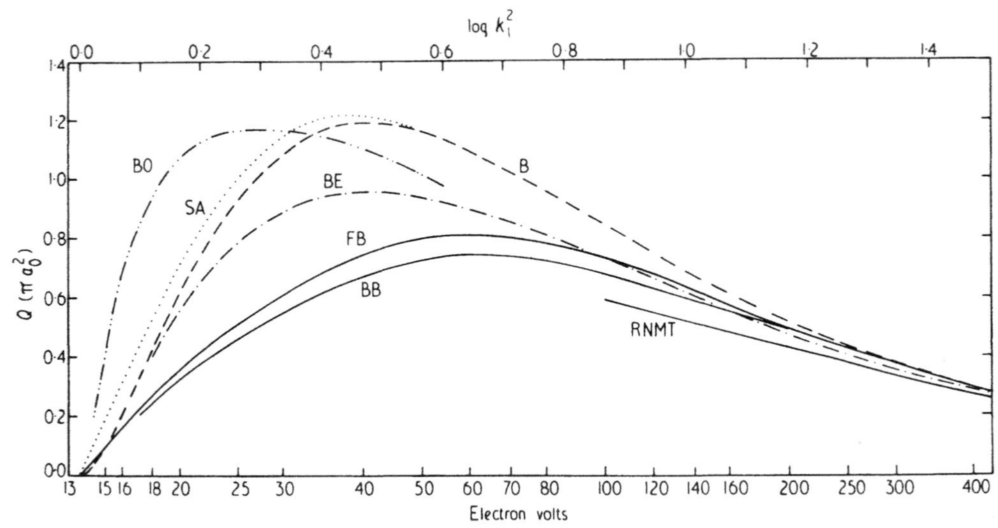


Fig. 4 Ionization of hydrogen by electron impact. The curves are as follows: BO, Born-Oppenheimer (Born approximation with some allowance for exchange); SA, spherically-averaged shielded Coulomb potential and Coulombic final state function; BE, Born approximation with exchange; B, Born approximation; FB, experimental curve of Fite and Brackmann; BB, experimental curve of Boyd and Boksenberg; RNMT, experimental curve of Rothe, Neynaber, Marino and Trujillo. (Reproduced by courtesy of Geltman, Ruege and Seaton,[57] the Institute of Physics and the Physical Society.)

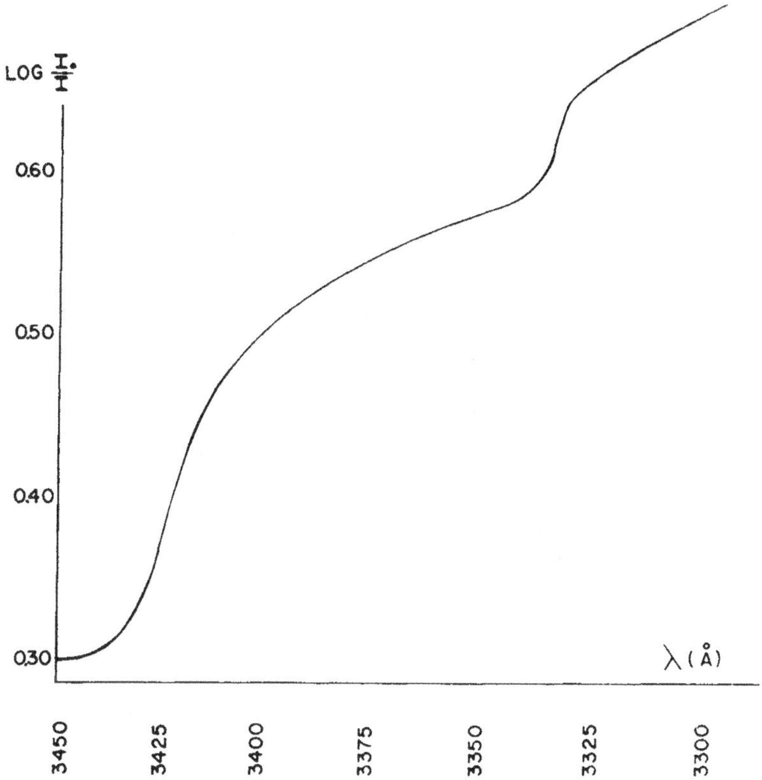


Fig. 5 Relative photodetachment cross section of $Cl^-$ in the region near threshold. The two breaks in the curve correspond to the two final states $^2P_{3/2}$ and $^2P_{1/2}$ available to the chlorine atom. The absolute cross section is about $1.5 \times 10^{-16}$ in the region 3350-3400 Å. (Reproduced by courtesy of Berry, Reimann and Spokes[10] and the Journal of Chemical Physics.)

# CHEMI-IONIZATION AND ION-MOLECULE REACTIONS IN GASES

J. L. Franklin,* M. S. B. Munson,† and F. H. Field‡

Humble Oil and Refining Company, Baytown, Texas

## Abstract

At elevated pressures (up to 500μ) in the ion source of a mass spectrometer, ions having different masses from those in the primary spectrum are observed, often in great abundance. These secondary ions result from reaction of primary ions or electronically excited atoms or molecules with the neutral, ground-state species present. Many second-order ion-molecule reactions are known and some reactions of higher order have been reported, a few being of order as high as six. Chemi-ionization reactions of the rare gases both with themselves and with other materials have been studied extensively and chemi-ionization reactions of a few simple gases are known. Both ion-molecule and chemi-ionization reactions usually exhibit high rates.

## Introduction

At the low pressures ($10^{-6}$mm) normally employed in the ion source of a mass spectrometer, the only ions observable are those having masses equal to or less than those of the molecules present in the source. As the pressure is increased, the intensity of the various primary ions increases proportionally. At sufficiently high pressures other ions begin to appear, usually with masses above that of the molecules present. These higher mass ions usually increase with pressure at a rate greater than the first power of pressure and thus are the result of collision processes in the source. Although it occasionally is possible to show that these result from thermal reactions on the filament or walls, this is usually not the case. In fact, such product ions in most instances result from reactions of the neutral species present with either ions or

---

Presented at the ARS Ions in Flames and Rockets Exhausts Conference, Palm Springs, Calif., October 10-12, 1962.

*Research Associate, Research and Development.

†Senior Research Chemist, Research and Development.

‡Senior Research Specialist, Research and Development.

electronically excited atoms or molecules. This paper will attempt to summarize recent developments in the knowledge of ion-molecule reactions and chemi-ionization processes.

## Ion-Molecule Reactions

Hogness and Harkness[1] in 1928 reported a brief study of secondary processes involving iodine and identified the following reactions:

$$I_2^+ + I_2 \rightarrow I_3^+ + I \quad (1)$$

$$I_2^- + I_2 \rightarrow I_3^- + I \quad (2)$$

Somewhat later, Mann, Hustrulid, and Tate,[2] in studying water in a mass spectrometer, observed the formation of $H_3O^+$ as a secondary product and attributed it to the reaction

$$H_2O^+ + H_2O \rightarrow H_3O^+ + OH \quad (3)$$

In addition, investigators studying ion mobilities have attributed certain of their results to ion-molecule reactions (see, for example, Loeb[3]). Recent studies by Tal'roze and Lyubimova,[4] Stevenson and Schissler,[5] Field, Franklin, and Lampe,[6] and Meisels, Hamill, and Williams,[7] demonstrating the existence of interesting and hitherto unsuspected ions, such as $CH_5^+$, $H_2Cl^+$, rare gas hydride ions, etc., have lead to increasing interest in ion-molecule reactions and initiated the now rather voluminous literature on the subject.

## Experimental Methods

With the appearance of an ion having a mass greater than that of the molecule present in the source, several problems arise if one is to understand the nature and source of the ion. First, it is necessary to establish whether the ion is indeed a secondary product, i.e., whether it exhibits a greater than first-order dependence upon pressure. This is done by varying the pressure and observing the variation in the intensity of the ion in question. It is readily shown that the number of ionizing collisions experienced by a beam of electrons in passing through a dilute gas is

$$I_i = I_{e_o}(1 - e^{-\sigma_i (M)\ell}) \quad (4)$$

where $I_i$ and $I_{e_o}$ are, respectively, the current of ions formed and the initial electron current, $\sigma_i$ is the cross section for formation of the ion, $\ell$ the path length, and (M) the contentration of molecules undergoing ionization. At sufficiently low pressures Eq. (4) reduces to

$$I_i \; + \; I_{e_o} \sigma_i (M) \ell \qquad (5)$$

and the ion intensity increases as the first power of the pressure.

For a secondary ion resulting from a reaction such as

$$P^+ + M \rightarrow S^+ + N \qquad (6)$$

one may write

$$d(S^+)/dt = k(M)(P^+) \qquad (7)$$

which integrates to

$$(S^+) = (P_o^+)(1-e^{-k(M)t}) \qquad (8)$$

Taking ion currents to be proportional to ion concentration, one obtains

$$I_{S^+} = I_{P_o^+} (1-e^{-k(M)t}) \qquad (9)$$

At low pressures, Eq. (9) reduces to

$$I_{S^+} = I_{P_o^+} k(M)t \qquad (10)$$

which, in view of Eq. (5), shows that $I_{S^+} \propto M^2$. Thus, the product of such an ion-molecule reaction will increase with the second power of the pressure. In fact, some product ions have been observed which exhibited as much as sixth-order dependence upon pressure.

In order to identify the reaction leading to a given secondary ion, principal dependence is usually placed upon determining the appearance potential of the secondary ion and matching it to appearance potentials of various primary ions in the mass spectrum of the compound present in the source. Usually a reasonably satisfactory match can be obtained, and from this one ordinarily can deduce a reasonable reaction. For example, with $CD_4$ present in the source, one finds secondary ions of great abundance at masses 22 and 34 corresponding to $CD_5^+$ and $C_2D_5^+$. Comparisons of the appearance potentials of these ions with those of the known primary ions of methane are given in Table 1. It is quite evident that $CD_4^+$ is the precursor of $CD_5^+$, and that $CD_3^+$ is the precursor of $C_2D_5^+$. The corresponding reactions are

$$CD_4^+ + CD_4 \rightarrow CD_5^+ + CD_3 \qquad (11)$$

$$CD_3^+ + CD_4 \rightarrow C_2D_5^+ + D_2 \qquad (12)$$

Stevenson and Schissler[5,8] in an early study varied the temperature in the mass spectrometer and showed that the several reactions investigated exhibited a slight negative temperature dependence, indicating that none involved an activation energy. Although this is the only study of the temperature dependence of ion-molecule reactions thus far reported, it is readily shown that those reactions observable in the mass spectrometer cannot involve significant amounts of activation energy. In fact, at the highest pressure yet used in studies of secondary reactions in mass spectrometer sources, one can easily show that, with the very small retention times and modest pressures (up to 500 μ), a reaction having an activation energy greater than 4 or 5 kcal/mole would be too slow to give observable products. As a result of Stevenson's studies and the very obvious limitations just discussed, it is now generally conceded that endothermic ion-molecule reactions simply do not occur, at least at the conditions normally existing in mass spectrometer sources. Further, for all of the now rather considerable number of such reactions that have been studied, there are no authentic instances of endothermic reactions. Thus, the heat of reaction has become a useful criterion of the possibility of a reaction's occurring.

## Some Observable Reactions

By this time several hundred ion-molecule reactions have been reported. Several reviews[9-12] of ion-molecule reactions have been published, and these can be referred to for complete and detailed reports. Several types of reactions have been observed, however, and a few examples of each are given in Table 2. It will be observed that there are reactions involving exchange of H atoms, protons, and $H^-$ ions, as well as symmetrical processes for which it is not possible to decide upon the entity transferred. There are also examples of reactions involving profound changes in the carbon skeleton. Although most of the reactions studied involve positive ions, there are a few examples of ion-molecule reactions of negative ions. It need not be implied that reactions of negative ions are less extensive than those of positive ions. Rather, they are more difficult to investigate, and only a limited number of studies have been made.

Table 2 shows no examples of reactions in which an ion and a neutral molecule simply condensed without decomposition. Such reactions would necessarily involve the formation of a product ion with excess energy and indeed with sufficient energy to decompose to the original reactants. For isolated systems such as one ordinarily encounters at the pressures employed in these studies, there is usually ample time for such an energetic complex to decompose, and thus such complexes usually are not observable. However, a few examples are known in which such complexes are observed. These usually have exhibited a pressure dependence of third order or higher and thus appear to result from stabilizing collisions between the complex and neutral species present in the source. For example, Field[13] has observed the reaction

$$C_2H_4^+ + 2C_2H_4 \rightarrow C_4H_8^+ + C_2H_4 \qquad (13)$$

Rather recently, however, some authentic examples of energetic complexes exhibiting second order pressure dependence have been reported by Pottie and Hamill[14] and by Henglein[15].

$$C_2H_5Br^+ + C_2H_5Br \rightarrow C_4H_{10}Br_2^+ \qquad (14)$$

$$C_3H_7I^+ + C_3H_7I \rightarrow C_6H_{14}I_2^+ \qquad (15)$$

$$C_3H_3N^+ + C_3H_3N \rightarrow C_6H_6N_2^+ \qquad (16)$$

$$C_6H_5^+ + C_6H_6 \rightarrow C_{12}H_{11}^+ \qquad (17)$$

It is evident that these result from reactions of rather complex systems. There are numerous degrees of freedom to which excess energies of reactions can be rapidly distributed.

## Kinetics

Since the mass spectrometer presents an opportunity for measuring the variation of product ions with variations in intensity of primary ions and in the concerntration of neutral reactants, it should be possible to measure the rate of reaction, and, indeed, the literature presents a number of such studies. For a simple second-order reactions, $P^+ + M \rightarrow S^+ + N$, it has been shown that

$$I_{S^+}/I_{P_o^+} = 1 - e^{-k(M)t} \qquad (18)$$

When only a small amount of primary ion disappears, this simplifies to

$$I_{S^+}/I_{P_o^+} = k(M)t \qquad (19)$$

This condition normally holds at ion source pressure in the order of 10 to 50 μ. A plot of the ratio of intensity of secondary to primary ions against concentration of neutral reactant will give a straight line of slope kt. If t can be determined, then k is readily calculated.

The electron beam passes through the ion source of a mass spectrometer in a narrow path parallel to the ion exit slit and to the repeller electrode. To a first approximation, it may be assumed that the thickness of the electron beam is negligible, and, consequently, that ions are formed in a plane parallel to both the exit slit and the repeller. Further, it may be assumed that the thermal velocities of the ions at the time of ionization are distributed randomly in direction. From these considerations one can readily deduce the average time spent by the primary ion traveling from the point of formation to the exit slit to be

$$\tau = (2Md/eE)^{1/2} \tag{20}$$

where $\tau$ is the average time of transit of the ion to the exit slit, d is the perpendicular distance from the center of the electron beam to the exit slit, e is the unit electric charge, and E is the field strength in volts/centimeter. Space charge and penetration of external fields through the slits usually can be neglected. When this is true, the field strength may be taken to be the voltage applied to the repeller electrode divided by the distance from the repeller electrode to the exit slit. Thus, from the dimensions of the source, the applied voltage, and the mass of the primary ion in question, the average residence time of the ion in the source can be calculated. In this source, this is usually in the order of $10^{-6}$ sec. Since the time can be calculated, one may calculate the rate constants for the secondary reaction, and this has been done in a number of instances. Table 3 presents several rate constants determined this way.

It should be pointed out that for many reactions the rate constant is a function of the mean relative velocity of the reacting ion and neutral molecule. The rate constant, in general, becomes smaller as the relative velocity increases. Certain classes of reactions appear to be independent of relative velocities, and these have been treated successfully by Gioumousis and Stevenson.[16] An empirical method of treating rate constants that are velocity-dependent has been proposed by Field, Franklin, and Lampe[6,10] and a rather satisfactory phenomenological treatment of such reactions has been developed by Boelrijk and Hamill.[17] It is beyond the scope of this paper to deal with these problems of rate phenomena in detail. In most instances, however, the variation of rate constants from thermal velocity to about 10 ev of energy is less than an order of magnitude.

Rate constants for most ion-molecule reactions tend to be in the order of $10^{-10}$ to $10^{-9}$ cc/molecule sec. The upper range of these values is such as to correspond to reaction cross sectionsin some instances 10 to 20 times that of normal collision cross sections.

There are many examples of reactions in which several different ionic products arise from the same set of reactants. For example,

$$C_2H_4^+ + C_2H_4 \rightarrow (C_4H_8^+) \rightarrow C_3H_5^+ + CH_3 \quad (21a)$$
$$\phantom{C_2H_4^+ + C_2H_4 \rightarrow (C_4H_8^+)} \rightarrow C_4H_7^+ + H \quad (21b)$$

$$C_2H_2^+ + C_2H_4 \rightarrow (C_4H_6^+) \rightarrow C_3H_3^+ + CH_3 \quad (22a)$$
$$\phantom{C_2H_2^+ + C_2H_4 \rightarrow (C_4H_6^+)} \rightarrow C_4H_5^+ + H \quad (22b)$$

$$C_4H_6^+ + C_4H_6 \rightarrow (C_8H_{12}^+) \rightarrow C_5H_6^+ + C_3H_6 \quad (23a)$$
$$\rightarrow C_5H_7^+ + C_3H_5 \quad (23b)$$
$$\rightarrow C_6H_6^+ + C_2H_6 \quad (23c)$$
$$\rightarrow C_6H_7^+ + C_2H_5 \quad (23d)$$
$$\rightarrow C_6H_8^+ + C_2H_4 \quad (23e)$$

In most instances the complex, as has been mentioned, is not detectable, and so the rate constant that one measures is limited by the rate of formation of the complex rather than by the rate of decomposition. When this is taken into account, one finds that, to a satisfactory approximation, the rate constant for formation of the complex is rather well predicted by the equation first developed by Eyring, Hirschfelder, and Taylor,[18]

$$k = 2e\pi(\alpha/\mu)^{1/2}$$

where e is the charge on the electron, $\alpha$ is the polarizability of the neutral reactant, and $\mu$ is the reduced mass of the complex.

In many instances, no doubt, the absence of higher order products suggests that the rate of decomposition of the complex is quite high, so that the complex decomposes before it can encounter a second neutral molecule. However, there are many examples, as were mentioned previously, in which substantial reactions are observed with third-order and higher dependence upon pressure. In some instances these reactions have been shown rather conclusively to result from further encounters of the complex with additional neutral species, rather than from reaction of the secondary product ion with other neutrals. Field[13], for example, has studied the behavior of ethylene at high pressures (up to 500 μ) and found several important ions to be produced by third-order reactions. He showed that $C_2H_4^+$ is the precursor of $C_4H_8^+$ and $C_5H_9^+$, both of which are formed by third-order reactions. The reactions could go by one or two paths.

Path 1:

$$C_2H_4^+ + C_2H_4 \rightarrow (C_4H_8^+) \begin{cases} \rightarrow C_3H_5^+ + CH_3 & (25a) \\ \rightarrow C_4H_7^+ + H & (25b) \end{cases}$$

$$C_3H_5^+ + C_2H_4 \rightarrow (C_5H_9^+) \begin{cases} \rightarrow C_4H_8^+ + CH & (25c) \\ \rightarrow C_5H_9^+ & (25d) \end{cases}$$

$$C_4H_7^+ + C_2H_4 \rightarrow (C_6H_{11}^+) \begin{cases} \rightarrow C_4H_8^+ + C_2H_3 & (25e) \\ \rightarrow C_5H_9^+ + CH_2 & (25f) \end{cases}$$

Path 2:

$$C_2H_4^+ + C_2H_4 \rightarrow (C_4H_8^+) \rightarrow C_3H_5^+, \text{ etc.} \qquad (26a)$$

$$(C_4H_8^+) + C_2H_4 \rightarrow (C_6H_{12}^+) \begin{cases} \rightarrow C_4H_8^+ + C_2H_4 & (26b) \\ \rightarrow C_5H_9^+ + CH_3 & (26c) \end{cases}$$

The reactions of $C_4H_7^+$ to form $C_4H_8^+$ and $C_5H_9^+$, and of $C_3H_5^+$ to form $C_4H_5^+$ are endothermic and would not occur. Further, the formation of $C_5H_9^+$ from $C_3H_5^+$ is a simple association reaction, and such reactions are seldom observed. Thus, the reaction by path 1 is quite improbable. On the other hand, the reactions that make up path 2 are exothermic and would be expected to occur. Consequently, Field concluded that path 2 represents the mechanism for the formation of $C_4H_8^+$ and $C_5H_9^+$ from ethylene. Similar deductions can be made concerning other ternary ions from ethylene.

With reactions of this kind it is possible to compute third-order rate constants. Without considering the detailed mechanism it is quite simple to compute an overall third-order constant, and Field[13] has done so for several of the reactions in the ethylene system. Rather surprisingly, these rate constants seem to be in the order of $10^{-27}$ cc$^2$/molecule$^2$ sec. These are very large rate constants, being in the order of $10^4$ to $10^5$ times greater than the rate constants for recombination of atoms such as nitrogen[19] or bromine.[20]

Thus far, only a few studies have been published of systems at comparatively high pressures in the mass spectrometer ion source. Such high pressure can be achieved only with excellent differential pumping so as to avoid scattering in the analyzer, and suitable instruments with such qualifications have become available only recently. Aside from the work of Field on ethylene, this laboratory has made several studies of light paraffin and olefin hydrocarbons.[21] Wexler[22] has reported extensive studies of methane at elevated pressures, and Melton and Rudolph[23] have studied cyanogen. At high pressures a number of problems arise. The simple linear dependence of the ratio of intensities of secondary to primary ions upon pressure no longer holds. In fact, one observes a rapid exponential decrease in intensity of primary reactant ions and a corresponding increase in secondary ions tending toward saturation. Figure 1 gives typical results. For a second-order (pseudo first-order) reaction such as Eq. (6)

$$I_{P^+}/I_{P_o^+} = e^{-k(M)t} \tag{27}$$

$$I_{S^+}/I_{P_o^+} = 1 - e^{-k(M)t} \tag{9}$$

These equations give a reasonable description of observed behavior as long as no extensive amount of tertiary and higher ion formation is experienced. However, as the pressure increases, the time between collisions in the source become smaller, and the opportunity for higher order reactions becomes greater. With the ethylene system, Field[13] has observed ions resulting from as many as five or six successive collisions. Table 4 tabulates reaction orders for several ions from ethylene.

## Rare Gases

One of the interesting results of studies of secondary reactions in mass spectrometric ion sources is the rather surprising ionic compounds that have been observed. In addition to $CH_5^+$, which was reported above, and $C_2H_7^+$, which has been observed recently,[21,22] the rare gases have been shown to undergo extensive chemical reactions and to form some strange and interesting compound ions. It has been known for some time that all of the rare gases form diatomic ions, and several years ago it was found by Stevenson and Schissler[5] that most of the rare gases form ionic compounds with hydrogen. In the latter case the reactions in most instances were

$$R^+ + H_2 \rightarrow RH^+ + H \tag{28}$$

where R is any rare gas. These reactions are similar to other ion-molecule reactions and apparently occur with comparable rates. It has been observed recently that rare gases undergo a number of ion-molecule reactions with both organic and inorganic materials, and some of these are indeed surprising. Thus, Field and Franklin,[24] and Field, Head, and Franklin[25] have reported extensive reactions of argon, krypton, and xenon with methane. It has been shown, for example, that all three of these rare gases react with methane to abstract a hydrogen, a $CH_2$, and a $CH_3$. Table 5 gives the various reactions and the approximate rate constants involved. The rates are one to three orders of magnitude less than that of most ion-molecule reactions. This, no doubt, is attributable to the preference of these systems to undergo charge exchange reactions. Thus, Head et al[25] found charge exchange reactions for argon and krypton ions with methane to have rate constants in the order of 1.1 to 1.2 x $10^{-9}$ cc/molecule sec in rather good agreement with the rates of most ion-molecule reactions. The formation of the compound ions is much smaller than that of the charge exchange reaction. In addition to the ions formed from methane, a number of other ions have been observed, although the nature of the reactions leading to some of them is not known. Thus, xenon is known to form compounds $XeI^+$, $XeO^+$, $XeC^+$, $XeOH^+$, etc. Argon and krypton also form a number of such compounds. Some, no doubt, involve reactions of electronically excited ions but these have not been studied sufficiently to determine the reactions.

## Chemi-Ionization Reactions

It has been known for some time that the rare gas diatomic ions exist. In 1951, Hornbeck and Molnar[26] reported the result of mass spectrometric studies of the formation of these ions. They measured the appearance potential of several of the rare gas diatomic ions and found them to be 0.7 to 1.4 ev below the ionization potential of the rare gas in question. This was interpreted as meaning that the diatomic ions were formed by the attack of highly excited neutral atoms upon atoms in the ground state with ejection of an electron, thus

$$Ar^* + Ar \rightarrow Ar_2^+ + e \tag{29}$$

Such reactions have come to be called chemi-ionization reactions. The appearance potentials of the diatomic ions are in all instances well above the metastable state of the atom in question, and occur at such levels that the large number of possible states prevent identification of the actual state reacting. Presumably, although it cannot be proved, the appearance potential corresponds to the lowest excited state capable of reaction, and, no doubt, any higher states that are present will participate also.

Recently several investigators have become interested once more in the rare gases and their chemi-ionization reactions. In this laboratory the appearance potentials of the diatomic ions have been remeasured[27] by the most modern techniques with results that are essentially confirmatory of those of Hornbeck and Molnar. Curran[28] has reported a measurement of the appearance potential of $He_2^+$, and Kaul and Taubert[29] have reported measurements of several of the rare gases. All are in essential agreement for the rare gas in question.

Ionization efficiency curves for $Kr^+$ and $Kr_2^+$ are given in Fig. 2. These are typical of the rare gases. It will be observed that the curve for $Kr^+$ rises sharply with increasing electron voltage and approaches a very broad plateau several volts above onset. On the other hand, the ionization efficiency curve for $Kr_2^+$ starts about 1 v below that of $Kr^+$, rises to a very sharp maximum only 1 or 2 v above onset, and then drops rather precipitously. This behavior is markedly different from that of the formation of a primary ion by electron impact or indeed of a secondary ion formed by ion-molecule reaction. The shape of the curve is very similar to that of a number of excitation processes[30] and indeed strongly supports the proposal by Hornbeck and Molnar that a chemi-ionization process involving an electronically excited rare gas atom is involved.

It is known now that the rare gases undergo a large number of chemi-ionization reactions with neutrals other than their own neutral atom. Thus, Field and Franklin[24] show that $XeCH_4^+$ is formed by a chemi-ionization reaction of electronically excited xenon and methane; Henglein and Muccini[31] report reactions of electronically excited argon and krypton with iodine; and Rudolph et al.[32] have observed the formation of $XeC_2H_2^+$, presumably by a chemi-ionization process. Table 6 presents a number of these known reactions and chemi-ionization products.

The ionic products resulting from chemi-ionization reactions are usually of rather low intensity, which suggests either that the reactions are slow or that the number of electronically excited reactants is comparatively small. Dahler, Franklin, Munson, and Field[33] have studied the rates of formation of the rare gas diatomic ions by chemi-ionization processes. These reactions do not lend themselves as readily to measurement of rate constants as do the ion-molecule reactions because no direct measurement can be made of the concentration of electronically excited reactant atoms in the source. Further, the problem in the case of the rare gases is complicated by the fact that the reaction rates need to be measured at moderately high pressures in order for the intensity of the diatomic ions to be sufficiently great to render measurement reliable. At these conditions some three-body recombination processes also may occur and must be avoided or accounted for. One may write as a reasonable reaction mechanism the following equations:

$$R + e \rightarrow R^+ + 2e \qquad k_i \tag{30}$$
$$R + e \rightarrow R^* + e \qquad k_e \tag{31}$$
$$R^* \rightarrow R + h\nu \qquad k_u = 1/\tau_u \tag{32}$$
$$R^* + R \rightarrow R_2^+ + e \qquad k_2 \tag{33}$$
$$R^+ + 2R \rightarrow R_2^+ + R \qquad k_3 \tag{34}$$

(The asterisk signifies excitation.)

Applying the steady-state treatment to the transient species R* one obtains

$$\frac{I_{R_2^+}}{I_{R^+}} = \frac{(R_2^+)}{(R^+)} = \frac{k_e}{k_i} \frac{k_2(R)\tau_u}{1 + k_2(R)\tau_u} + k_3(R)^2\tau_i \tag{35}$$

Here $k_u$ is the rate constant for unproductive loss of energy from excited atoms by radiation, or diffusion to the walls, and $\tau_i$ is the residence time of the monatomic ions in the source.

In order to permit measurement of the rate of the chemi-ionization process, the three-body reaction was suppressed by carrying out the studies at low electron energies, as near as possible to the maximum in the ionization efficiency curve, and at high repeller field strength to minimize the residence time of primary ions in the source and reduce the rate constant for the third-order (ion-molecule) reaction to negligible values. At these conditions the intensity ratio was given by the curves in Fig. 3. It is observed from these curves that the ratio of diatomic to monatomic ions tends to approach saturation. If the ternary process is completely suppressed, one may invert Eq. (35)

$$I_{R^+}/I_{R_2^+} = k_i/k_e + k_i/k_e\, k_2\tau_u(R) \tag{36}$$

Thus a plot of the ratio of the intensities of monatomic to diatomic ions against the reciprocal of the rare gas pressure should give a straight line of intercept $k_i/k_e$, and slope $k_i/k_e k_2 \tau_u$. In Fig. 4 such plots have been made, and it is evident that reasonably straight lines are obtained. The results are given in Table 7. Unfortunately $\tau_u$ in this case represents the lifetime of the excited atom and is not directly measurable. However, as an upper bound the value of about $10^{-6}$ can be taken, this value being approximately the time required for diffusion to the walls where, no doubt, the energy would be discharged. As a lower bound, the time of radiation can probably be selected. This time is not known, although values in the order of $10^{-8}$ to $10^{-7}$ sec would appear to be reasonable in view of the fact that in no cases are metastable states involved. It is of interest, therefore, to consider in the case of the argon the range of rate constants which would result from the choice of the two limits of $\tau_u$. From Table 7, $k_2\tau_u$ is $3.6 \times 10^{-16}$ cc/molecule; the choice of $10^{-6}$ sec as the upper bound for $\tau_u$ leads to a value for k of $3.6 \times 10^{-10}$ in reasonable agreement with the rate constants for ion-molecule reactions. On the other hand, choice of the lower value of $\tau_u$, $10^{-8}$, leads to a rate constant of $3.6 \times 10^{-8}$ cc/molecule sec which if true would be the fastest secondary reaction that has thus far been reported. This higher value for $k_2$ corresponds to a reaction cross section of about 5200 Å$^2$. This is, of course, a very large cross section. However, if one assumes that the orbital radius increases as the square of the principal quantum number, one would expect that, with states about three principal quantum numbers above the ground state, the cross section of the excited atom would indeed be very large, and the reaction cross section would probably be in the order of several thousand square angstroms, in qualitative agreement with the higher value of the reaction cross section computed on the assumption that the lifetime of the excited species is $10^{-8}$ sec. Similar considerations, of course, apply to other chemi-ionization processes of the rare gases.

One expects diatomic rare gas ions to be formed by three-body processes as well as by chemi-ionization, and indeed the literature contains some evidence for these recombination reactions. Dahler et al[33] in their studies of rates adjusted conditions in the ion source so as to minimize the chemi-ionization reactions and to maximize the opportunities for ion-molecule reactions. This was done by operating at 70 ev electron energy and at comparatively low repeller field strength. This has the effect of reducing proportionally the electronically excited atoms, increasing markedly the number of atomic ions, maximizing the rate constant for the ion-molecule reactions and increasing the residence time for the atomic ions in the source. Under these conditions one still expects a certain amount of chemi-ionization reaction, and so the total production of diatomic ions will include contributions from both chemi-ionization and ion-molecule reactions. In Fig. 5 the ratio of the intensities of diatomic to monatomic ions is plotted against pressure at conditions favorable to maximum ion-molecule reactions. Both the argon and helium ions have a general upward curvature showing a predominance of third-order reactions. At the lower pressures neon shows a tendency to saturate, as might be expected if chemi-ionization predominates. Above the 150 μ the curve turns upward along a path that shows an approximate third-order dependence upon pressure. Obviously these curves, especially for argon and neon, express the results of second- and third-order processes. In order to determine the rate constant for the third-order process, it is necessary to deduct the contribution from the chemi-ionization reactions. Since the proportion of excited species at high electron voltage or the rate constant for reaction of the species present at high energies is not known, it was necessary to redetermine the chemi-ionization rates at conditions of high electron energies by employing high field strengths so as to minimize ion-molecule reactions. Fig. 6 shows the plot of intensity ratio against pressure when employing 70 v electrons and a 50 v/cm field strength. At the lower pressures both the neon and argon curves show a general second-order dependence, as would be expected if the chemi-ionization process was predominant. Helium, on the other hand, shows a monatomic upward trend indicating predominantly third-order behavior. Because of this behavior of helium, its kinetics will be discussed separately. Plotting the data from the lower pressure region for argon and neon in the same manner as was done at the lower electron energy, it was possible to obtain the ratio of electronically excited atoms to monatomic ions and the kτ values for neon and argon. These results are given in Table 7.

Employing the results thus obtained, it is possible to calculate the number of diatomic ions produced by chemi-ionization in neon and argon at high electron energies and low field strengths (since the chemi-ionization reaction is independent of field strength) and to deduct this from the total curve for the intensity of diatomic ions. In Fig. 5, the line marked H.M. represents the Hornbeck-Molnar or chemi-ionization contribution calculated in this way. The difference between these curves and the corresponding total product of diatomic ions will represent the contribution of the third-order ion-molecule reaction. A plot of this difference against the square of the pressure gives a straight line, the slope of which is $k\tau$, and since, in this case $\tau$ is the residence time of the monatomic ion, k is readily computed. The results are given in Table 8.

The problem of determining the nature of the third-order process leading to $He_2^+$ at higher electron energy is a baffling one. The helium diatomic ion formation is independent of field strength even at 300 μ pressure at which conditions argon and neon exhibit definite dependence upon field strength. If an ion-molecule reaction were involved in the formation of $He_2^+$, an increase in the field strength should serve to reduce the rate constant and to reduce the residence time, both of which would have the effect of reducing the intensity of $He_2^+$. Since this is not observed, it appears that an ion-molecule reaction is not involved in the formation of the helium diatomic ion. This is rather surprising and indeed rather difficult to accept, especially in view of the fact that at the higher electron energies the formation of $He_2^+$ appears to be definitely third order. Conceivably this behavior might be attributable to a third-order chemi-ionization process, but the nature of this reaction is not understood.

It is of interest to consider the third-order rate constants determined for argon and neon, namely, about $2 \times 10^{-28}$ and $2 \times 10^{-29}$ $cc^2/molecule^2$ sec, respectively. These values are extraordinarily large for reactions involving three atoms. The rates of recombination of nitrogen atoms in the presence of $N_2$ for comparison are approximately $10^{-32}$ $cc^2/molecule^2$ sec,[19] and thus the third-order ion-molecule reactions are in the order of $10^3$ to $10^4$ times that of systems of neutral species of comparable complexity. If one considers the detailed mechanism for this three-body formation of a diatomic rare gas ion, the reaction may be written as follows:

$$R^+ + R \underset{2}{\overset{1}{\rightleftarrows}} R_2^{+*} \tag{37}$$

$$R_2^{+*} + R \overset{3}{\rightarrow} R_2^+ + R \tag{38}$$

where the asterisk indicates excitation. Standard kinetic treatment leads to the equation

$$(R_2^+)/(R^+) = k_1 k_3 (R)^2 t/(k_2 + k_3(R)) \tag{39}$$

At the conditions of these studies $k_2 \gg k_3(R)$, so Eq. (39) becomes

$$(R_2^+)/(R^+) = k_1k_3/k_2(R)^2t \quad (40)$$

and the overall third-order rate constant is $k_1k_3/k_2$. If a typical value of 2 x $10^{-10}$ cc/molecule sec is taken for $k_1$, then $k_3/k_2 = 10^{-18}$. If it is assumed that $k_3$ is about $10^{-10}$, then $k_2$ is about $10^8$ sec. This represents an unbelievably long lifetime of $10^{-8}$ sec for a diatomic ion having sufficient energy to decompose. Comparable lifetimes in the case of atom recombinations prove to be in the order of $10^{-12}$ to $10^{-13}$ sec, i.e., approximately the time of a single vibration. On the other hand, if the value of $k_2$ is taken to be about $10^{11}$ $sec^{-1}$, which would correspond to a not inconceivable value of the time of a single vibration, then one calculates $k_3$ to be about $10^{-7}$ cc/molecule sec, which is a tremendously fast reaction. Perhaps this is acceptable in view of the fact that highly excited species can be expected to undergo more rapid reaction than species in their ground state. In any event, the probable values of rate constants for chemi-ionization reactions appear to bear this out.

Munson, Field, and Franklin[34] have extended the study of chemi-ionization to nitrogen and carbon monoxide and mixtures of these compounds with the rare gases. At about 150 μ pressure, the nitrogen mass spectrum includes an ion at mass 56, which is no doubt $N_4^+$. At the same conditions carbon monoxide also shows a dimeric ion, $C_2O_2^+$. At pressures below 100 μ neither of these ions is observable. An ionization efficiency curve of $N_4^+$ is given in Fig. 7. This ion has an appearance potential of about 1 ev below the ionization potential of $N_2$, and the shape of the curve resembles that of the diatomic ions of the rare gases and of electronically excited species generally. These facts strongly suggest that $N_4^+$ is formed by chemi-ionization process, which can be written

$$N_2^* + N_2 \rightarrow N_4^+ + e \quad (41)$$

Carbon monoxide exhibits similar behavior, so the reaction to form $C_2O_2^+$ can be written

$$CO^* + CO \rightarrow C_2O_2^+ + e \quad (42)$$

At pressures up to 150 μ the intensity of $N_4^+$ and $C_2O_2^+$ shows the reactions to be second order. If, however, the pressure is raised above about 250 μ in the source, breaks appear in the ionization efficiency curves at about the ionization potential of $N_2$ and CO, respectively, indicating that a third-order process begins to occur at the higher pressures.

In this respect the formation of $N_4^+$ and $C_2O_2^+$ is directly analogous to the production of rare gas diatomic ions exhibiting both chemi-ionization reactions and third-order ion-molecule recombination reactions. The principal difference lies in the fact that the $N_4$ and $C_2O_2^+$ ions occur in considerably smaller intensitites than do those of diatomic ions of rare gases.

It should be mentioned that these results disagree with the observations of Saporoschenko[35] who found the appearance potential of $N_4^+$ to be very close to the ionization potential of $N_2$ and the pressure dependence to be second order, so that the reaction is, in effect, a long-lived collision of $N_2^+$ and $N_2$. It seems highly improbable that such a simple system would persist for so long, and, in view of the more recent results from this laboratory, it appears that Saporoschenko's results are in error.

In this same series of experiments argon was introduced into the source when nitrogen was present at a partial pressure of about 80 μ. At these conditions there was no $N_4^+$ observed with nitrogen alone in the source. However, as soon as argon was introduced, a rather surprisingly large $N_4^+$ peak appeared and increased in intensity with addition of further argon. This was a rather surprising result, and Munson et al went to some length to develop an explanation. In addition to the $N_4^+$ ion, the argon diatomic ion was, of course, observed, and $ArN_2^+$ ion appeared. The latter had previously been observed by Kaul and Fuchs.[36] The ionization efficiency curves for these ions were determined, and the appearance potentials of all were found to be very nearly the same (Figure 8). Further, all passed through a sharp miximum about 5 v above onset and thus appeared to result directly from chemi-ionization reactions or to have chemi-ionization reactions as precursors. A pressure study of the system is given in Figs. 9 and 10 and shows, as would be expected, that $Ar_2^+$ depends upon the second power of argon, that $ArN_2^+$ depends upon the first power of both argon and nitrogen, and that $N_4^+$ is dependent upon the first power of argon and the second power of nitrogen. In fact, the $N_4^+$ intensity tends to saturate, suggesting that this ion can be decomposed by collision. The dependence of these various species upon field strength was examined, and the results given in Table 9 show that the formation of $N_4^+$ is strongly dependent upon field strength, but that the formation of $Ar_2^+$ and $ArN_2^+$ are relatively little dependent upon field strength. Of course, the formation of the argon diatomic ion at the conditions employed results mainly from chemi-ionization reactions and should be independent of field strength. The formation of $ArN_2^+$ apparently also involves a chemi-ionization process. The strong dependence of $N_4^+$ upon repeller field proclaims that this ion results from an ion-molecule reaction. Its ionization efficiency curve, however, shows that an electronically excited species was the precursor of the ion and the identity of the appearance potential of all three ions, $N_4^+$, $ArN_2^+$, and $Ar_2^+$, suggests an electronically excited argon atom. From these various observations it appeared reasonable to write the following set of equations representing the mechanism for the formation of the various ions present.

$$Ar + e \rightarrow Ar^+ + 2e \qquad k_i \qquad (43)$$
$$Ar + e \rightarrow Ar^* + e \qquad k_e \qquad (44)$$
$$N_2 + e \rightarrow N_2^+ + 2e \qquad k_i' \qquad (45)$$
$$Ar^* \rightarrow Ar + h\nu \qquad k_u \qquad (46)$$
$$Ar^* + Ar \rightarrow Ar_2^+ + e \qquad k_1 \qquad (47)$$
$$Ar^* + N_2 \rightarrow ArN_2^+ + e \qquad k_2 \qquad (48)$$
$$ArN_2^+ + N_2 \rightarrow N_4^+ + Ar \qquad k_3 \qquad (49)$$
$$N_4^{+2} + Ar \rightarrow ArN_2^+ + N_2 \qquad k_4 \qquad (50)$$

By applying the usual steady-state criterion to transient species approximate rate equations can be developed for the various ions considered. Perhaps of greatest interest is the result one obtains by taking the ratio of $ArN_2^+$ and the $N_4^+$ ions to $Ar_2^+$. The results for comparatively low pressures are given in Eqs. (51) and (52).

$$(ArN_2^+)/Ar_2^+ = k_2N_2/k_1R \qquad (51)$$

$$(N_4^+)/(Ar_2^+) + k_2k_3(N_2)^2t/2k_1(Ar) \qquad (52)$$

Table 10 shows the ratio of $k_1$ to $k_2$ and the value of $k_3$ obtained from pressure studies. It is of interest for the argon-nitrogen system that the ratio of $k_1$ to $k_2$ is approximately unity, and $k_3$, which is the rate constant for an ion-molecule reaction, has a value quite similar to that of other ion-molecule reactions, about $10^{-10}$ cc/molecule sec. With minor differences the same criteria apply to the system krypton-carbon monoxide where the appearance potentials of the $Kr_2^+$, $KrCO^+$, and $C_2O_2^+$ agree approximately, and the dependence upon pressure and field strength is quite analogous to the behavior of the argon-nitrogen system. Table 10 shows that the $k_1$ to $k_2$ ratio is 2.5 rather than 1 and that $k_3$ is perhaps slightly greater than that of the corresponding reaction in the argon-nitrogen system.

The formation of $C_2O_2^+$ is promoted by the presence of argon, and superficially, this system is quite similar to the argon-nitrogen and to the krypton-carbon monoxide systems. However, the appearance potential of $C_2O_2^+$ is too far below that of the other chemi-ionization products to permit treatment of this system in accordance with the mechanism outlined above for argon-nitrogen. Apparently other states of argon are present to transfer energy to the lower states of carbon monoxide below the energies at which $Ar_2^+$ and $ArCO^+$ can be formed. This being the case, it is not possible to obtain a meaningful calculation of $k_3$. It is, however, possible from the kinetic evidence to obtain a ratio of $k_1$ to $k_2$ for this system and the result, 1.7, is not unreasonable in view of the values obtained from the other systems.

It is of interest also that values of $k_1/k_2$ could be determined from the systems krypton-nitrogen, xenon-nitrogen, senon-nitrogen, xenon-CO, and xenon-methane and these results are given in Table 10. None of these systems bring about the formation of $N_4^+$ or $C_2O_2^+$, and, rather as one would expect, the $k_1/k_2$ values for these systems are much larger than those already discovered. This is perhaps to be expected in view of the rather considerable differences in ionization potential of the principal reactants, although ions, of course, are not involved. The existence of states of energy appropriate for chemi-ionization generally appear to be greater the closer the ionization potential of the two reactants, and these results are in accordance with these observations.

## Reactions of Excited Ions

No doubt many ionic systems include excited ions that undergo characteristic secondary reactions; however, it is comparatively difficult to identify most of these processes, and so the literature contains relatively little information concerning excited ions and their reactions. Henglein[31,37] has identified several of these especially in reactions of carbon disulfide, some aromatics, and rare gases. The formation of $N_3^+$ is now known to result from the reaction of an electronically excited nitrogen ion

$$N_2^{+*} + N_2 \rightarrow N_3^+ + N \tag{53}$$

The formation of $C_2O^+$ in carbon monoxide involves a similar reaction involving electronically excited $CO^+$. Table 11 lists a number of these reactions. Of course, the manner in which the nature of these reactions is identified is similar to that of identifying other reactions. In the case of $N_3^+$, for example, Kaul and Fuchs[36] found the appearance potential of $N_3^+$ to be about 20 ev. This is well above the ionization potential of nitrogen and below the appearance potential of $N^+$. The ion shows a definite dependence upon field strength in the source and consequently must involve the reaction of $N_2^+$ ion excited to several volts above the ground state ionization potential.

Munson et al.[34] studied the formation of $N_3^+$ and other ions resulting from the reaction of electronically excited nitrogen ions with various rare gases. Helium and neon formed no measurable amounts of compound ions, but argon, krypton, and xenon all formed the compound $RN^+$ in addition to $N_3^+$, which was formed in the presence of all rare gases. Although the appearance potentials were not measured accurately, in all cases they were about the same as that of $N_3^+$. Further, pressure studies showed that the $N_3^+$ ion was depleted by the addition of argon, krypton, and xenon. In all cases $N_3^+$ showed a second-power dependence upon nitrogen pressure, and $RN^+$ was dependent upon the first power of the pressure of the rare gas and nitrogen. Since the appearance potential of $RN^+$ and $N_3^+$ were rather similar in all cases, it seemed reasonable that $RN^+$ and $N_3^+$ arose from the same precursor, namely, the excited nitrogen ion. The following mechanism describes the system adequately:

$$N_2 + e \rightarrow N_2^+ + 2e \qquad k_i \qquad (54)$$
$$N_2 + e \rightarrow N_2^{+*} + 2e \qquad k_i^* \qquad (55)$$
$$N_2^{+*} + N_2 \rightarrow N_3^+ + N \qquad k_a \qquad (56)$$
$$N_2^{+*} + R \rightarrow RN^+ + N \qquad k_b \qquad (57)$$
$$H_2^{+*} \rightarrow N_2^+ + h\nu \qquad k_u \qquad (58)$$

This system may be treated kinetically in a manner analogous to that employed in the study of kinetics of chemi-ionization reactions, it being necessary to employ such treatment in view of the fact that there is no direct measure of the number of excited nitrogen ions involved in the reaction. Assuming a steady state concentration for the excited $N_2^+$ ion, one readily deduces the following for formation of the $RN^+$ and $N_3^+$ ions:

$$(RN^+)/(N_2^+) = k_i^*/k_i \; [k_b(R)/(k_a(N_2) + k_b(R) + k_u)] \qquad (59)$$

$$(N_3^+)/(N_2^+) = k_i^*/k_i \; [k_a(N_2)/(k_a(N_2) + R_b(R) + k_u)] \qquad (60)$$

From Eqs. (59) and (60) one readily obtains

$$N_3^+/(RN^+) = k_a(N_2)/k_b(R) \qquad (61)$$

which enables one to obtain the ratio of rate constants from the pressure dependence of $(N_3^+)/(RN^+)$ or its inverse. Inverting Eq. (60) one obtains

$$(N_2^+)/(N_3^+) = k_i/k_i^* (1 + k_u/k_a(N_2) + (k_i/k_i^*)(k_b/k_a)(R)/(N_2) \qquad (62)$$

From the slope of the plot of $(N_2^+)/(N_3^+)$ vs rare gas pressure and the values of $k_b/k_a$, one can obtain $k_i/k_i^*$, the ratio of normal molecule ions to excited ions. Table 12 gives results of this kinetic analysis for the various rare gases with nitrogen. The concentration of excited nitrogen ions is obviously quite small. The relative rates of the reaction of nitrogen with the rare gases vary as one would expect, being greater with xenon than with argon, but the variation is not that predicted by Eq. (24), relating k to the square root of the polarizability.

If one considers the case of the formation of $N_3^+$ in the nitrogen system alone, one can readily obtain Eq. (63) from Eq. (62) by letting (R) = 0.

$$(N_2^+)/(N_3^+) = k_i/k_i^*(1 + k_u/k_a(N_2)) \qquad (63)$$

This is, of course, directly comparable to the equation for the formation of the rare gas diatomic ions by chemi-ionization. Taking the values if $k_i/k_i^*$ of about 200 from Table 12 and a nitrogen pressure of about 80 μ, $k_a\tau_u$ is found to be about $2 \times 10^{-15}$ cc/molecule sec, where $\tau_u = 1/k_u$ refers to the lifetime of the excited molecule ion. This value is quite similar to those obtained for the excited atoms of argon and neon, namely, about $4 \times 10^{-16}$ and $12 \times 10^{-16}$, respectively. Although there is no way of ascertaining the lifetime $\tau_u$ of the electronically excited ion, it is believed that the criterion applied to the rare gases probably is equally applicable to the electronically excited nitrogen ion. If $\tau$ is determined by the residence time in the source, then k is found to be about $2 \times 10^{-9}$ cc/molecule sec. If, however, $\tau$ is the time required for radiation of the excited ion, it may be as small as $10^{-8}$ sec, in which case k would have the extremely large value of $10^{-7}$ cc/molecule sec. Although one cannot be sure it is correct, it is suspected that the higher value for the rate constant is probably the proper one.

With carbon monoxide alone or in the presence of rare gases, $C_2O^+$ was always formed, and the appearance potential suggested that electronically excited CO ion was the precursor. Consequently, the following reaction is involved:

$$CO^{+*} + CO \rightarrow C_2O^+ + O \qquad (64)$$

With argon, krypton, and xenon, small but definite amounts of the $RC^+$ ions are always observed. All of these had high appearance potentials, indicating that the precursors were electronically excited ions; however, the $C_2O^+$ concentration was not decreased by addition of any of the rare gases, and so it was formed from species other than those forming $RC^+$. A study of the kinetics of these processes was not undetaken.

In ordinary third-order reactions it is often true that the ternary product ion arises from collision of the primary complex with a neutral molecule. This was discussed briefly in consideration of some of the higher order ions from ethylene as studied by Field.[13] The following equations indicate the nature of the system:

$$P^+ + M \rightarrow PM^+ \qquad k_1 \tag{65}$$
$$PM^+ \rightarrow S^+ + N \qquad k_2 \tag{66}$$
$$PM^+ + M \rightarrow T^+ + F \qquad k_3 \tag{67}$$

Now from these equations it is evident that $PM^+$ will be highly excited in many cases, and so reaction (67) to form the ternary will be that of an excited ion. Applying the steady-state treatment, one finds for $PM^+$ at modest conversion

$$(PM^+) = k_1(M)(P^+)/(k_2 + k_3(M)) \cong k_1(M)(P^+)/k_2 \tag{68}$$

since $(PM^+)$ and $k_3(M)$ are quite small. Similarly for the tertiary ion

$$\frac{(T^+)}{(P_o^+)} = \frac{k_3(M)}{k_2 + k_3(M)}(1 - e^{-k_1(M)t}) \cong \frac{k_1 k_3 (M)^2 t}{k_2} \tag{69}$$

The observed third-order constants $K = k_1k_3/k_2$ are in the order of $10^{-26}$ for some hydrocarbon reactions, and, taking a reasonable value for $k_1$, $10^{-10}$,

$$k_3/k_2 = 10^{-16} \tag{70}$$

Since no $PM^+$ is observed,

$$(PM^+)/(P_o^+) = k_1(M)/k_2 \lessdot 10^{-4} \tag{71}$$

With $(M) = 10^{15}$ molecules/cc and $k_1 = 10^{-10}$, $k_2 > 10^9$. Using this value of $k_2$ in Eq. (70), one obtains a value of $k_3 \cong 10^{-7}$ cc/molecule sec. From these considerations the reaction rate of the excited complex is estimated to be a factor of $10^2$ to $10^3$ faster than an ordinary ion-molecule reaction.

A large number of ternary reactions involving hydrocarbon have been observed in this laboratory, and many of them have rate constants in the order of $10^{-25}$ to $10^{-27}$ $cc^2/molecule^2$ sec. From the foregoing results it is evident that all of these must involve very fast reactions on the part of the complex with the neutral species. It will be recalled that somewhat similar criteria were applied to the ternary reactions of the rare gas ions forming diatomic ions, indicating that the excited complexes react very rapidly. For the reactions of electronically excited nitrogen ion and electronically excited rare gas atoms values of $k\tau$ were computed, and, associating $\tau$ with a radiation time of $10^{-8}$ sec, it was deduced that k would again be very large. Thus, although rigorous proof cannot be adduced, in many instances the weight of evidence seems to be increasingly convincing that excited atoms, ions, and molecules undergo reactions at much higher rates than do species in their ground states. Perhaps this should not be surprising, but heretofore there has been comparatively little evidence upon which to base any generalizations. It is the authors' opinion now that this will be generally ture and that future studies probably will continue to extend the evidence in support of this conclusion.

References

[1] Hogness, R. R. and Harkness, R. W., "The ionization processes of iodine interpreted by the mass-spectrograph," Phys. Rev. 32, 784-790 (1928).

[2] Mann, M. M., Hustrulid, A., and Tate, J. T., "The ionization and dissociation of water vapor and ammonia by electron impact," Phys. Rev. 58, 340-347 (1940).

[3] Loeb, L. B., Basic Processes of Gaseous Electronics (University of California Press, Berkeley and Los Angeles, California , 1955).

[4] Tal'roze, V. L. and Lyubimova, A. K., Dokl. Akad. Nauk. SSSR 86, 909 (1952).

[5] Stevenson, D. P. and Schissler, D. O., "Rate of the gaseous reactions, $X^+ + YH = XH^+ + Y$," J. Chem. Phys. 23, 1353-1354 (1955); "Reactions of gaseous molecule ions with gaseous molecules. II," J. Chem. Phys. 24, 926-28 (1956).

6 Field, F. H., Franklin, J. L., and Lampe, F. W., "Reactions of gaseous ions. I. Methane and ethylene," J. Am. Chem. Soc. 79, 2419-2429 (1957).

7 Meisels, G. G., Hamill, W. H., and Williams, R. R., Jr., "Ion-molecule reactions in radiation chemistry," J. Chem. Phys. 25, 790 (1956).

8 Stevenson, D. P. and Schissler, D. O., "Reactions of gaseous molecule ions with gaseous molecules. IV. Experimental method and results," J. Chem. Phys. 29, 282-294 (1958).

9 Lampe, F. W. and Field, F. H., "Reactions of ions with molecules in the gas phase," Tetrahedron 7, 189-205 (1959).

10 Lampe, F. W., Franklin, J. L., and Field, F. H., "Kinetics of the reactions of ions with molecules," Progress in Reaction Kinetics (Pergamon Press, London, 1961), Vol. 1, pp. 69-103.

11 Rosenstock, H. M., Mueller, C. R., Wallenstein, M. B., Vestal, M. L., Tory, A., Rivers, D., and Johnston, W. H., "Ion-molecule reactions," William H. Johnston Labs., Inc. Rep. JLI-650-3-7 UC-23 (October 1959).

12 Stevenson, D. P. and Schissler, D. O., "Mass spectrometry and radiation chemistry," The Chemical and Biological Reactions of Radiations, edited by M. Haissinsky (Academic Press, London, 1961), Vol. V, pp. 167-271.

13 Field, F. H., "Reactions of gaseous ions. VIII. Multiple order ion-molecule reactions and the ultra-high pressure mass spectrum of ethylene," J. Am. Chem. Soc. 83, 1523-1534 (1961).

14 Pottie, R. F. and Hamill, W. H., "Persistent ion-molecule collision complexes of alkyl halides," J. Phys. Chem. 63, 877-879 (1959).

15 Henglein, A., "Mass spectrometric observation of persistent ion-molecule complexes of acrylonitrile and of benzene," Z. Naturforsch. 17a, 44-46 (1962).

16 Gioumousis, A. and Stevenson, D. P., "Reactions of gaseous molecule ions with gaseous molecules. V. Theory," J. Chem. Phys. 29, 294 -299 (1958).

[17] Boelrijk, N. and Hamill, W. H., "Effects of relative velocity upon gaseous ion-molecule reactions; charge transfer to the neopentane molecule," J. Am. Chem. Soc. 84, 730-734(1962).

[18] Eyring, H., Hirschfelder, J. O., and Taylor, H. S., "The theoretical treatment of chemical reactions produced by ionization processes. Part I. The orth-para hydrogen conversion by Alpha-particles," J. Chem. Phys. 4, 479-491 (1936).

[19] Herron, J. T., Franklin, J. L., Bradt, P., and Dibeler, V. H., "Kinetics of nitrogen atom recombination," J. Chem. Phys. 30, 879-885 (1959).

[20] Rabinowitch, E., "The recombination-velocity of free atoms," Trans. Faraday Soc. 33, 283-293 (1937).

[21] Field, F. H., Franklin, J. L., and Munson, M. S. B., "Mass spectra of light paraffins and olefins at elevated pressures," presented at 142nd Meeting of American Chemical Society before the Division of Physical Chemistry, Atlantic City (1962).

[22] Wexler, S. and Jesse, N., "Consecutive ion-molecule reactions in methane," J. Am. Chem. Soc. 84, 3425-3432 (1962).

[23] Melton, C. E. and Rudolph, P. S., "Transient species in the radiolytic polymerization of cyanogen," J. Chem. Phys. 33, 1594-95 (1960).

[24] Field, F. H. and Franklin, J. L., "Reactions of gaseous ions. X. Ionic reactions in xenon-methane mixtures," J. Am. Chem. Soc. 83, 4509-15 (1961).

[25] Field, F. H., Head, H. N., and Franklin, J. L., "Reactions of gaseous ions. XI. Ionic reactions in krypton-methane and argon-methane mixtures," J. Am. Chem. Soc. 84, 1118-20 (1962).

[26] Hornbeck, J. A. and Molnar, J. P., "Mass spectrometric studies of molecular ions in the noble gases," Phys. Rev. 84, 621-25 (1951).

[27] Munson, M. S. B., Franklin, J. L., and Field, F. H., unpublished data .

[28] Curran, R. K., "Negative ion formation in various gases at pressures up to 0.5 mm Hg," Mass Spectrometry Conference, ASTM Committee E-14, New Orleans, La., June 3-8, 1962, pp. 324-332.

[29] Kaul, W. and Taubert, R., "Sedundarreaktionen in edelgas-edelgasund edelgas-stickstoff-gerischen," Naturforsch. 17a, 88-89 (1962).

[30] See for example, Massey, H. S. W. and Burhop, E. H. S., Electronic and Ionic Impact Phenomena (Oxford University Press, London, 1952).

[31] Henglein, A. and Muccini, G. A., "Mass spectrometric studies of ion-molecule reactions in mixtures of methane, methanol, water, argon and krypton with iodine: participation of excited ions and atoms and some radiation chemical considerations," Z. Naturforsch. 15a, 584-593 (1960).

[32] Rudolph, P. A., Lind, S. C., and Melton, C. E., "Ionic complexes of Xe and $C_2H_2$ produced by the radiolysis of these gases," J. Chem. Phys. 36, 1031-37 (1962).

[33] Dahler, J. S., Franklin, J. L., Munson, M. S. B., and Field, F. H., "Rare-gas molecule-ion formation by mass spectrometry. Kinetics of $Ar_2^+$, $Ne_2^+$, and $He_2^+$ formation by second- and third-order processes," J. Chem. Phys. 36, 3332-3344 (1962).

[34] Munson, M. S. B., Field, F. H., and Franklin, J. L., "High-pressure mass spectrometric study of reactions of rare gases with $N_2$ and CO," J. Chem. Phys. 37, 1790-1799 (1962).

[35] Saporoschenko, M., "Ions in nitrogen," Phys. Rev. 111, 1550-1553 (1958).

[36] Kaul, W., and Fuchs, R., "Mass spectrometric study of $A\text{-}N_2$ mixtures," Z. Naturforsch. 15a, 326-30 (1960).

[37] Henglein, A., "Mass spectrometric observations on reactions of excited ions in carbon disulfide and some aromatic compounds," Z. Naturforsch. 17a, 37-43 (1962).

Table 1 Appearance potentials of ions from methane

| | Ap(ev) | | Ap(ev) |
|---|---|---|---|
| $CD^+$ | 23.0 | $CD_4^+$ | 13.1 |
| $CD_2^+$ | 15.5 | $CD_5^+$ | 13.1 |
| $CD_3^+$ | 14.4 | $C_2D_5^+$ | 14.3 |

Table 2 Some known ion-molecule reactions

| | |
|---|---|
| $D_2^+ + D_2 \rightarrow D_3^+ + D$ | $C_2H_4^+ + C_3H_8 \rightarrow C_3H_7^+ + C_2H_5$ |
| $A^+ + H_2 \rightarrow AH^+ + H$ | $C_2H_4^+ + C_2H_4 \rightarrow C_3H_5^+ + CH_3$ |
| $H_2O^+ + H_2O \rightarrow H_3O^+ + OH$ | $C_2H_2^+ + C_2H_2 \rightarrow C_4H_2^+ + H_2$ |
| $NH_3^+ + NH_3 \rightarrow NH_4^+ + NH_2$ | $C_4H_6^+ + C_4H_6 \rightarrow C_6H_6^+ + C_2H_6$ |
| $O_2^+ + H_2 \rightarrow H_2O^+ + O$ | $CH_3^+ + H_2S \rightarrow CH_3S^+ + H_2$ |
| $CD_4^+ + H_2S \rightarrow CD_4H^+ + HS$ | $I^+ + C_2H_5I \rightarrow HI_2^+ + C_2H_4$ |
| $C_2H_3^+ + C_2H_4 \rightarrow C_2H_5^+ + C_2H_2$ | $H^- + H_2O \rightarrow OH^- + H_2$ |
| $C_3H_8^+ + C_2H_4 \rightarrow C_2H_5^+ + C_3H_7$ | $O^- + I_2 \rightarrow OI^- + I$ |
| $CH_3^+ + C_2H_6 \rightarrow C_2H_5^+ + CH_4$ | $SO^- + SO_2 \rightarrow SO_2^- + SO$ |

Table 3 Rate constants for several second-order reactions[10]

| | $10^{10}$ k, cc/molecule sec |
|---|---|
| $D_2^+ + D_2 \rightarrow D_3^+ + D$ | 14.5 |
| $Kr^+ + H_2 \rightarrow KrH^+ + H$ | 5 |
| $CH_4 + CH_4 \rightarrow CH_5^+ + CH_3$ | 8.5 |
| $H_2O^+ + H_2O \rightarrow H_3O^+ + OH$ | 12.8 |
| $CH_3OH^+ + CH_3OH \rightarrow CH_3OH_2^+ + CH_3O$ | 112.0 |
| $HI^+ + CH_3I \rightarrow CH_3I_2^+ + H$ | 3.6 |
| $CH_3^+ + CH_4 \rightarrow C_2H_5^+ + H_2$ | 8.5 |
| $C_2H_2^+ + C_2H_4 \rightarrow C_3H_3^+ + CH_3$ | 2.3 |
| $C_3H_5^+ + C_3H_6 \rightarrow C_4H_7^+ + C_2H_4$ | 21.2 |
| $O^- + CH_3I \rightarrow OI^- + CH_3$ | 21.2 |
| $O^-$ x $CH_3NO_2 \rightarrow CHNO_2^- + H_2O$ | 22.8 |

Table 4 Ions produced by high-order processes in ethylene[13]

| Ion | Reaction Order | Ion | Reaction Order |
|---|---|---|---|
| $C_2H_5^+$ | 2 | $C_5H_9^+$ | 3 and 4 |
| $C_3H_3^+$ | 2 | $C_7H_7^+$ | 4 |
| $C_3H_5^+$ | 2 and 3 | $C_8H_7^+$ | 4 |
| $C_4H_6^+$ | 3 | $C_8H_9^+$ | 5 |
| $C_4H_8^+$ | 3 | $C_6H_9^+$ | 5 |
| $C_5H_7^+$ | 3 | $C_7H_{13}^+$ | 6 |

Table 5 Rates of reaction of rare gases with methane[24, 25]

| Reaction | | $10^{11}$ k, cc/molecule sec |
|---|---|---|
| $Ar^+ + CH_4$ | $\rightarrow ArH^+ + CH_3$ | 2.0 |
| | $\rightarrow ArCH_3^+ + H$ | 0.03 |
| | $\rightarrow ArCh_2^+ + H_2$ | 0.08 |
| | $\rightarrow CH_2^+ + H_2 + Ar$ | 23.0 |
| | $\rightarrow CH_3^+ + H + Ar$ | 91.0 |
| $Kr^+ + CH_4$ | $\rightarrow KrH^+ + CH_3$ | 1.2 |
| | $\rightarrow KrCH_3^+ + H$ | 0.3 |
| | $\rightarrow KrCH_2^+ + H_2$ | 0.1 |
| | $\rightarrow CH_3^+ + H + Kr$ | 49.0 |
| | $\rightarrow CH_4^+ + Kr$ | 74.0 |
| $Xe^+ + CH_4$ | $\rightarrow XeH^+ + CH_3$ | 3.8 |
| | $\rightarrow XeCH_3^+ + H$ | 0.22 |
| | $\rightarrow XeCH_2^+ + H_2$ | 0.045 |

Table 6 Some chemi-ionization reactions

| | |
|---|---|
| $Xe^* + CH_4 \rightarrow XeCH_4^+ + e$ | $Ar^* + I^+ \rightarrow ArI^+ + I + e$ |
| $(Xe^* + C_2H_2)$ / $(Xe + C_2H_2^*)$ $\rightarrow XeC_2H_2^+ + e$ | $Ar^* + N_2 \rightarrow ArN_2^+ + e$ |
| $(Xe^* + O_2)$ / $(Xe + O_2^*)$ $\rightarrow XeO_2^+ + e$ | $Kr^* + CO \rightarrow KrCO^+ + e$ |
| | $N_2^* + N_2 \rightarrow N_4^+ + e$ |
| $Kr^* + I_2 \rightarrow KrI^+ + I + e$ | $CO^* + CO \rightarrow C_2O_2^+ + e$ |

Table 7 Rates of chemi-ionization reactions in Ar, Ne, and He at 50 v/cm field strength[33]

| Rare gas | Electron volts | $k_e/k_i$ | $10^{16}\ k_2\tau_u$, cc/molecule |
|---|---|---|---|
| Ar | 15 | 0.055 | 3.6 |
| | 70 | 0.014 | 2.7 |
| Ne | 20 | 0.010 | 11.6 |
| | 70 | 0.0045 | 1.1 |
| He | 22 | 0.064 | 0.58 |

Table 8 Termolecular rate constants[33, a]

| Rare gas | k, cc$^2$/molecule$^2$/sec |
|---|---|
| Ar | $2.1 \times 10^{-28}$ |
| Ne | $2.0 \times 10^{-29}$ |
| He | $9.8 \times 10^{-29}$[b] |

[a] Electron energy = 70 v; field strength = 12.5 v/cm.
[b] Computed as if an ion-molecule reaction were involved.

Table 9 Field strength dependence of dimeric ions in the Ar-$N_2$ system [34]

| $P(Ar)$, μ | Field strength, v/cm | | |
|---|---|---|---|
| | 12.5 | 30 | 50 |
| | $N_4^+/Ar_2^+$ | | |
| 60 | 0.35 | 0.22 | 0.14 |
| 70 | 0.28 | 0.18 | 0.11 |
| 80 | 0.23 | 0.13 | 0.09 |
| 90 | 0.19 | 0.12 | 0.08 |
| | $ArN_2^+/Ar_2^+$ | | |
| 60 | 1.7 | 1.5 | 1.35 |
| 70 | 1.5 | 1.3 | 1.15 |
| 80 | 1.3 | 1.1 | 1.01 |
| 90 | 1.1 | 1.0 | 0.87 |

Table 10 Kinetic data from rare gas + $N_2$ or CO systems [34]

| R | Y | $k_1/k_2$ | $k_3$, cc/molecule sec, at FS = 50 v/cm |
|---|---|---|---|
| Ar | $N_2$ | 0.9 | $1 \times 10^{-10}$ |
| Ar | CO | 1.7 | |
| Kr | CO | 2.5 | $3 \times 10^{-10}$ |
| Kr | $N_2$ | 10 | |
| Xe | CO | 13 | |
| Xe | $N_2$ | >15 | |
| Xe | $CH_4$ | 9 | |

Table 11 Some reactions of excited ions

| | Reference |
|---|---|
| $CS_2^{+*} + CS_2 \rightarrow CS_3^+ + CS$ | 37 |
| $CS_2^{+*} + CS_2 \rightarrow C_2S_2 + S_2$ | 37 |
| $CS_2^{+*} + CS_2 \rightarrow C_2S_3^+ + S$ | 37 |
| $CS_2^{+*} + H_2O \rightarrow H_2OS^+ + CS$ | 37 |
| $CS_2^{+*} + I_2 \rightarrow CSI^+ + SI$ | 37 |
| $CS_2^{+*} + I_2 \rightarrow SI^+ + CSI$ | 37 |
| $C_6H_5CN^{+*} + C_6H_5CN \rightarrow C_{12}H_9CN^+ + HCN$ | 37 |
| $C_6H_5Cl^{+*} + C_6H_5Cl \rightarrow C_{12}H_9^+ + HCl + Cl$ | 37 |
| $C_6H_5Br^{+*} + C_6H_5Br \rightarrow C_{12}H_{10}^+ + Br_2$ | 37 |
| $I_2^{+*} + H_2O \rightarrow H_2OI^+ + I$ | 37 |
| $O_2^{+*} + O_2 \rightarrow O_3^+ + O$ | 37 |
| $N_2^{+*} + N_2 \rightarrow N_3^+ + N$ | 36, 34 |
| $N_2^{+*} + Ar \rightarrow ArN^+ + N$ | 36, 34 |
| $CO^{+*} + CO \rightarrow C_2O^+ + O$ | 34 |

Table 12 Kinetic data for rare gas-nitrogen systems[34,a]

| a | b | $k_b/k_a$ | $k_i/k_i^*$ |
|---|---|---|---|
| $N_2$ | Ar | 0.54 | 111 |
| $N_2$ | Kr | 0.85 | 245 |
| $N_2$ | Xe | 2.5 | 216 |

[a] EV = 30 ev; FS = 12.5 v/cm.

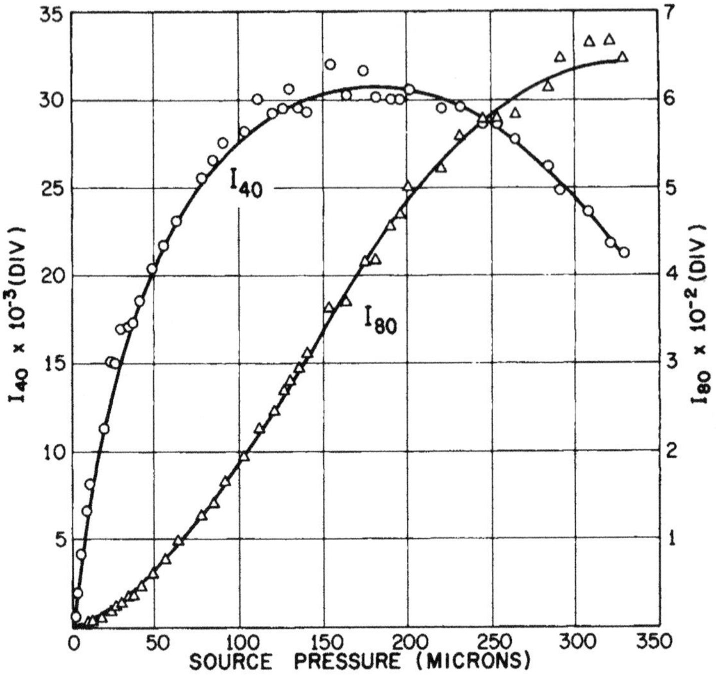


Fig. 1 Ion currents in Ar to high pressure

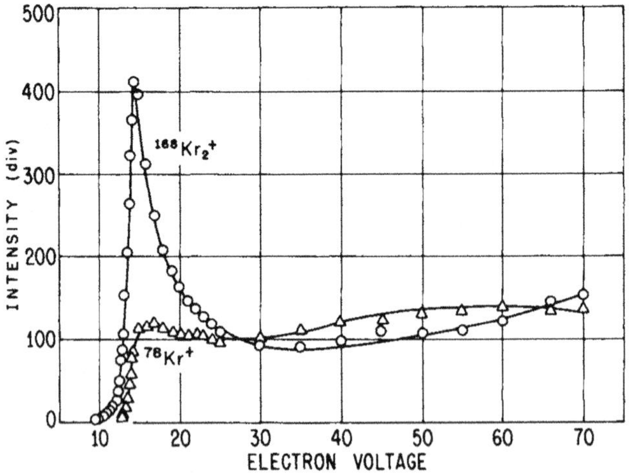


Fig. 2 Krypton ionization efficiency curves; monoisotopic $I_{Kr^+}$ = 282 $I_{^{78}Kr^+}$, monoisotopic $I_{Kr_2^+}$ = 2.75$I_{^{168}Kr_2^+}$

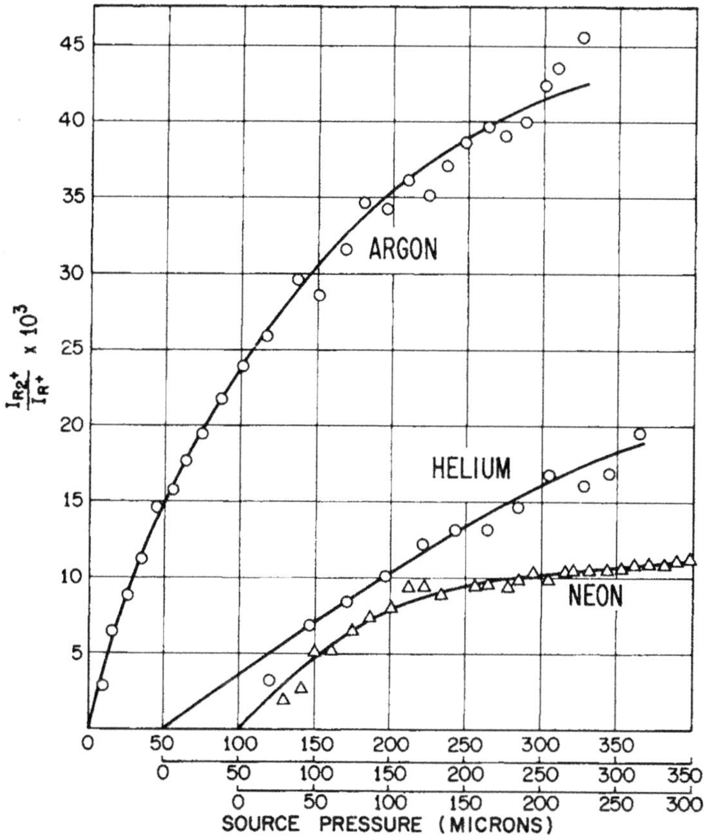


Fig. 3 $I_{R_2}+/I_R+$ vs source pressure in Ar, Ne and He; EV = 15v (Ar), 20v (Ne), 22v (He); FS = 50 v cm in all cases

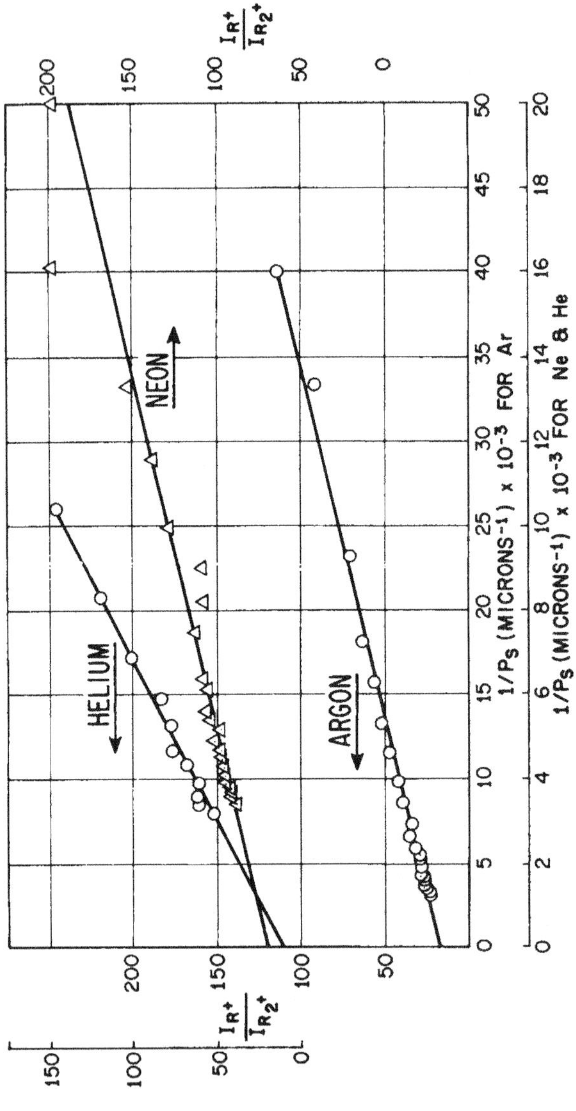


Fig. 4 $I_{R^+}/I_{R_2^+}$ vs $1/P_S$, conditions as in Fig. 3

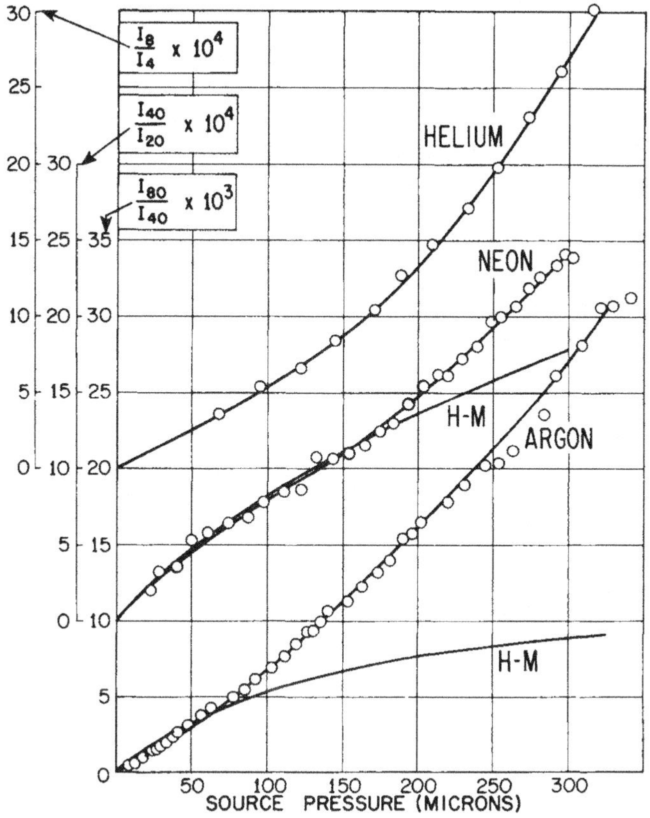


Fig. 5 $I_{R_2}{}^+/I_R{}^+$ vs source pressure in Ar, Ne, and He; EV = 70 v; FS = 12.5 v/cm; "H-M" stands for calculated Hornbeck-Molnar contribution

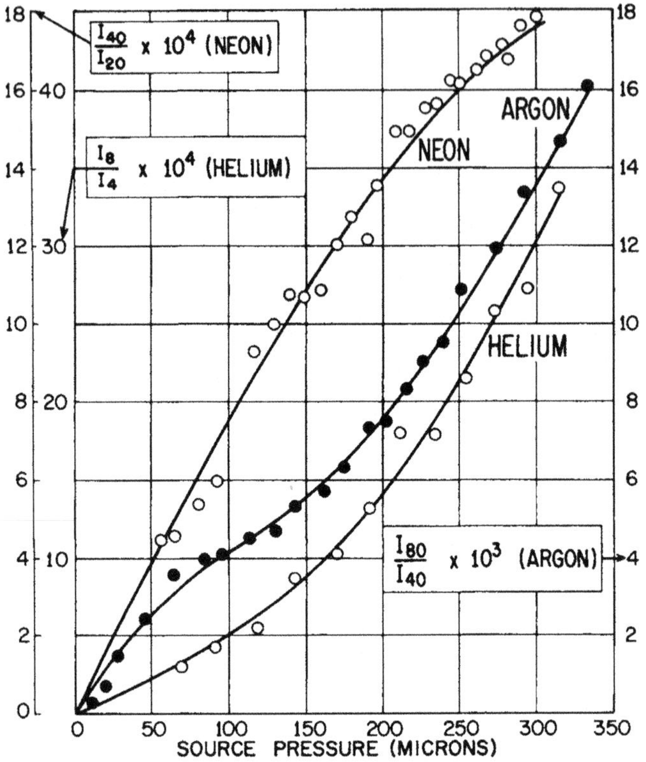


Fig. 6 $I_{R_2}+/I_R+$ vs source pressure in Ar, Ne, and He; $EV = 70$ v, $FS = 50$ v/cm

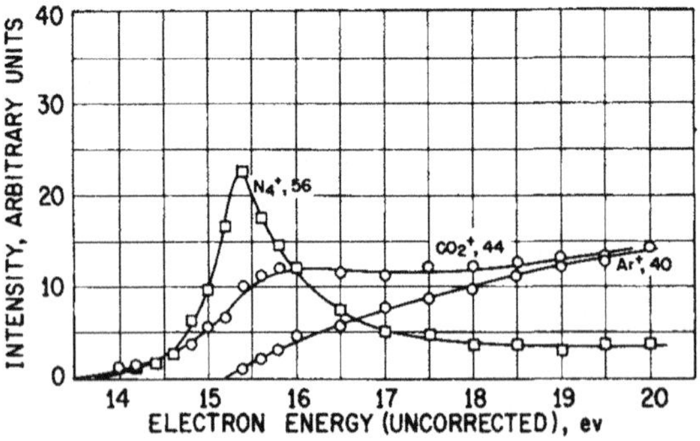


Fig. 7 Ionization efficiency curve for $N_4^+$ from $N_2$; $P(N_2) = 154\mu$; $FS = 12.5$ v/cm; $EC = 20\mu a$

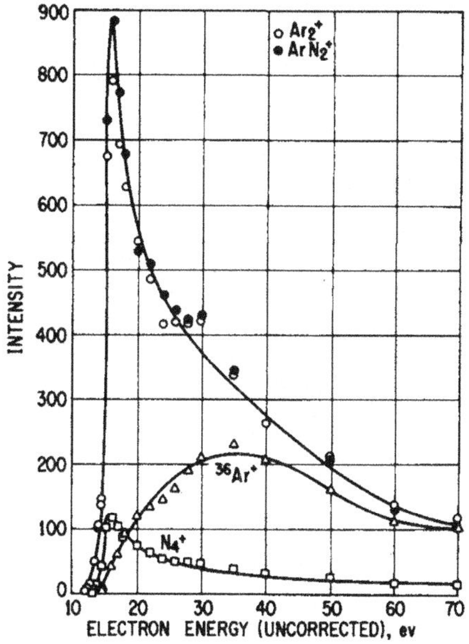


Fig. 8 Ar - $N_2$ ionization efficiency curves; P = 160μ(Ar ≅ $N_2$); FS = 12.5 v/cm

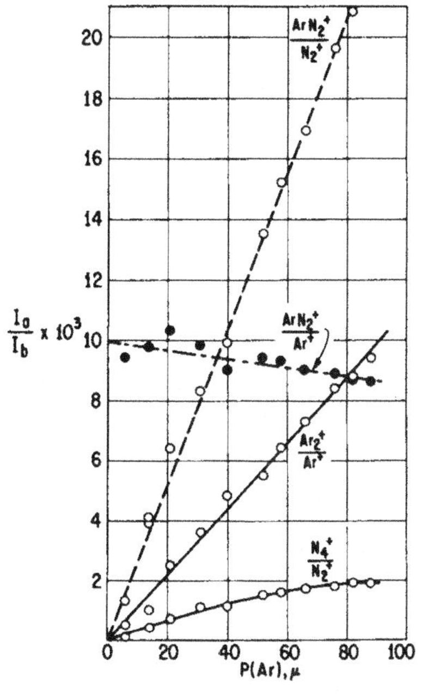


Fig. 9 Addition of Ar to $N_2$; P($N_2$)= 81μ, E ≙ 15 ev, FS = 50 v/cm, EC = 0.5μa

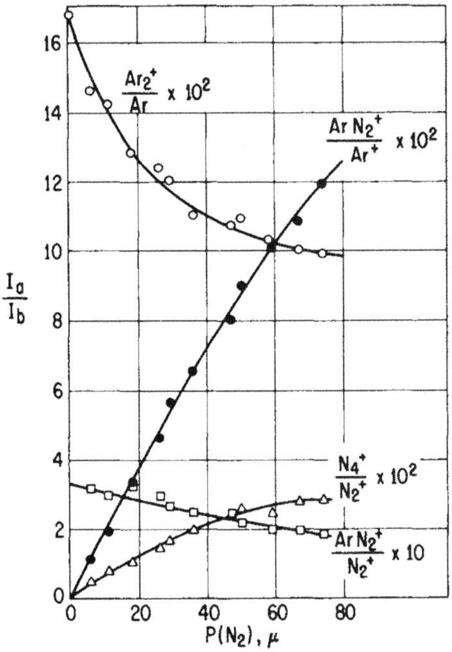


Fig. 10 Addition of $N_2$ to Ar; P(Ar) = 85μ, E ≅ 15 ev, FS = 12.5 v/cm; EC = 0.5 μa

# II

# IONIZATION IN FLAMES

# NONEQUILIBRIUM IONIZATION IN FLAMES

H. F. Calcote*

AeroChem Research Laboratories, Inc., Princeton, N. J.

## Abstract

This paper reviews the efforts of the author on "ionization in flames" from 1946 to 1962. A literature study led to the concept of chemi-ionization. Experiments on the deflection of flames in both transverse and longitudinal d.c. electric fields were interpreted by an "electric wind" theory from which it was concluded that high concentrations of ions occurred in a thin reacting zone of the flame. These conclusions were verified by Langmuir probe studies which were used to measure the rate of ion recombination in hydrocarbon-air flames. The probe studies also led to a proposed mechanism of ion formation, charge transfer, ion recombination and ambipolar diffusion which essentially accounts for the dominant ion processes occurring in hydrocarbon-air or oxygen flames. A mass spectrometric

---

Presented at the ARS Ions in Flames and Rocket Exhausts Conference, Palm Spring, Calif., October 10-12, 1962. This report was prepared under contract Nonr-3809(00). The original work was sponsored by the Office of Naval Research via Project SQUID. The development of the mass spectrometer was supported by AF-BMD, and work is continuing with support from the Naval Bureau of Weapons, Air Force Combridge Research Center and the Office of Naval Research. Several people have participated in these programs throughout the years whose assistance is acknowledged in the references. I am particularly indebted to the following who reviewed the manuscript and are currently extending this work: A. Fontijn, S. C. Kurzius and H. Silla.

*Director of Research.

technique for obtaining ion profiles in low pressure flames was developed and used to identify the ions and to study their production and decay. The information on mechanisms of ion formation and decay in flames is being applied to the problem of radar attenuation in rocket exhausts.

## Introduction

The acceptance of nonequilibrium ionization in flames is a very recent phenomenon although the necessary experimental observations have been in the literature since the early part of the century. Evaluation of the early literature, flame deflection experiments, ion profile measurements with Langmuir probes, microwave measurements, and mass spectrometric studies of ionization in flames have persuaded the scientific community that ion formation in flames is a reaction rate phenomenon and not a thermodynamic equilibrium phenomenon due to impurities, free radicals, or incipient carbon particles. There is general agreement that the formation of ions is due to reactions of specific species of well-known flame front constituents.[1] There is still some disagreement as to exactly what these reactant species are.[2] We have reached the point, however, in our understanding where flames may be used as a medium for studying both elementary ion-molecule reactions and ion recombination reactions at temperatures and pressures which do not seem to be available in any other experimental apparatus. This, then, is an opportune time to review the observations and arguments which have led to this state of affairs.

Because the early literature is so often neglected in the attempt to keep up with new developments, a "literature review and analysis" made in 1948[3] of the early literature on the electrical properties of flames will be summarized briefly. The information in this early literature made it evident that ionization in flames is an interesting subject and led to the concept of chemi-ionization. Experimental studies on the deflection of flames in both transverse and longitudinal electric fields therefore were undertaken.[4,5] This work confirmed the author's convictions about chemi-ionization but indicated a need for detailed ion profile studies through the flame front. The only tool he could conceive of which might give absolute values of ion and electron concentrations through the flame front with sufficient spatial resolution to be meaningful † was the Langmuir probe.[6] This tool gave absolute ion and electron concentrations in the flame front and led to measured values of the rate of ion recombination as well as estimates of the rate of ion formation.[7] The ions present still needed to be identified because a rather large number of ion producing reactions meeting the requirements for chemi-ionization could be written down, and it seemed that identification of the ions present in the flame would resolve the problem. Thus began a program to develop a mass spectrometric technique of identifying the ions present and to study their profiles in low pressure flames.[8] To the disappointment of all, the mass spectrometric results did not resolve the problem but increased it because of the very large number of ion species observed. Thus, the bits of information from all sources had to be sifted to arrive at the specific ion forming reactions and subsequent ion-molecule reactions which must occur. Of course, involvement in studies of ions in flames usually involves one in the very important problem of electromagnetic wave attenuation in rocket exhausts.[9]

The author's personal efforts on the subject of "ions in flames" will be outlined below. Of course the work of others has had a strong interaction with these efforts, but space does not permit tracing through all of these here.

## An Assessment of Early Literature - Chemi-Ionization

Reviews of early work on "ions in flames" can be found by H. A. Wilson, 1912,[10] and again in 1931,[11] by Thomson and Thomson, 1928,[12] and by Becker, 1929.[13] The author reviewed the subject[3] in 1948 and had to state then (p. 1), "Ions and their contingent phenomena have [therefore] been eliminated from modern discussions on combustion." The purpose of that review as recorded was "to emphasize the fact that the experimental results thus far obtained demand more consideration of this neglected phase of the combustion problem." Some of the observations which led to the above statement

† Estimates of flame thickness varied from $10^{-4}$ to $2 \times 10^{-2}$ cm.

and to the interpretation of the ion forming step as a chemi-ionization reaction (similar to a chemiluminescent reaction) rather than a thermal process will be presented.

Chemi-ionization (Ref. 3, p. 49) was defined as

$$A + B \rightarrow \text{activated complex} \rightarrow C + D^+ + e^-$$

$$\text{or} \quad \rightarrow C^+ + D^-$$

$$\text{or} \quad \rightarrow C + D^* \xrightarrow{M} D^+ + e^-$$

The species A, B, C, D, and M were not specifically identified but allowed to be atoms, radicals, or molecules.

Experiments by several investigators[14] as early as 1906 had demonstrated that in a Bunsen flame the conductivity in the inner cone exceeded the conductivity in the outer cone by several orders of magnitude. This observation alone was sufficient to eliminate thermal ionization of impurities which do not change their identity as they move through the flame front because, if anything, the temperature of the outer cone exceeds that of the inner cone. A rough estimate of $4 \times 10^{10}$ ions/cc was obtained for the ion concentration in a typical inner Bunsen cone by using the ratio of 100 to 1 between conductivity in the inner cone and outer cone observed by Tufts[14] in 1906 and the measurements of flame conductivity and electron mobility of a number of investigators to calculate the electron concentration in the outer cone.[3] Because of its general historical interest, Table 1 of the author's review is reproduced (Table 1). A value for the flame conductivity of $\lambda \approx 5 \times 10^{-7}$ mhos/cm and $k_e \approx 8 \times 10^3$ cm/sec/v /cm was chosen which led to an estimate of $n_e \approx 5 \times 10^8$ ions/cc in the outer cone.

Such a relatively high value of electron concentration made it difficult to account for the ions by a thermal process because either very high concentrations of the species being ionized or impossibly low values of the ionization potential would be required. This argument was further strengthened by the measurements of the temperature variation of ionization[15,16] which indicated an ionization potential, using the thermal theory, of about‡ 2 eV. A similar value recently has been obtained by van Tiggelen and associates,[18] and they interpret it as an activation energy for ionization.

The argument that Bunsen burner flame ionization is due to particles being thermally ionized[19] should have been rejected because of Tuft's observation[14] in 1906 that in diffusion flames the maximum electrical conductivity occurs in the lower part of the flame and not in the upper luminous part. The observation of ions in lean flames should also have made the argument involving thermal ionization of particles questionable.

A series of interesting experiments by Garner, Saunders, and associates[20] were reported in the late 1920's in which they measured the conductivity of explosions in closed vessels. For hydrogen-oxygen explosions they estimated $n_e \sim 10^{13}$ electrons/cc. Nevertheless, they concluded ionization was a thermal process. In measurements made over a range of compositions they found:

| Fuel with oxygen | Maximum current (mamp) |
|---|---|
| CO | 0.006 |
| $H_2$ | 4 |
| $CH_4$ | 4 |
| $C_2H_2$ | 200 |

Today it is known that ionization in hydrogen-oxygen flames is negligibly small when hydrocarbon impurities are removed.§ [21]

A rather extensive series of studies by various individuals[22] in the 1920's and 1930's on the effect of additives on flame ionization demonstrated that additives produce changes in the degree of ionization, which emphasizes the importance of chemistry in the ionization process. Most of this work was related

‡ Lusby's data[17] reported in 1911 on flame current variation with temperature can be interpreted to give $V_i \approx 1.7$ eV (Ref. 3, p. 13).

§ Widespread use of Ionization Flame Detectors in Gas Chromotography using hydrogen-air flames.

to the "knock" problem in reciprocating engines. Many other observations in the old literature beckon one to look further. Several other interesting and pertinent clues could be quoted. For example, Rossikhin and Timkovskii, 1940,[23] found that in a 25% $C_2H_2$ air flame the intensity of the $C_2$ bands (5635 Å, 4364 Å) and the CH band (4314 Å) did not change in a high frequency field (25 and 34 mc/sec), but the intensity of the CH band (3900 Å) was increased.

The question of whether ions and electrons play a role in flame propagation also was pursued with vigor prior to the mid 1930's. A large number of deflection and extinction experiments were carried out in electric and magnetic fields. Unfortunately these do not lend themselves to a systematic analysis or interpretation. The experiments can be used to argue with equal vigor for either a chemical effect, i.e., ions and/or electrons are important in flame propagation; or a mechanical effect, i.e., the electric field moves the flame because of a mechanical force, electric wind, exerted on the ions which transmit this force to the bulk gas.

In assessing the older literature it is difficult to conclude anything other than that the ions are produced by chemi-ionization. Yet the discussion and argument continued unabated until very recently for thermal ionization of impurities or carbon particles.[7]

## Flame Deflection in Electric Fields

Discordant results and interpretations reported in the literature concerning the effect of electric fields on flame speeds and other flame properties prompted an investigation of the phenomena. The principle object in these experiments was to clarify the mechanism by which flames are deflected in electric fields. As a by-product, an estimate of the ion concentration in the flame front was obtained. Three types of experiments were undertaken with dc electric fields. Bunsen burner type flames were exposed to both transverse fields, i.e., the field approximately normal to the unburned gas flow lines, and longitudinal fields, i.e., the field approximately parallel to the unburned gas flow lines. Experiments also were performed with dc electric fields applied longitudinally to the direction of flame propagation in a tube.

## Transverse Fields on Burner Flames

A Bunsen-type flame burning on a rectangular duct was exposed to a dc electric field applied between two plates 6 cm apart.[4] The deflection of the flow lines was determined by means of aluminum particle tracks, Fig. 1. An analysis was made based upon the "electric wind" concept, i. e., the concept that the flame front experiences a force due to the drag of the positive ions, which are accelerated toward the negative electrode, on the rest of the gas. Because of the light mass of the electrons, assuming negligible electron attachment, the force on the flame is toward the negative electrode. Such an analysis is reasonable only when the thickness of the ionized region is less than the distance the electrons must travel to reach a steady state velocity. When the steady state velocity is attained, the momentum transfer might be expected to be the same from both electrons and ions, and no ion wind would exist unless the electron velocity in the steady state is so high that nonelastic collisions dominate. The subject of electric winds and flame deflection deserves more attention.

The analysis for a n-butane-air flame at 1 atm and an equivalence ratio of 0.82 produced a value for the multiple of the positive ion concentration $n_+$ and the thickness $L$ of the ionized region:

$$n_+L = 7.7 \times 10^{10} \text{ ion/cc/cm}^{-1}$$

A major difficulty in this analysis was estimating the field strength in the flame zone because large potential drops would be expected between the combustion zone and the electrodes where there was no conducting gas. This was verified by the use of probes which also were used to estimate the field strength in the flame. Such measurements are seldom very satisfactory however.

The above value could be used to estimate the ion concentration in the combustion zone if the thickness of the ionized flame were known. However, at the time this experiment was done there were no measurements of ionized flame thicknesses.

Therefore, a value of the ion concentration was estimated by adding sodium to the flame and deducing that if the addition of sodium to give $n_+ = 2 \times 10^{12}$ ions/cc by Saha's equation did not cause an increase in deflection, then the ion concentration in the unseeded flame must be at least this value. There are apparently several things wrong with this argument: a) even alkali metal ions may give an excess of ionization in the combustion zone; b) the ionization zone in a seeded flame is much thicker than in a nonseeded flame so not only $n_+$ is increased but L is also increased. Yet the value of $n_+$ from the deflection experiments is the value most frequently quoted for the ion concentration in a hydrocarbon flame front at 1 atm. This is probably because it has been dignified by being quoted in a widely-read book.[24]

With the value of $n_+$ "known," it was possible to calculate a value of L. Thus,

$$L \lesssim 8 \times 10^{-2} \text{ cm}$$

This seemed reasonable at the time because it was in the range of flame thickness which had been estimated by other means, i.e., $L = 10^{-4}$ to $2 \times 10^{-2}$ cm. It is interesting to compare this rather crudely derived value for L with one determined by measuring the ion concentration through the flame front with Langmuir probes.[6] For a propane-air flame at an equivalence ratio of 0.8 the measured thickness was $2 \times 10^{-2}$ cm.

The high value of $n_+$ was interpreted as evidence for chemi-ionization.[4]

<u>Longitudinal Fields on Burner Flames</u>

A longitudinal electric field was applied to Bunsen burner flames[5] by placing a quartz tube of 2.5-cm internal diameter over a burner (Fig. 2) with an internal diameter of 0.66 cm such that the flame burned inside the quartz tube. The dc electric field was applied between a platinum ring 1.5 cm wide placed on the outside of the quartz tube and 4.2 cm above the burner. This required that any current must flow through the quartz tube so that the actual voltage gradient in the flame must have been very low due to the voltage drop through the quartz. The large electrical resistance of the quartz tube also prevented electrical breakdown in the flame. Flame pressures were determined with a manometer filled with petroleum ether and mounted at an angle of $1^{o}40'$ to the horizontal. When the metal ring was positive with respect to the burner so that positive ions would be accelerated toward the burner and the unburned gas, the flame stability was greatly increased. Figure 3 shows that both the blow-off limits and maximum total flow velocity were increased by the field. Simultaneously the dead space, i.e., the distance between the base of the flame and the burner, was decreased. The flame pressure as indicated on the sensitive manometer was increased.

With the reverse field direction the results were not so clearly defined. At low applied electric fields (total 8 kv) the stability limits were decreased, the dead space increased, and the flame pressure decreased. However, at greater applied voltages an increase in stability was observed apparently due to local electrical breakdown around the burner rim. This led to an interesting change in flame appearance with increasing electric field (Fig. 4). The flame first became polyhedral; then, after a strong vibration, it would suddenly change to a "derby" shape.

The performance of these flames in electric fields was also interpreted in terms of an electric wind. It was thus possible from the measurements of change in flame pressure to estimate $n_+L$. At a positive field (ring positive) of 10-kv total voltage the current was ~ 20 mamps and the flame pressure about 5 dynes/sq cm. This yielded a value for $n_+ > 1 \times 10^{10}$. The main uncertainty was in estimating the actual field strength to which the flame front was exposed. This value for $n_+$ is consistent with that previously obtained from transverse electric fields.

## Role of Ions in Flame Propagation

The literature abounds in experimental observations which have been interpreted either in terms of an important role of ions in the mechanism of flame propagation or in terms of mechanical effects when the ions are exposed to an electric field. The results are not sufficiently clear-cut to base a decision in favor of one or the other interpretations. In the work just reported a mechanical model seemed adequate although the quantitative character of the experiments left something to be desired.

The initial suggestion of J. J. Thomson[25] in 1910 that electrons play a significant role in flame propagation by rapidly diffusing ahead of the flame front sparked the initial studies on the deflection and extinction of flames in both electric and magnetic fields. This suggestion reminds one of more recent theories of flame propagation based upon the rapid diffusion coefficients of hydrogen atoms in flames. Knowledge of the nature of the parent ion and the recombination process may lead to further speculation as to mechanisms by which ions in flames could participate in the process of flame propagation. Ambipolar diffusion of the dominant $H_3O^+$ ion ahead of the flame front could be rather rapid because of a large ion concentration gradient. Once ahead of the flame the ion could recombine with the electrons to produce several free radicals by, e. g.,

$$H_3O^+ + e^- \rightarrow OH + 2H \qquad (1)$$

Radicals and atoms so produced would initiate the combustion reactions common to other concepts of flame propagation. Such speculation is tempered by the relatively low concentrations of flame ions; notwithstanding, the exercise of placing the argument on a quantitative basis is suggested as being informative.

In an attempt to determine the effect of an electric field on the burning velocity of a Bunsen burner flame, the flame deflection experiments in transverse fields were further analyzed.[4] The particle tracks in Fig. 1 were interpreted to give the burning velocity $S$ at any position in the flame front. Two common techniques were used. The local burning velocity was computed from the measured angle $\alpha$ which the flow lines make with the unburned side of the visible flame front. The local burning velocity also was calculated by measuring the change in diameter of a stream tube as it leaves the burner port and as it enters the flame $dr/dl$. The results are presented in Fig. 5. The no-field case, represented by the solid curve, is normal. With the application of an electric field (positive plate on the left and negative plate on the right) the results are scattered, but an average curve through them (dashed curve in Fig. 5) indicates about a 20% increase on the positive electrode side where positive ions would be accelerated into the unburned gas and electrons into the burned gas away from the flame front, and a similar decrease on the negative electrode side where the opposite phenomena occur. The observation is indicative of an important role for the positive ions or possibly for an inhibitory role for electrons. The experiment was conceived to separate out the electric wind mechanism, which bends the flame, from the changes in local burning velocity. Although the results indicate a positive effect, they should be accepted with caution because of the nature of the experiment. The work should be repeated more carefully.

Experiments of a different nature also were performed to gain further insight into the nature of the interaction of electric fields with flames.[26] In this experiment a flame was made to propagate downward through a Pyrex tube 150 cm long and 2.5 cm in internal diameter. The upper part of the tube was connected to a 5-liter flask through a large stopcock which was opened just prior to igniting the combustible mixture in the tube with a spark or hot wire. The flask initially was filled with air to equal the pressure in the tube. The flame propagation velocity down the tube was measured with rf flame ionization detectors[27] placed at a known distance apart. A longitudinal electric field was applied to the tube by two external metal rings as in Fig. 2.

The results are shown in Fig. 6. At some critical applied voltage the flame speed was suddenly increased, and then very little change in speed of propagation was realized with further increase in voltage. The result was qualitatively independent of direction of the electric field although it was somewhat greater when the first electode was positive. The flame front was observed to be wrinkled after entering the electric field and the increased velocity persisted beyond the second electrode, i. e., beyond the electric field. These results are consistent with a mechanical interpretation of the effect of the electric field although they do not eliminate the possibility of a chemical effect.

## Studies with Langmuir Probes

The studies already outlined indicated the need for detailed studies of the ion profiles through the flame front. Early literature on flame conductivity had shown that the degree of ionization was greatest in the flame front, and interpretation of the flame deflection experiments in electric fields indicated that the thickness of the highly ionized zone was very thin and of the same magnitude as estimates of the thickness of the combustion wave. Thus, to study chemi-ionization a very small probe would be required. The conductivity experiments previously made[3,11] used fairly large electrodes so they were quickly eliminated from consideration. Microwave techniques are only able to resolve a region of the flame comparable to or greater than one wavelength (at least several millimeters in the very best case) so they also were eliminated from consideration. Heumann[28] had shown in 1940 that a small wire probe inserted in a flame in which the current was collected as a function of the applied voltage behaved as a Langmuir probe. In spite of the complications of inserting a probe into the flame, this device was chosen as the tool for studying ionization in the flame front of a large number of flames.

### Ion Concentrations

Initial work[6] gave low results for ion concentrations because of neglect of a pressure correction which subsequently has been taken into account.[29-31] In fact, the careful use of small cooled probes with appropriate considerations in interpreting the data, including corrections for ambipolar diffusion in the flame, allows the determination of accurate values of the positive ion concentration $n_+$, the electron concentration $n_e$, and the electron temperature $T_e$. Proper use of the probe is fraught with pitfalls, and, like any other tool, it must be used with care in spite of its apparent simplicity. The probe has given valuable data on ion concentrations and rates of ion recombination.

The first use of the probe in 1955[6] was to demonstrate that in lean flames with temperatures as low as 1300°K the ion concentrations were far too great to be accounted for by thermal ionization of impurities, carbon particles, or free radicals, and that the ions must be due to chemi-ionization. Some of the results are summarized in Table 2, where the results are now corrected for the pressure effect on the probe current as referred to above. It is interesting to compare the maximum ion concentration for the flame at an equivalence ratio of 0.835 with that deduced for a butane-air flame (equivalence ratio of 0.82) from flame deflection experiments. The values are $2 \times 10^{11}$ and $10^{12}$ ions/cc, respectively. The propane air flame was diluted with 59.6 mole % nitrogen. It also was shown that neither hydrogen sulfide nor carbon disulfide flames produce ions to any detectable extent although their temperatures exceed 2000°K.

## Ion Recombination Coefficients

The maximum ion concentration occurs just at the downstream edge of the luminous cone, Fig. 7, in hydrocarbon-air or oxygen flames. It was shown in 1957[7] that the decay of ionization behind the flame front behaved as a second-order process with respect to the positive ion concentration. This was interpreted as an ion recombination process by assuming that the production of ions effectively ceased at the downstream edge of the luminous zone. The rate of recombination is given by

$$\frac{dn_+}{dt} = \alpha\, n_+ n_- = \alpha\, n_+^2$$

when $n_+ \approx n_-$, i.e., the concentrations of positive and negative species recombining are equal. The recombination coefficient $\alpha$ can be obtained by integration of the above to give

$$\frac{1}{n_+(t)} = \alpha t + \frac{1}{n_+(0)}$$

where $n_+(t)$ and $n_+(0)$ represent the ion concentration at times t and 0, respectively. A plot of $1/n_+(t)$ against time gave a straight line. The slope yielded $\alpha$. The distance scale through the flame was converted to time via the known gas flow velocity. A value of $\alpha = 2 \times 10^{-7}$ cc/sec thus was obtained for a propane-air flame at 1 atm. This value subsequently was confirmed by more careful studies[30] and shown to be independent of pressure when appropriate corrections were made at lower pressure for ambipolar diffusion. King reported[32] in 1957 a value close to this but failed to confirm the pressure independence[33] because he neglected the correction for ambipolar diffusion. Green and Sugden in 1962[21] also found the same value for $\alpha_+$ using a mass spectrometer and following the decay of the $H_3O^+$ peak. Because the dominant ion is $H_3O^+$,[8,34,35] the mass spectrometer and the Langmuir probe give the same result.

## Rate of Ion Formation

It was argued[7] further that, if the process for recombination remains the same through the flame front, the maximum ion concentration in the flame occurs at the position at which the rate of ion recombination equals the rate of ion formation. Thus, the rate of ion formation $dn_+/dt$ was found to be $2 \times 10^{15}$ ions/cc/sec.

From a kinetic theory estimate of the rate constant for ion formation from free radicals with no activation energy, $k = 5 \times 10^{-11}$ to $5 \times 10^{-10}$ cc/sec was obtained″ at 2000°K.[7,30] The value of k from kinetic theory and the measured rate of ion formation were used to estimate the concentrations of reactants in the flame producing ions, and it was argued that these concentrations were reasonable for the reactants considered for chemi-ionization in 1957. This compares with Green and Sugden's estimate[21] of $k = 3 \times 10^{-12}$ cc/sec.

## Mechanism of Ion Formation

Reference 7 was a detailed evaluation of the various mechanisms which had been proposed for the formation of ions in

″There was a numerical mistake in the original paper, and k was incorrectly given as $8 \times 10^{-15}$ cc/sec.

flames. These mechanisms were:

1) thermal ionization of impurities,
2) thermal ionization of reaction products and intermediates with low ionization potentials,
3) thermal ionization of carbon particles,
4) ionization via nonequilibrated translational energy,
5) cumulative excitation, and
6) chemi-ionization.

It was shown that only cumulative excitation and chemi-ionization fit the experimental observations. Some possible reactions were proposed for the last two mechanisms, based mostly on arguments of energetics. The basic problem this work posed was that of the energy distribution between reaction products in an energetic reaction.

At the Seventh Combustion Symposium both Sugden and van Tiggelen presented mass spectrometric data to show that the dominant ion in hydrocarbon flames is $H_3O^+$ with many other ions present in smaller concentrations. This at first presented a problem in that no simple reactions could be written to produce $H_3O^+$ from neutral species and thus the chemi-ionization mechanism seemed in jeopardy. However, it was shown[36] that ion-molecule proton transfer reactions could be written which would explain the dominant $H_3O^+$ ion as being produced from ions formed by chemi-ionization or cumulative excitation reactions. Mechanisms also were presented to explain the large number of other ions observed, particularly large hydrocarbon ions.

In 1959[37] a review of the various relaxation processes in flames such as ion recombination, electron attachment, ion-molecule charge transfer, ambipolar diffusion, and electron temperature decay was prepared. The ion forming reaction

$$CH(^2\Sigma^+) + O \rightarrow CHO^+ + e^- \qquad (2)$$

was suggested. This reaction is exothermic to the extent# of 71 kcal/mole which might easily account for the high electron temperatures observed in flames. The ground state reaction[30]

$$CH + O \rightarrow CHO^+ + e^- \qquad (3)$$

is endothermic (19 kcal/mole) so that it could not account for the high electron temperatures observed. On the other hand, Green and Sugden[21] used 203 kcal/mole for the heat of formation of $CHO^+$ and find reaction (3) to be thermoneutral. The low value of 203 kcal/mole is from Field and Franklin[38] and the high value the author has chosen, 220 kcal/mole, is from a recent review by Bernecker and Long.[39] The whole subject of electron temperature and the thermochemistry needs further clarification.

The theoretical electron temperature decay times are of the order of $10^{-7}$ sec for electron-molecule collisions and milliseconds for coulombic collisions. Yet the experimental electron temperatures persist for some distance, i.e., milliseconds, downstream of the flame front. This is presumably the result of their high diffusion velocity, but this problem still requires quantitative description.[31]

Ion-Molecule Charge Transfer

It was also shown[37] that the $CHO^+$ ion would be expected to be short lived in the environment of a flame, and it thus would transfer its charge rapidly via

$$CHO^+ + H_2O \rightarrow H_3O^+ + CO \qquad (4)$$

A rate constant for this reaction was estimated to be between $10^{-9}$ and $10^{-8}$ cc/sec. In the hydrocarbon flames studied by mass spectrometry by Knewstubb and Sugden,[35] Deckers and van Tiggelen,[40] and Calcote and Reuter[8] the relaxation time

#The original thermochemistry has been revised using data based upon the heat of vaporization of carbon being 170 kcal/mole (see Table 3).

for the $CHO^+$ ion thus would be $10^{-10}$ to $10^{-7}$ sec. This small relaxation time was suggested as the reason for the $CHO^+$ ion's not being originally reported in mass spectrometric studies.[35,40] Deckers did, however, add some data to his original paper at the Seventh Combustion Symposium[41] that showed the $CHO^+$ ion to be present to the extent of only about 1% of $H_3O^+$ and $C_3H_3^+$. The presence of $CHO^+$ in small concentrations subsequently was verified in both Sugden's laboratory[21] and the author's.[8]

## Dissociative Recombination

The recombination process was then suggested[30,37] to be the dissociative recombination

$$H_3O^+ + OH^- \rightarrow 2H_2O$$

$$\text{or} \qquad \rightarrow H_2O + OH + H \qquad (5)$$

or

$$H_3O^+ + e^- \rightarrow H_2O + H$$

$$\text{or} \qquad \rightarrow OH + 2H \qquad (6)$$

The fact that the recombination is second order and that the recombination coefficient $\alpha$ is independent of pressure already has been mentioned.

The pressure independence of $\alpha$ was obtained only after correcting the ion decay rates for ambipolar diffusion. Without this correction the recombination rate was inversely dependent upon pressure to about the 0.6 power. Below about 30-60 mm Hg the ion decay process was shown to be first order with respect to positive ion concentration. This was interpreted as due to ambipolar diffusion, and diffusion coefficients were deduced from the data. When these values were reduced to 1 mm Hg and 298°K, the value for stoichiometric propane-air ($\phi = 1.0$), obtained at 33 mm Hg, was 90 $cm^2/sec$, and when the nitrogen of the air was replaced by helium, the value was 260 $cm^2/sec$. The reasonableness of these values gave us confidence in the interpretation of the first-order decay at low pressures as being due to ambipolar diffusion. The success in obtaining pressure-independent recombination coefficients when correcting for ambipolar diffusion - this must be so for a second-order process - further increased our confidence in the interpretation of the low-pressure decay as ambipolar diffusion.

## Further Considerations of Ion Formation and Recombination

The method of treating the Langmuir probe data and the corrections for ambipolar diffusion were discussed in more detail at the Eighth Combustion Symposium.[30] Thirteen possible chemi-ionization reactions were tabulated. These are repeated in Table 3. Although reaction (4) has gained rather wide support[21,42,43] and is certainly the major source of ions in flames, there seems no basic reason why some of the other reactions may not also contribute to ion formation in flames. In fact, it would not be surprising if chemi-ionization turned out to be a rather common phenomenon. It is interesting in this regard to note that Brewer[44] in 1924 observed ion formation in every gas phase reaction studied although the temperature was far too low for thermal ionization. The choice of possible reactions presented in Table 3 was determined by the requirements that

1) the reaction should be exothermic or nearly thermoneutral; 2) the reactants should be simple chemical species known to be present in flames; 3) the reactants must involve species from the fuel and the oxidizer; 4) the products should be simple ions.

The observation that the maximum mole fraction of ions in a flame is independent of pressure, coupled with the observation that ion recombination is second order with respect to positive ion concentration, indicates that the mole fractions of ion precursors must be independent of total pressure.

Further considerations in using the Langmuir probe were outlined at the Ninth Combustion Symposium[31] where it was shown how the probe might be used for determining electron concentrations as well as positive ion concentrations. An internal self-consistency check was used by which the experimentally observed wall potential was compared with the wall potential calculated from the experimentally measured $n_+$, $n_e$, and $T_e$. The agreement was in general very good, and the principle limitation on accuracy appeared to be the electron collision cross section which is required to reduce the data.

Some results are reported in Fig. 7 in which the positive ion concentration demonstrates the usual decay. However, rather typically the electron concentration is less than the positive ion and rather constant through the flame front, showing only a very slow decay. This low concentration of electrons indicates rather large concentrations of negative ions, perhaps $OH^-$. Actually this concentration of $OH^-$ would not be unreasonable in view of the excess concentration of OH radicals in hydrocarbon flame fronts and the OH radicals profile in hydrocarbon-air flames.[45] The convergence of positive ion and electron concentrations downstream from the flame front tends to support the validity of the measurement. Further justification for large concentrations of negative ions can be found in the observation that measured recombination coefficients of alkali metal ions and lead ions in flames give values of about $10^{-9}$ cc/sec (King[46] and Padley and Sugden.[47]) This is consistent with a Thomson three-body process involving negative ions, but not so for a three-body process involving electrons that gives $\alpha \approx 10^{-11}$ cc/sec. Mass spectrometric analysis,[1,2,47] however, has shown the concentration of $OH^-$ to be only 1/100th that of the electrons.

Both low electron concentrations and the constancy of electron concentration with distance through the flame indicate that recombination is via reaction (5) rather than by reaction (6). It must be admitted, of course, that the accuracy of the Langmuir probe and the collision cross-section data that go into data reduction are such that further verification of the ratio of positive ions to electrons is highly desirable.

The electron temperature in a propane-air flame was found to exceed the adiabatic flame temperature by several hundred degrees, (Fig. 7). The electron temperature also exhibits a slower relaxation than theory would predict and, as has already been suggested, this may be due to a high diffusion coefficient. The electron temperature measurements always show more scatter than $n_+$ or $n_e$, but it always falls above the adiabatic flame temperature. It thus far has received less attention than $n_+$ and will be considered more carefully in the future. Confidence in the validity of $T_e$ depends upon the internal agreement obtained between the observed wall potential and the wall potential computed from the observed $n_+$, $n_e$, and $T_e$.

## Nonequilibrium Ionization of Alkali Metals

The nonequilibrium ionization of potassium when added to a hydrocarbon-air flame also has been observed[26] (Fig. 8). This is probably due to the reaction

$$H_3O^+ + K \rightarrow K^+ + H_2O + H \tag{7}$$

When small quantities of alkali metals, potassium or sodium, are added to hydrocarbon-air flames, the mass spectrometric studies[8] show that the alkali metal ion appears and all other ions are suppressed.

## Mass Spectrometric Studies

All of the foregoing work with probes gives total concentrations only. To determine the identity of ions present, it is necessary to use another tool. A mass spectrometer is ideal for this purpose. The first object of mass spectrometric studies was obviously to identify the ions present and thus to determine which of the possible chemi-ionization reactions proposed[30] are responsible for ions in flames. The development of a mass spectrometric technique for studying ions in low-pressure flames began at AeroChem in 1957. In this same year both van Tiggelen[40] and Sugden[35] reported their first results of mass spectrometric studies of ions in flames. The results laid the way for a much deeper insight into the subject but were somewhat disappointing, to the author at least, in that the dominant ion was not one of the ions proposed but $H_3O^+$. Many other ions were also present. Of course this also left much more interesting work to be done, and some of this has been discussed already in the previous section. The author's first results were not reported until

1961[8] and were in agreement with observations of both Sugden's and van Tiggelen's groups.

The mass spectrometric studies of the three groups mentioned have raised many problems, particularly the determination of the mechanisms and rates of formation of the various ions observed. Most important, however, they have demonstrated which ions are present and to what extent.

In mass spectrometer work, Reuter and the author[48] have concentrated on obtaining detailed ion profiles through the flame front, (Fig. 9). The rate of appearance and decay of the various ions eventually should permit the calculation of ion formation, charge transfer, and ion recombination rates. However, before this can be done, the results must be corrected for diffusion, and the concentration of the non-ionized reactant must be known. Thus it is not possible to interpret the appearance of the $C_3H_3^+$ ion peak ahead of the $H_3O^+$ peak as evidence that $C_3H_3^+$ is produced first. In a rapid reaction,

$$H_3O^+ + M \rightleftarrows C_3H_3^+ + \cdots \tag{8}$$

the concentration of M may decay rapidly through the flame so that $C_3H_3^+$ could appear before $H_3O^+$ although $H_3O^+$ is produced in the reaction sequence before $C_3H_3^+$. In the two mechanisms thus far considered[21,31] for the formation of $C_3H_3^+$, the reactant M is $C_3H_2$ or $CH_2O$ and $C_2H_2$ (a two-step reaction.) Both of these would meet the above criteria.

The $CHO^+$ ion (Fig. 7) is present in a very small concentration, and its profile certainly does not indicate it to be the first ion produced. This conclusion must be arrived at by more profound arguments.[21,30,42] The small concentration is explained by the rapid ion-molecule reaction (4) to produce $H_3O^+$.

## Microwave Attenuation in Rocket Exhausts

Recent efforts at AeroChem have been directed to the application of the understanding of ion formation and loss processes in flames to the problem of radar attenuation in rocket exhausts.[9] The free electrons in flames attenuate, reflect, and refract electromagnetic waves which are used for communication with rocket vehicles. Ionization in the rocket exhausts thus may work havoc with the communications system. The principles involved in this problem are largely known but some of the details, particularly reaction rate constants, are not known with sufficient accuracy to combine them with electromagnetic theory and fluid dynamics to give accurate numerical answers. Fundamental laboratory studies such as described above someday will yield sufficient understanding to specify the system completely. Nevertheless, the present understanding of ions in flames is sufficient to give much useful information. For example, the recombination coefficients measured in laboratory flames, not only for chemi-ions but for thermal ions, can be used to predict the rate of electron concentration decay in the rocket nozzle and at what point the electron concentration freezes.[49]

The attenuation in decibels, db, of a radar wave propagating through an ionized flame is given approximately by

$$db = -1.5 \times 10^{-10} \frac{\omega_p^2 \nu}{\nu^2 + \omega^2} \cdot x$$

where

$x$ = distance through ionized flame, cm

$\nu$ = electron collision frequency with the gas molecules calculable from collision cross sections

$\omega$ = radar or electromagnetic wave frequency

$\omega_p$ = plasma frequency

$\omega_p^2 = \left(\frac{4\pi e^2}{m_e}\right) n_e$

$e$ = charge on electron

$m_e$ = mass of electron

The more exact relationship contains exactly the same quantities so that attenuation is dependent only on x, $\nu$, $\omega$, and $n_e$. It is thus easy to see how the essential problem reduces to an understanding of the mechanism of electron formation and electron losses in flames.

Some of the complexities of the problem can be seen in Fig. 10 which shows the attenuation as a function of the microwave frequency. Because the three frequencies - $\nu$, $\omega$, and $\omega_p$ are of the same order of magnitude, and $\nu$ and $\omega_p$ have different altitude dependencies, the attenuation can change in very different fashions with altitude. The distance x also has a different dependence on altitude as determined by fluid dynamics and mixing of the exhaust with the ambient. The importance of the interaction of these parameters with altitude can be determined readily by a study of Fig. 10. The futility of ground testing or scale testing of rockets for prediction of attenuation at high altitudes without some understanding of the problem and a somewhat sophisticated interpretation of the results is thus apparent.

This point has been made to demonstrate the importance of fundamental studies of ions in flames to the radar attenuation problem in rocket exhausts.

## Current Research

Since this is a personal review of efforts in the field of "ions in flames," a proper closing requires a summary of current efforts at AeroChem. These efforts include: 1) study of ionization in the combustion of solid propellants and relation of this work to attenuation of radar waves in the exhaust of solid propellant rockets, 2) determination of the effect of chemical additives on flame ionization, 3) detailed profile studies in low pressure flames with an rf mass spectrometer and Langmuir probes and deduction from these measurements of the fundamental rate constants, 4) a fundamental study of the CH + O reaction and related reactions in an electrical discharge flow system, and 5) detailed study of the formation of negative ions in flames (planned for the near future).

Table 1 Flame conductivities, electron mobilities, and electron concentrations in the outer cone of bunsen flames.

| Flame Conductivity $10^{-7}$mhos/cm | Electron Mobility cm/sec/volt/cm | Electron Concentration $10^{8}$electrons/cc | Source | Ref. |
|---|---|---|---|---|
| 72 | 10,000 | 4.5 | E. Gold and B. S. Lacy | a |
| 15 | (1,000) | 100 | F. Haber | b |
| 5-8 | ... | ... | E. Zachman | c |
| ... | 8,000 | ... | B. T. Barnet | d |
| 3.2 | 2,700 | 7.4 | J. A. J. Bennett | e |
| 10 | 2,500 | 25 | H. A. Wilson | f |
| 2.4 | ... | -- | H. Ullmann | g |
| 4.1 | ... | -- | W. Kisselman and A. Becker | h |

a. Gold, E., "The velocity of the negative ions in flames," Proc. Roy. Soc. (London) A79, 43-66 (1907).

b. Haber, F. and Lacy, B. S., "Uber den innenkegel der bunsenflamme," Z. Physik. Chem. 68, 726-752 (1909). (assumed mobility-1000)

c. Zachmann, E., "Untersuchungen uber die electrische leitfahigkeit metalldampfhaltiger flammen mit einfuhrung einwandfreier feldmessungen," Ann. Phys. 74, 461-517 (1924).

d. Barnes, B. T., "The ionization of caesium vapor and the mobility of electrons in the bunsen flame," Phys. Rev. 23, 178-188 (1924).

e. Bennett, J. A. J., Electricity in Flames Philosophical Magazine and Journal of Science 53, 127-146 (1927).

f. Wilson, H. A., "Electrical conductivity of flames," Rev. Mod. Phys. 3, 156-189 (1931).

g. Ullmann, H., "Vergleichende messungen der leitfahigkeit der bunsenflamme mit Gleich und Wechselstrom," Z. Physik 97, 496-510 (1935).

h. Kisselman, W. and Becker, A., "Die elektrische leitfahigkeit der alkalimetallflammen," Ann. Phys. 25, 49-73 (1936).

Table 2 Ion concentration in lean flames diluted with nitrogen.

| Equivalence ratio | Nitrogen added to air, Mole % | Flame temperature, °K | Maximum ion concentration, $10^{10}$ ions/cc |
|---|---|---|---|
| 0.467 | 0.0 | 1330 | 2.1 |
| 0.521 | 0.0 | 1520 | 2.5[a] |
| 0.540 | 12.7 | 1390 | 3.3 |
| 0.635 | 31.7 | 1420 | 4.3 |
| 0.765 | 51.4 | 1490 | 8.3 |
| 0.835 | 59.6 | 1540 | 17.0 |

a. The double probe of E. O. Johnson and L. Malter ["A floating double probe method for measurements in gas discharges," Phys. Rev. 80, 58-68 (1950)] was used for this experiment.

Table 3 Possible chemi-ionization reactions in flames.[a,b]

| | Reaction | Heat of reaction |
|---|---|---|
| 1. | $C_2(^1\Pi_g) + OH \rightarrow CH^+ + CO + e^-$ | |
| | 16.2   0.39   17.3   -1.15 | -0.5 |
| 2. | $C_2(^3\Pi_g) + O_2 \rightarrow CO^+ + CO + e^-$ | |
| | 12.6   0   12.85   -1.15 | -0.9 |
| 3. | $C_2 + HO_2 \rightarrow C_2O_2H^+ + e^-$ | |
| | 10.2   0.2   ~ 7.0[a.6.] | ~ -3.4 |
| 4. | $CH + O \rightarrow CHO^+ + e^-$ | |
| | 6.16   2.56   9.6 | 0.9 |
| 5. | $CH(^2\Sigma^+) + O \rightarrow CHO^+ + e^-$ | |
| | 10.1   2.56   9.6 | -3.1 |
| 6. | $CH(^2\Sigma^+) + O_2 \rightarrow CHO_2^+ + e^-$ | |
| | 10.1   0   7.1 | -3.0 |
| 7. | $CH(^2\Sigma^+) + HO_2 \rightarrow CHO^+ + OH^-$ | |
| | 10.1   0.2   9.6 | -0.7 |
| 8. | $CH(^2\Sigma^+) + HO_2 \rightarrow CHO_2^+ + H + e^-$ | |
| | 10.1   0.2   7.1   2.27 | -0.9 |
| 9. | $CH + O_2(^3\Sigma_u^-) \rightarrow CHO_2^+ + e^-$ | |
| | 6.16   6.19   7.1 | -5.3 |

Table 3 continued on next page.

| | | | | |
|---|---|---|---|---|
| 10. | $C_2H + O_2(^3\Sigma_u^-)$ | → | $CH^+ + CO_2 + e^-$ | |
| | 5.0 6.19 | | 17.29 -4.08 | 2.0 |
| 11. | $C_2H + O_2(^3\Sigma_u^-)$ | → | $C_2O_2H^+ + e^-$ | |
| | 5.0 6.19 | | ~7.0[a. 6.] | ~ -4.2 |
| 12. | $C_2H + O_2(^1\Sigma_g^+)$ | → | $C_2O_2H^+ + e^-$ | |
| | 5.0 1.64 | | ~7.0[a. 6.] | ~ 0.4 |
| 13. | $CO(^1\Pi) + O$ | → | $CO_2^+ + e^-$ | |
| | 6.43 2.56 | | 9.71 | 0.5 |

a. From Calcote, H. F., Eighth Symposium on Combustion (Williams and Wilkins Company Baltimore, Md., 1962), p. 184. The heats of reaction have been recalculated using data from: 1) "Selected values of chemical thermodynamic properties," Nat. Bur. Standards Circ. 500 (February 1952, reprinted July 1961); 2) Bernecker, R.R. and Long, F. Z., "Heats of formation of some organic positive ions and their parent radicals and molecules," J. Phys. Chem. 65, 1565-1569 (1961); 3) Laidler, K. J. and Shuler, K. E., "Kinetics of combustion," Ind. Eng. Chem. 43, 2758-2761 (1951); 4) Herzberg, G., Molecular Spectra and Molecular Structure I. Spectra of Diatomic Molecules (D. Van Nostrand Co., Inc., Toronto, Canada, 1951), p. 454 - for $C^2$ excited states; 5) Branscomb, L. M., "Photodetachment," Atomic and Molecular Processes (Academic Press, New York, 1962), p. 100 - for electron affinity of $OH^-$ of 1.78 eV. This is probably a low value; 6) Estimated from bond energies.

b. Heats of formation and reaction are given in electron volts.

## References

[1]See other papers in this symposium and the papers and discussion at the Ninth Symposium (International) on Combustion, Cornell University, Ithaca, N.Y., August 27 to September 1, 1962 (Academic Press, Inc., New York, 1963).

[2]See Ref. 1, discusssions by van Tiggelen and Kinbara.

[3]Calcote, H. F., "The electrical properties of flames," Part I, Project SQUID Tech. Memo. No. PR-6, Tech. Paper No. 37, Princeton University, ASTIA No. 148-558, (June 1948).

[4]Calcote, H. F., "Electrical properties of flames-burner flames in transverse electric fields," Third Symposium on Combustion, Flame, and Explosion Phenomena (Williams and Wilkins Publishers, Baltimore, Md., 1949), p. 245.

[5]Calcote, H. F. and Pease, R. N., "Electrical properties of flames-burner flames in longitudinal electric fields," Ind. Eng. Chem. 43, 2726-2731 (1951).

[6]Calcote, H. F. and King, I. R., "Studies of ionization in flames by means of Langmuir probes," Fifth Symposium on Combustion (Reinhold Publishing Corporation, N. Y., 1955), p. 423.

[7]Calcote, H. F., "Mechanisms for the formation of ions in flames," Combustion and Flame 1, 385-403 (1957).

[8]Calcote H. F., and Reuter, J. L., "Mass spectrometric study of ion profiles in low pressure flames," Instrument Society of America Fall Instrument-Automation Conference, Preprint No. 69-LA-61, Los Angeles (September 11-15, 1961).

[9]Calcote, H. F. and Silla, H., "Radar attenuation in solid propellant rocket exhausts," Bulletin of the 18th Meeting JANAF-ARPA-NASA Solid Propellant Group (June 1962), Vol. III, p. 3 (Confidential).

[10]Wilson, H. A., Electrical Properties of Flames and Incandescent Solids (University of London Press, London, England, 1912).

[11]Wilson, H. A., "Electrical conductivity of flames," Rev. Mod. Phys. 3, 156-189 (1931).

[12]Thomson, J. J. and Thomson, G. P., Conductivity of Electricity Through Gases (Cambridge University Press, Cambridge, England, 1928), 3rd ed., Vol. 1.

[13]Becker, A., "Die elektrischen eigenschaften der flamme," Handbuck der Experimental physik (Akademische Verlagsejsellschaft M. B. H., Leipsig, 1929), Vol. 13.

[14]Tufts, F. L., "The phenomena of ionization in flame gases and vapors," Phys. Rev. 22, 193-220 (1906); Davidson, J. F., "Bemerkungen uber die elektrische leitfahigkeit von flammen," Physik Z. 7, 108-112, (1906); Haber, F. and Lacy, B. S., "Uber den innenkegel der bunsenflamme," A. Physik. Chem. 68,

726-752 (1909); Epstein, F. and Krassa, P., "Uber die leitfahigkeit des innern konus gespaltener flammen, nebst bemerkung die explosibilitat von gasmischungen," Z. Physik. Chem. 71, 28-46 (1910); Vogt, K., "Untersuchung uber die elektronenbewegung in flammen," Ann. Phys. 12, 433-476 (1932).

[15]Bennett, J. A. J., "Electricity in Flames", Phil. Mag. Seventh Series, 3, 127-146 (1927).

[16]Kisselmann, W. and Becker, A., "Die elektrische leitfahigkeit der alkalimetallflammen," Ann. Phys. 25, 49-73 (1936).

[17]Lusby, S. G., "The mobility of the positive ion in flames," Phil. Mag., Sixth Series 22, 775-791 (1911).

[18]Poncelet, J., Berendsen, R. and van Tiggelen, A., "Comparative study of ionization in acetylene-oxygen and acetylene-nitrous oxide flames," Seventh Symposium on Combustion (Butterworths Scientific Publications. London, England, 1959), p. 256.

[19]Lewis, B. and von Elbe, G., Combustion, Flames and Explosions in Gases (Cambridge, England, 1961), p. 558.

[20]Garner, W. E. and Saunders, S. W., "Ionization in gas explosions," Trans. Faraday Soc. 22, 281-288 (1926); Saunders, S. W., "Investigations on gaseous explosions. Part IV Ionization in methane-oxygen and acetylene-oxygen explosions," ibid. 23, 256-266 (1927); Saunders, S. W. and Sato, K., "Investigations on gaseous explosions; Part III Ionization in explosions of carbon monoxide and oxygen," ibid. 23, 248-256 (1927); Semenov, N., Nalbandjan, A. and Dubovizkii, F., "The mechanism of the upper limit of inflammation of electrolytic gas mixture," ibid. 29, 606-611 (1933).

[21]Green, J. A. and Sugden, T. M., "Some observations on the mechanisms of ionization in flames containing hydrocarbons," Ninth Symposium on Combustion, Cornell University. August 27 to September 1, 1962; Academic Press, Inc., New York, 1963, p. 607.

[22]See, for example; Garner, W. E. and Roffey, F., "The radiation from explosions of carbon monoxide and oxygen to

which hydrogen has been added," Nature 121, 56-57 (1928); Garner, W. E. and Johnson, C. J., "The effect of catalysts on the speed of flames, infrared emission, and ionization during the combustion of carbon monoxide and oxygen," J. Chem. Soc. 131, 280-298 (1928); Wendt, G. L. and Grimm, F. V., "A suggested mechanism for antiknock action," Ind. Eng. Chem. 16, 890-893 (1924); Garner, W. E. and Saunders, S. W., "Investigations on gaseous explosions Part I Ionization in hydrogen and oxygen explosions," Trans. Faraday Soc. 22, 324-337 (1926); Bennett, J. A. J., "Ionization in flames of various organic substances," ibid. 23, 307-311 (1927).

[23]Rossikhin, V. S. and Timkovskii, V. P., "Spectral investigation of an acetylene-air flame in an ultrahigh frequency field," U. S. S. R. J. Exp. Theoret. Phys. 10, 1446-1449 (1940).

[24]Gaydon, A. G. and Wolfhard, H. G., Flames, Their Structure, Radiation and Temperature (Chapman and Hall, Ltd., London, England, 1953).

[25]Thomson, J. J., "Report on gaseous combustion; discussion," Rpt. Brit. Assoc. Adv. Sci. 80th V. 80, 501 (1910).

[26]Calcote, H. F., unpublished experiments.

[27]Calcote, H. F., "Ionization flame detectors," Rev. Sci. Instr. 20, 349-352 (1949).

[28]Heumann, T., "Uber stromspannungskurven in isonisierten flammengasen," Spectrochim. Acta. 1, 293-318 (1940).

[29]King, I. R. and Calcote, H. F., "Effect of probe size on ion concentration measurements in flames," J. Chem. Phys. 23, 2203-2204 (1955).

[30]Calcote, H. F., "Ion production and recombination in flames," Eighth Symposium on Combustion (Williams and Wilkins Company, Baltimore, Md., 1962), p. 184.

[31]Calcote, H. F., "Ion and electron profiles in flames," Ninth Symposium on Combustion, Cornell University, Ithaca, N. Y., August 27 to September 1, 1962; (Academic Press, Inc., New York, 1963), p. 622.

[32]King, I. R., "Ion recombination rates in methane-air flames," J. Chem. Phys. 27, 817-818 (1957).

[33]King, I. R., "Recombination of ions in flames," J. Chem. Phys. 37, 74-80 (1962).

[34]DeJaegers, S., Deckers, J. and van Tiggelen, A.,"Identity of the most abundant ions in some flames," Eighth Symposium on Combustion (Williams and Wilkins Company, Baltimore, Md., 1962), p. 155.

[35]Knewstubb, P. F. and Sugden, T. M., "Mass spectrometry of the ions present in hydrocarbon flames," Seventh Symposium on Combustion (Butterworths Scientific Publications, London, England, 1959), p. 247.

[36]Calcote, H. F., Seventh Symposium on Combustion (Butterworths Scientific Publications, London, England, 1959), discussion, p. 284

[37]Calcote, H. F., "Relaxation processes in plasma," Dynamics of Conducting Gases, Proceedings of the Third Biennial Gas Dynamics Symposium August 1959 (Northwestern Univ. Press, 1960), p. 36.

[38]Field, F. H. and Franklin, J. L., Electron Impact Phenomena (Academic Press Inc., N. Y., 1957).

[39]Bernecker, R. R. and Long, F. A., "Heats of formation of some organic positive ions and their parent radicals and molecules," J. Phys. Chem. 65, 1565-1569 (1961).

[40]Deckers, J. and van Tiggelen, A., "Ion identification in flames," Seventh Symposium on Combustion (Butterworths Scientific Publications, London, England, 1959), p. 254.

[41]Deckers, J., Seventh Symposium on Combustion (Butterworths Scientific Publications, London, England, 1959), discussion, p. 283.

[42]Bulewicz, E. M. and Padley, P. I., "A cyclotron resonance study of ionization in low pressure flames," Ninth Symposium on Combustion, Cornell University, Ithaca, New York, Aug. 27 to Sept. 1, 1962; (Academic Press, Inc., N. Y., 1963), p. 638.

[43]Sugden, T. M., "Excited species in flames," Annual Review of Physical Chemistry (Annual Reviews Inc., Palo Alto, Calif., 1962), Vol. 13, p. 369.

[44]Brewer, A. K., "A study of ionization produced in certain gaseous reactions", J. Am. Chem. Soc. 46, 1403-1419 (1924).

[45]Westenberg, A. A. and Fristrom, R. M., "Methane-oxygen flame structure IV. Chemical kinetic considerations," J. Phys. Chem. 65, 591-601 (1961).

[46]King, I. R., "Recombination rates of alkali metal ions," J. Chem. Phys. 36, 553-554 (1962).

[47]Padley, P. J. and Sugden, T. M., "Some observations on the production and recombination of ions and electrons from metallic additives in hydrogen and hydrocarbon flames." Eighth Symposium on Combustion (Williams and Wilkins Company, Baltimore, Md., 1962), p. 164.

[48]Calcote, H. F. and Reuter, J. L., "Mass spectrometric study of ion profiles in low pressure flames." J. Chem. Phys. 38, 310-317 (1963).

[49]Rosner, D. E., "Estimation of electrical conductivity at rocket nozzle exit sections." ARS J. 32, 1602-1605 (1962).

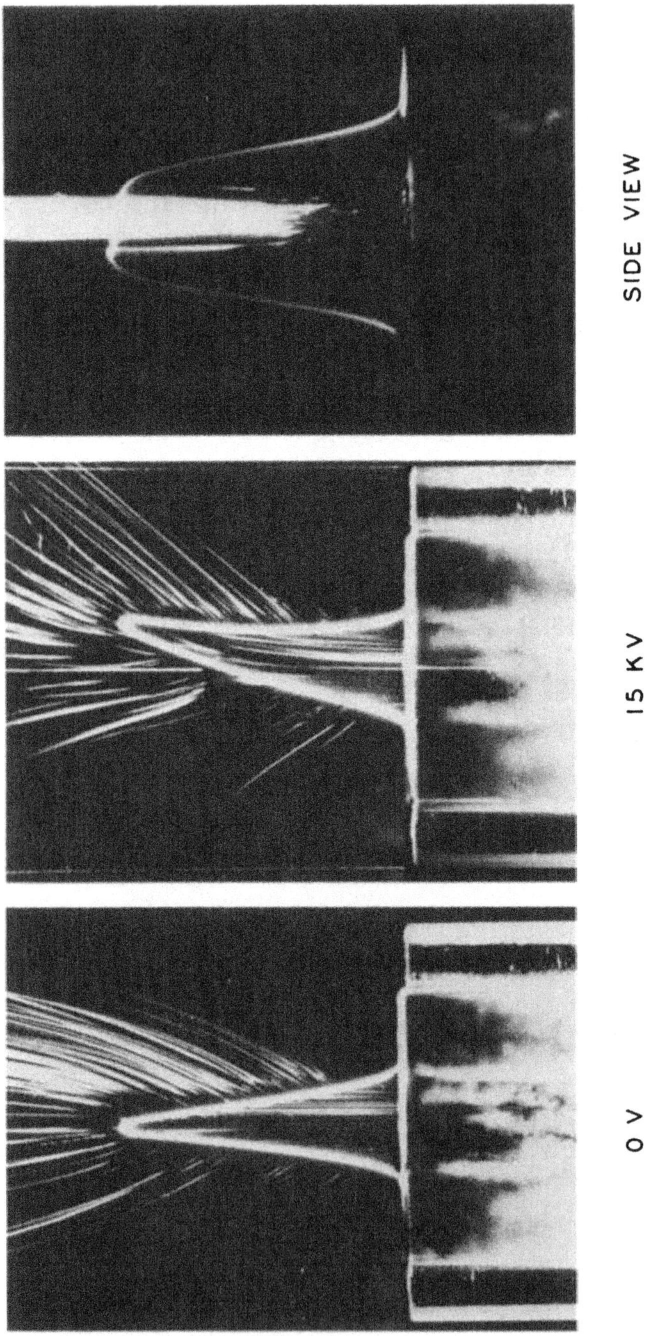


Fig. 1 Effect of a transverse electric field on flame deflection; n-butane-air, equivalence ratio = 0.95; total gas flow = 190 cc/sec; aluminum dust particles irradiated from side; rectangular burner: 1.70 x 0.750 cm; electrodes: 21 x 24 cm, 6 cm apart

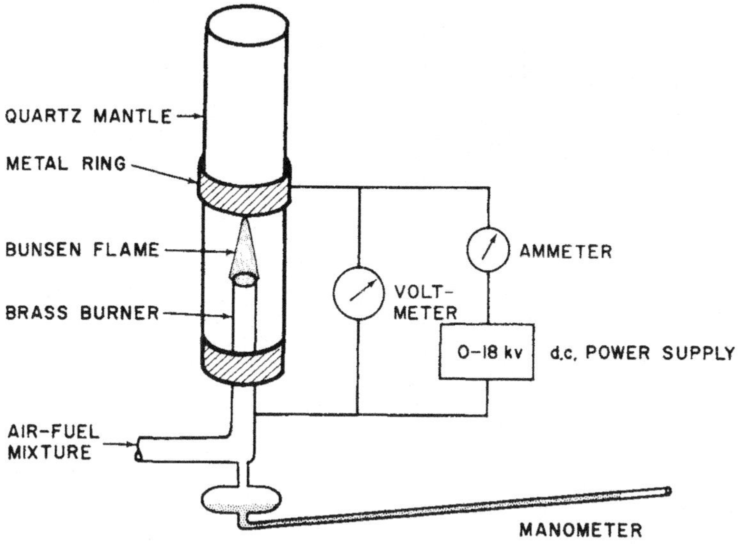


Fig. 2 Burner for application of longitudinal electric field

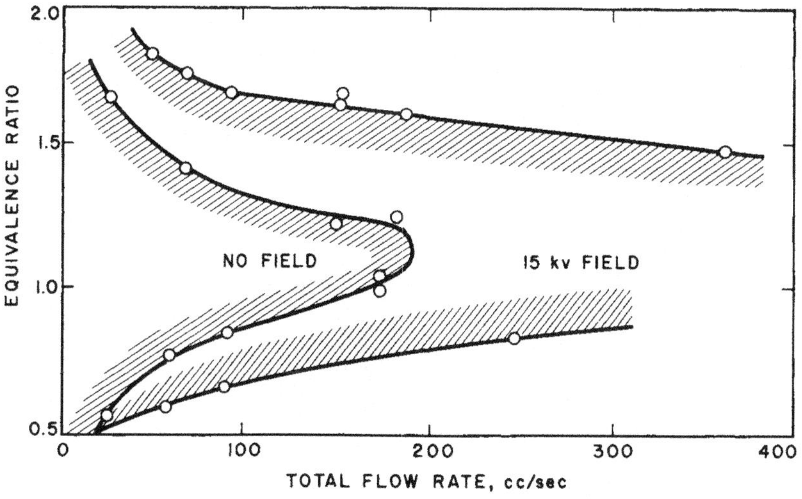


Fig. 3 Increase in flame stability by application of dc electric field; n-butane-air flames, 15 kv electric field with ring (Fig. 2) positive, burner diameter = 0.86 cm

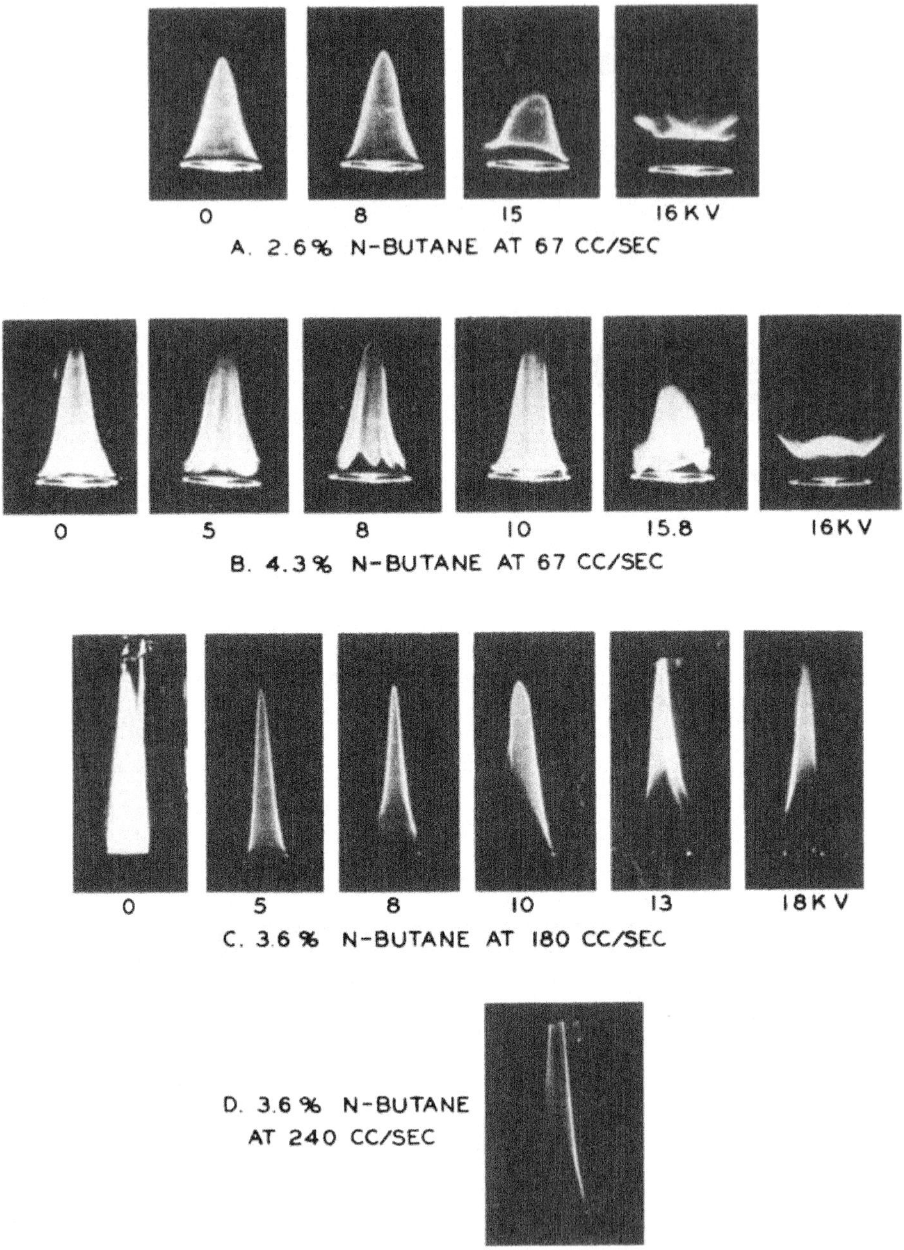


Fig. 4 Effect of negative longitudinal electric fields on the inner cone of a bunsen burner flame; n-butane-air, total flow = 67 cc/sec; burner diamter = 0.86 cm, total applied voltage in kv indicated under each photograph

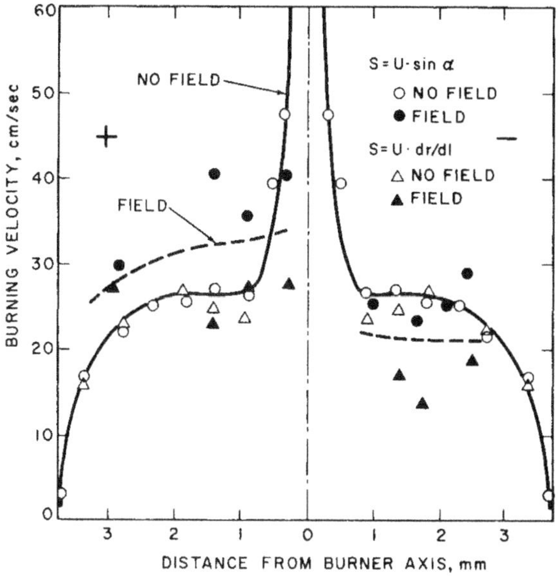


Fig. 5 Effect of transverse electric fields on burning velocity; n-butane-air, equivalence ratio = 0.95; total flow velocity = 191 cc/sec; pyrex burner diameter = 0.80 cm

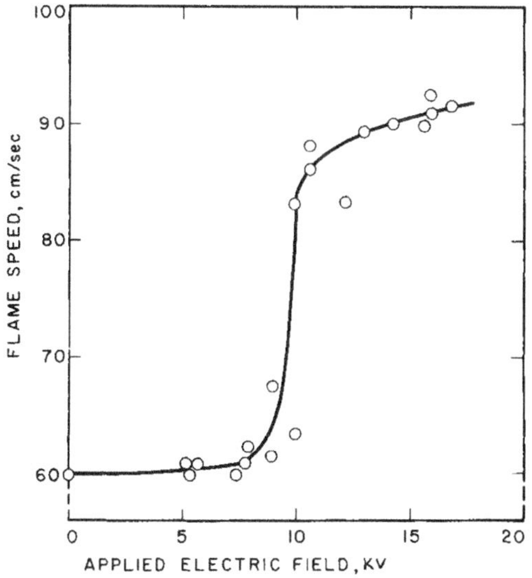


Fig. 6 Effect of dc longitudinal electric field on flame speed; n-butane-air, equivalence ratio = 1.2; pressure = 200 mm Hg; propagation in tube 150 cm long and 2.5 cm internal diameter

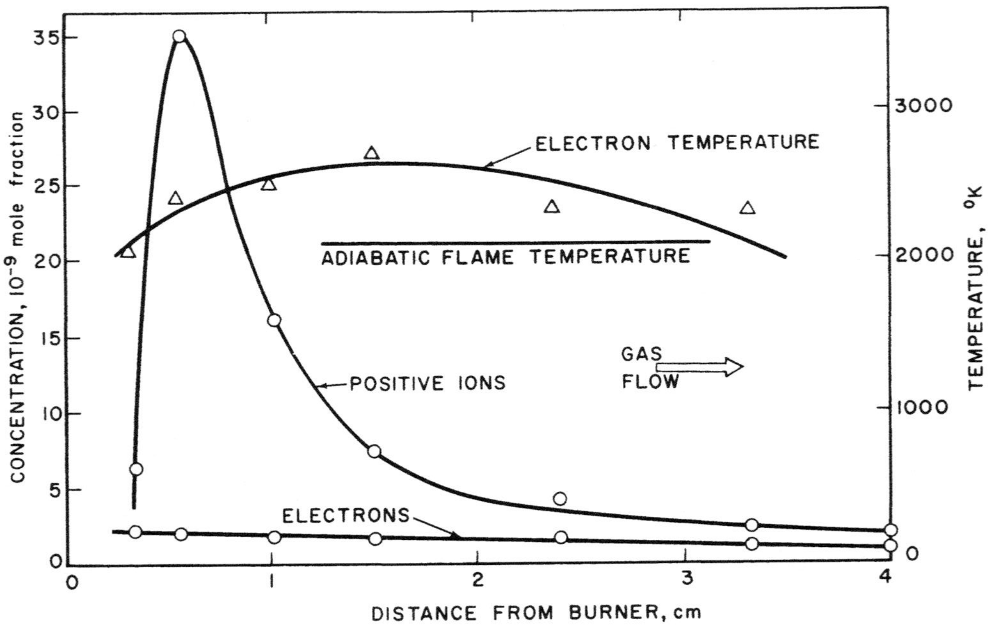


Fig. 7 Ion, electron and electron temperature profiles in a low pressure flame; propane-air, equivalence ratio = 0.88; pressure = 33 mm Hg; probe: Pt-40% Rh, diameter = 0.025 cm, length = 0.50 cm; total flow velocity = 180 cc/sec (STP): burner diameter = 10 cm

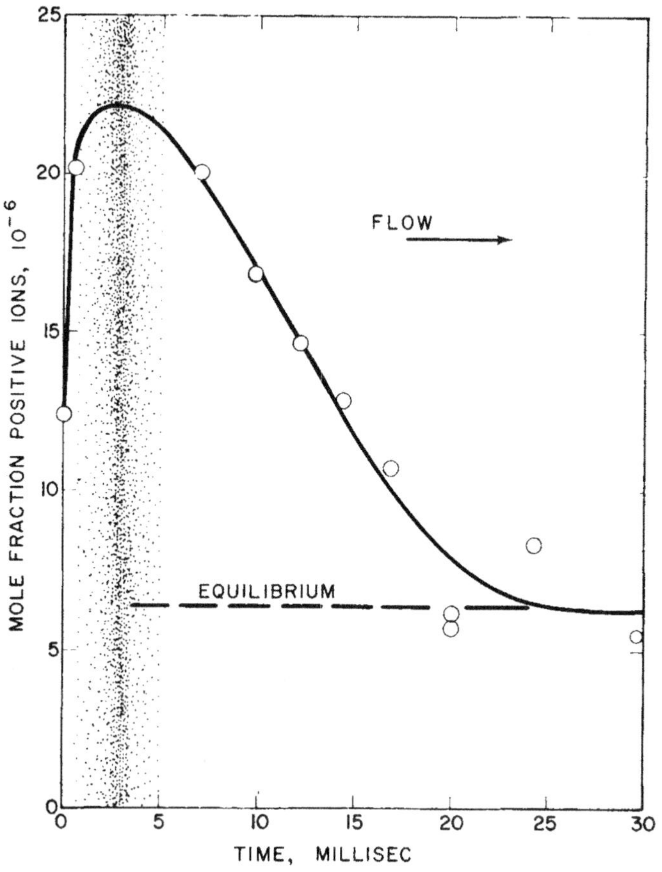


Fig. 8 Nonequilibrium ionization of potassium in a propane-air flame; pressure = 66 mm Hg; equivalence ratio = 1.2; potassium chloride added = 3.3 x 10 gm/sec; adiabatic flame temperature = 2140°K; probe: Pt-40% Rh, diameter = 0.015 cm stretched completely across flame

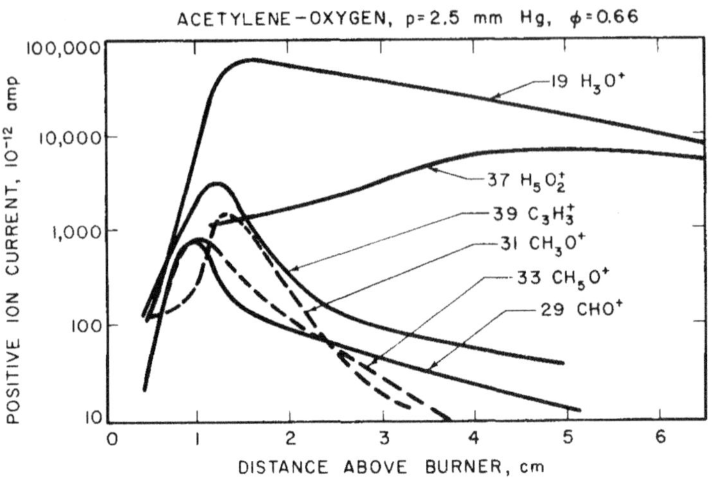


Fig. 9 Mass profiles for an acetylene-oxygen flame; pressure = 2.5 mm Hg, equivalence ratio = 0.66, total flow = 61 cc/sec (STP), burner diameter = 10 cm

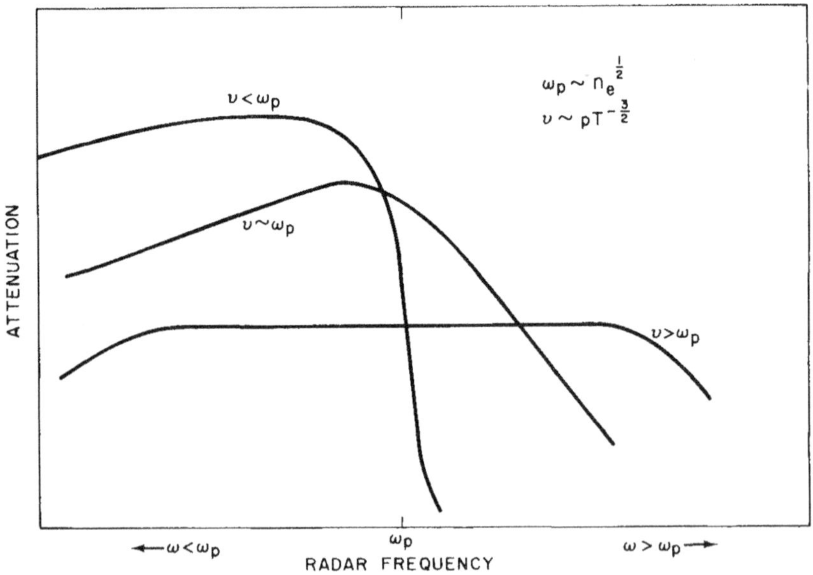


Fig. 10 Radar attenuation in rocket exhausts

# A SURVEY OF FLAME IONIZATION WORK AT THE UNIVERSITY OF CAMBRIDGE

T. M. Sugden*

University of Cambridge, Cambridge, England

Abstract

An account is given of the development of work on ionization in flames at the Department of Physical Chemistry in the University of Cambridge over the past fifteen years. The opportunity is taken to re-assess some of the earlier work in the light of new information about chemical reactions in flames. The paper begins with an account of the experimental methods which have been used for measuring electrical conductivity of flames, the microwave and resonant cavity techniques, and the mass spectrometric system which has been produced for the separation and analysis of positive ions in flames. Brief discussions of equilibrium ionization and of the fundamentals of flame chemistry precede the description of the results. The "natural" ionization of hydrocarbon flames is interpreted in terms of the mechanism suggested by Calcote, in which the primary step is the reaction of CH radicals with O atoms to give $CHO^+$ and a free electron. New information in support of this theory is given. The discussion of this kind of chemi-ionization is followed by an account of transfer of such ionization from molecular ions to give certain metallic atomic ions, thereby retarding the recombination process. The final section deals with the ionization of alkali elements in flames, and short accounts are given of the conditions under which equilibrium ionization is achieved, the rate at which it is set up, and the anomalous disturbances in electron concentration induced by the addition of small amounts of halogen in the presence of alkali metals.

---

Presented at the ARS Ions in Flames and Rocket Exhausts Conference, Palm Springs, Calif., October 10–12, 1962.

* Reader in Physical Chemistry, Department of Physical Chemistry.

## 1. Introduction

The purpose of this paper is to set before an audience concerned with ionization in flames and rocket exhausts a picture of the work carried out by the author and his collaborators in this field over the past fifteen years. It is hoped that this will be useful since much of the work is otherwise scattered over diverse journals, and since rapid developments in the knowledge of the physical chemistry of flames has necessitated re-interpretation of much of the earlier work.

Fundamental work about the nature of ionization in flames is being done mainly by three groups : that of Calcote in the United States, that of van Tiggelen in Belgium, and that of the author in England. Contributions from all of these are appearing in the present meeting. On the basis of this work it is probably true to say that the fundamental problems are well on the way to being solved, and that it is possible to offer meaningful advice to those concerned with the practical aspects.

Broadly these practical aspects are twofold - firstly, that of generation of electric power by magnetohydrodynamic methods, and secondly, that of attenuation or modification of radio signals by ionized gases in rocket exhausts. In the former one is mainly concerned with raising the level of ionization, and in the latter with lowering it.

In general it is possible to distinguish between two kinds of ionizing system - on the one hand the "natural" ionization of flame gases without special additives or impurities, in which case the ionization is usually confined to the reaction zone of the flame, and on the other hand, the ionization of flames with traces of readily ionizable additives or impurities, such as the alkali metals, where the ionization may persist into the burnt gases.

## 2. Experimental Methods

Two basic kinds of measurements present themselves. The electrical conductivity of flames derives from the presence of free electrons on account of their small mass and consequent high mobility, except in the presence of large excesses of heavier ions. Measurements of electrical conductivity can be made to lead to values of electron concentration. Such measurements at Cambridge have been made exclusively at microwave or radio frequencies. The other kind of measurement is mass-spectroscopic in which the ionized flame gases are sampled, and the positive (or negative) ions identified, and their relative and absolute concentrations obtained

Given this information over suitable ranges of the variables concerned, such as temperature, composition, amount, and nature of additive, etc., it is possible to make tests of suggested mechanisms and equilibria of ionization.

## a. Measurement of Electron Concentration

According to the simplest electrodynamical theory, the electrical conductivity $\sigma$ and the dielectric constant $\epsilon$ of an ionized gas containing a small number of electrons relative to neutral molecules are given by

$$\sigma = \frac{ne^2}{m} \cdot \frac{\omega_1}{\omega^2 + \omega_1^2} \tag{2.1}$$

$$\epsilon = 1 - \frac{4\pi ne^2}{m} \cdot \frac{1}{\omega^2 + \omega_1^2} \tag{2.2}$$

where $n$ = number of electrons $cm^{-3}$
$e$ = electronic charge
$m$ = electronic mass
$\omega$ = angular frequency of measuring radiation
$\omega_1$ = mean frequency of collision of electrons with molecules.

Similar terms for heavier ions generally can be ignored. The uncertain term in these expressions is the collision frequency, and this is open to experimental or theoretical determination.

Two methods are available for the measurement of $\sigma$ at microwave frequencies - the direct attenuation of a beam of microwaves by the ionized gas, and the modification of the characteristics of a resonant cavity by the presence of the gas.

The first reported measurements using direct attenuation of microwaves are those of Andrew, Axford, and Sugden[1], in which the electrical conductivity of the secondary flash of a gun was measured and compared satisfactorily with the value obtained by a crude d.c. method. The electron concentration was derived by calculating a collision frequency assuming point electrons and ordinary collision diameters of molecules, using simple kinetic theory. This collision theory probably estimated the frequency ($\sim 10^{11}$ $sec^{-1}$) low by a factor of about 3 in the light of recent measurements on flames.[2,3] It showed, however, that collision frequencies and microwave frequencies, at atmospheric pressure, are commensurate.

Further work using the direct attenuation method has been

reported by Smith and Sugden[4,5] and by Belcher and Sugden[6,7]. The last two papers dealt with the ionization of alkali metals in coal-gas/air flames, and the first two similarly with hydrogen/air flames. In Ref. (6) a direct measurement of collision frequency was attempted by finding $\sigma$ at four widely differing wavelengths (10 cm, 3 cm, $1\frac{1}{4}$ cm, and 0.8 cm). This value was also probably low by about threefold because the flame was rather small for use with 10-cm radiation and because techniques for the shortest wavelength were at that time in their infancy. Technically, all of these measurements suffered from the fact that the zone of flame gases examined by direct transmission of a beam of microwaves between opposed horns relatively is large and has ill-defined boundaries. In no sense can the method be used for detailed probing of a flame except of the crudest kind.[8] A limitation is that only relatively large electron concentrations ( > about $10^{11}$ $cm^{-3}$) can be obtained with laboratory flames ; it is, however, a very useful method for large practical systems.

A cylindrical microwave cavity with a coaxial column of burnt gases offers the advantage that the sample of gases being examined has well-defined boundaries. An X-band (3-cm) tunable cavity was used by Sugden and Thrush[9] to make observations in the burnt gases of acetylene/air flames containing soot, and by Sugden and Wheeler[10] to measure ionization in the burnt gases of hydrogen /air flames containing alkaline earth metals. In principle the presence of ionized gas in a resonant cavity affects both its selectivity (Q) and its frequency of resonance, and a measurement of both effects would provide two results, from which both the electron concentration and the collision frequency could be determined. In practice, however, the change in frequency is small and has proved difficult to measure, while the change in Q is large and overwhelmingly dependent on the conductivity. Hence these measurements rely on assumed collision frequencies. An advantage is that the method is more sensitive than the direct attenuation method, and can be used down to at least $10^9$ electrons $cm^{-3}$ at atmospheric pressure.

These early pieces of work had the disadvantage that the height of the flame column sampled was still rather large (about 2 cm), because of the choice of resonant mode in the cavity, which was in turn determined by the use of a tunable end-plate. This cavity also had the drawback that relatively large holes had to be sealed against escape of radiation by open-work metal gauzes made of platinum or Kanthal wire. Such gauzes tend to distort at high temperatures, and thus affect the performance of the cavity. Allo of these disadvantages were overcome, at the expense of a slight loss in sensitivity, in the cavity described and used by Padley and Sugden.[11] This is a cylindrical cavity operating at about 10 cm wavelength, so that gauzeless holes nearly 2 cm in diam could be put in the end-plates to allow passage of the flame. By sacrificing the tuning requirement, which in any case can be surmounted by tuning the oscillator, it was possible to choose a mode with frequency independent of the length of the cylinder, which was thus reduced to 3 mm, giving an essentially disk-shaped cavity about 8 cm in diam. The calibration difficulty was overcome in this case to a large extent by adding a known small amount of caesium to a hot flame where complete ionization to $Cs^+$ and electrons could be expected. Changes in the maximum current at resonance, which is simply related to Q, were measured.

Another method which has been developed at Cambridge for measurement of electron concentrations is based on radio-frequency techniques in the 10–100 Mc region. Here Equations (2.1) and (2.2) are simplified, since $\omega \ll \omega_1$ at atmospheric pressure. A simple resonant circuit of a condenser, between the plates of which the flame is placed, and an external inductance coil is coupled to a signal generator, and the current flowing in it is picked up by a loop containing a crystal rectifier. Thus there is a complete analogy with the microwave resonant circuit method, and the sensitivity is similar. As originally used by Smith and Sugden[4] parallel plates were placed on either side of the flame, but a more sophisticated system, of essentially cylindrical symmetry, has been described and used by Knewstubb and Sugden[12] to study, inter alia, ionization of alkali metals in hydrogen flames. It also has been used by Williams[13] to measure ionization in low-pressure flames. Again, it has the disadvantage of a rather large and ill-defined sample.

Perhaps the strongest criticism which can be made of all this work is the relative neglect of the collision frequency factor in relating conductivities to electron concentrations. This affects the absolute values obtained, and also the comparison of relative values from different flames. Under normal conditions of observation, however, with small microwave or radio-frequency powers and with small collision frequencies, the use of a mean collision frequency based on a Maxwellian distribution of velocities is reasonable, as comparison with more elaborate theories of distribution shows[14]. Calcote[20], from work using d.c. probe systems, has obtained some evidence that the mean electron temperature may be a good deal in excess of the general gas temperature in low-pressure flames ; this is not likely to be true of flames at atmospheric pressure. Further, the general interpretation of probe results for electrons is still rather uncertain. A very promising development is the application of cyclotron resonance techniques[2] at microwave frequencies. This method, suggested by Ingram[15] has been applied with considerable success by Bulewicz and Padley[2,16]. It leads to reliable values both of collision frequency and electron concentration. It has been shown that the molecules $H_2O$ and $CO_2$, which are the only triatomic molecules in the burnt gases of hydrogen and hydrocarbon flames, are mainly instrumental in determining the collision frequency. It has also largely justified the comparisons made between different flames in the earlier work described previously, since collision frequency turns out to be a not very sensitive function of flame gas composition.

It will be obvious from the foregoing that microwave and radiofrequency methods have little application to the narrow reaction zones of premixed flames at atmospheric pressure, and none of the Cambridge work has been directed towards obtaining electron concentrations in this region. It is probable that probe measurements, or d.c. methods such as that described by Poncelet, Berendsen and van Tiggelen[17] will be more valuable here.

## b · Mass-Spectrometric Measurements

Undoubtedly a great impetus has been given to the understanding of ionization in flames - particularly that in the reaction zones of hydrocarbon flames without additives - by the application of mass-spectrometric analysis. Instruments for sampling ionized flame gases, the separation of the ions from the neutral gases, and the subsequent analysis of the ions have been described by Knewstubb and Sugden[18] for atmospheric pressure flames, by Deckers and van Tiggelen[19] for flames at about 1/10 atm, and by Calcote[20] for flames at a few millimeters of mercury.

In the Cambridge instrument a small flame plays on a water-cooled platinum diaphragm about 0.002 in. thick containing a hole about 0.002 in. diameter. Vigorous pumping behind the hole reduces the pressure to about $10^{-3}$ mm Hg, while the ions are concentrated by ion-lenses to fall on a slit leading to a second chamber at about $10^{-5}$ mm Hg, followed by a third chamber and a conventional magnetic analyser at about $10^{-6}$ mm Hg. Ions are either counted directly or measured as an electrometer current. Calibration for ion concentrations can be made by the caesium method described for electron concentrations[11]. An ion concentration of about $10^{11}$ $cm^{-3}$ in the flame corresponds with a count of about $10^7$ ions $sec^{-1}$, so that concentrations down to $10^5$ $cm^{-3}$ can be estimated.

The difficulties here are concerned largely with interpretation of the plethora of results which can be obtained. Some of these will be discussed in a later chapter. One of interest here, however, is the problem associated with the boundary layer which forms between the flame and the platinum diaphragm, The presence of this layer can result in modification of the types and patterns of ion observed. Thus, $H_3O^+$ ions from the flame gases may attach molecules of water to make hydrated ions in this region, or they may react with ammonia molecules produced catalytically at the surface of the platinum to give $NH_4^+$ ions. It is important to be able to distinguish between such ions and genuine "flame" ions in the absence of apliccable methods for removal of the layer. This has been achieved to a great extent by studying the effects of varying hole size[21,22], since a small hole relatively enhances the effects of the boundary layer, while a large one permits direct passage of most of the genuine flame ions without interference.

### 3. Equilibrium and Non-equilibrium Ionization

Any discussion of experimental results on ionization in flames must take into account the extent to which the level of ionization permitted by thermodynamic equilibrium is reached. In the event of a system showing this degree of ionization, it is useless to expect to be able to obtain information about the mechanism of ionization. In the event of incomplete ionization, or of excess ionization (often known as chemi-ionization), such information can, in principle, be obtained.

In any gaseous system the equilibrium of an ionization reaction

$$X \rightleftharpoons X^+ + e^- \qquad (3.1)$$

can be expresses in the form of the thermodynamic equation of Saha[23]

$$\log_{10}K = -\frac{5050\,V}{T} + \frac{5}{2}\log_{10}T - 6.5 + \log_{10}\frac{g_{X^+}\cdot g_{e^-}}{g_X} \qquad (3.2)$$

where $K = (p_{X^+}\cdot p_{e^-})/p_X$, with partial pressures p in atmospheres, T in $^{\circ}K$, V is the ionization potential of X in electron-volts, and the g's are the statistical (electronic) weights of the various species. Thus for sodium, $g_{Na} = 2$ (ground state $^2S_{\frac{1}{2}}$), $g_{Na^+} = 1$ (ground state is $^1S_0$), while $g_{e^-} = 2$. This equation is of course subject to modification if appreciable amounts of electronically excited X and $X^+$ are present, but the temperatures concerned in flames are in general rather too low for this to be important.

In full equilibrium each species will conform to Eq. (3.2). One may have, in addition, the possibility of negative ion formation

$$X^- \rightleftharpoons X + e^- \qquad (3.3)$$

in which case an equation like (3.2) also holds, with substitution of E, the electron affinity of X, for the ionization potential and with insertion of the appropriate statistical weights. Thus for Cl $(^2P_{3/2})$, $g = 4$, and for $Cl^-$ $(^1S_0)$, $g = 1$.

These equations represent a norm to which the actually observed ionization in a known gas mixture at known temperature can be referred. Excellent tabulations exist of V for metals[24], and for the molecules and many of the free radicals concerned in flame reactions[25]. Data on electron affinities are somewhat more sparse and unreliable, but, with the exceptions of the halogens, and possibly OH and CN, negative ions do not seem to be very important in flames because of the low values of the electron affinity.

The burnt gases of hydrogen and hydrocarbon fuels with air or oxygen have, as their normal constituents, the species $H_2O$, $CO_2$, CO, $H_2$, $N_2$, $O_2$, together with some OH, H, and O. The amounts of these are such, and their ionization potentials so high, that the they should not give rise to electron concentrations large enough to be measurable by the methods described in Sect. 2a, according to the Saha equation. This is well known to be the case. Even in the reaction zones of hydrogen flames, where the concentration of free radicals OH, H and O exceeds equilibrium values in some cases by two or three orders of magnitude, the amount of ionization would appear to be negligible.[22,26] Ions observed mass-spectrometrically in flames of cylinder hydrogen[18] have been shown[22] to derive from trace impurities of hydrocarbon. Hydrogen flames thus form a useful medium in which to study the ionization of additives. Descriptions of work of this latter kind, based mainly on measurements of electron concentration, will be given in Sect. 7.

The best-known example of "natural" - i.e., nonadditive - ionization occurs in the reaction zones of premixed flames of hydrocarbons with air or oxygen. Knowledge of the composition of these zones, and of the chemical reactions involved, is much less complete than for hydrogen flames, but small amounts of radicals such as CH and $C_2$ have been detected, and there is little doubt that others are present.[2,7]. The concentrations of these intermediates, although small, vastly exceed those for thermodynamic equilibrium. Nevertheless, what is known of the ionization potentials of these radical species leads one to believe that the observed ionization in the reaction zone does not derive from thermal ionization of them, and the true explanation must be sought in some other kind of ionization mechanism.

## 4. Chemistry of Hydrogen and Hydrocarbon Flames

This is not the place to go into a detailed account of the chemistry of the reactions of combustion in hydrogen and hydrocarbon flames, the systems on which most of the ionization work to be discussed has been done. Excellent surveys of existing knowledge are to be found in standard monographs[27,28]. At the same time, since many important aspects of ionization have been found to be contingent on the main flame processes, some brief account of the relevant considerations must be made at this point.

The oxidation of hydrogen in a flame takes place by the branched chain process embodied in the forward reactions of the reversible processes

$$H + O_2 \rightleftharpoons OH + O \tag{4.1}$$

$$O + H_2 \rightleftharpoons OH + H \tag{4.2}$$

$$OH + H_2 \rightleftharpoons H_2O + H \tag{4.3}$$

These result in the accummulation of free radicals, which begin to recombine by ternary processes, principally

$$H + H + M \rightarrow H_2 + M \tag{4.4}$$

$$OH + H + M \rightarrow H_2O + M \tag{4.5}$$

These latter reactions, unlike the first three, generate a great deal of heat, which further speeds up the reactions (4.1–4.3), including the reverse steps, so that, emerging from the reaction zone (about 0.1 mm thick at atmospheric pressure and traversed in about 50 $\mu$sec)are gases in which there are excess free radicals, recombining by reactions (4.4) and (4.5), subject to the maintenance of the balanced reactions (4.1) to (4.3), which determine certain relationships between the concentrations of the species concerned.

These excess concentrations persist several centimeters downstream from the reaction zone,[26] a distance corresponding with a few milliseconds in time. The situation on the burnt gases of nonsoot-forming hydrocarbon flames is very similar, with addition of the water gas equilibrium

$$H_2 + CO_2 \rightleftharpoons H_2O + CO \tag{4.6}$$

The species CO and $CO_2$ are additional third bodies M for the recombination reactions.

The reaction zones of hydrocarbon flames are characterized by the presence of many other free radicals, as described in the last section. These, unlike H, OH, and O, do not persist beyond the reaction zone, and unlike those three radicals, do not lend themselves to accurate determination of their concentrations. Thus there is room for considerable speculation about them, and their role in flame ionization. There has been no lack of such speculation.

## 5. "Natural" Ionization in Hydrocarbon Flames

An excellent review of early views on this subject has been given by Calcote.[29] The position, however, has been made much clearer by the advent of mass spectrometry, and the early results of Knewstubb and Sugden,[30,31] as well as those of van Tiggelen and co-workers[32,33], established the principal positive ions to be $H_3O^+$ and $C_3H_3^+$, although many other positive ions were observed The former of these persists further downstream than any of the others. Further results have been obtained by Calcote,[20] who has formulated the following mechanism for the main reactions:

$$CH + O \rightarrow CHO^+ + e^- \qquad (5.1)$$

$$CHO^+ + H_2O \rightarrow H_3O^+ + CO \qquad (5.2)$$

$$H_3O^+ + e^- \rightarrow H_2O + H \qquad (5.3)$$

The Cambridge group is strongly in favour of this mechanism, since it recently has done a great deal of work which supports it. This work[21,22,34] deals with mass-spectrometric studies of the ionization from addition of controlled amounts of hydrocarbons (up to 1%) - principally acetylene - to the fuel of a hydrogen/air flame. This system has the following advantages: 1) different hydrocarbons may be compared under similar conditions; 2) the amount of hydrocarbon becomes a controllable variable under approximately constant conditions; and 3) the concentration of atomic oxygen in the reaction zone of a hydrogen/air flame can be estimated with fair accuracy on the basis of flame photometric measurements,[35,36,37] and is almost unchanged[38] by addition of small quantities of hydrocarbon. This work has shown the following :

1) The rate constant of reaction (5.3) is $(2.2 \pm 1.0) \times 10^{-7}$ $cm^3$ $sec^{-1}$, in agreement with Calcote's work.[20]

2) The rate constant of reaction (5.2) is $7 \times 10^{-9}$ $cm^3 sec^{-1}$;

to within a factor of about 2.[22]

3) The application of steady-state considerations to the reactive intermediate $CHO^+$ ion, and to the maximum value of the amount of $H_3O^+$, with these rate constants, and the known concentration of $H_2O$ molecules, leads to the observed ratio of $[H_3O^+]/[CHO^+]$, and its variation with concentration. For 1% of acetylene, this ratio is of order of magnitude $10^5$, which shows the advantage of the wide range of the mass spectrometer.

4) Taking the flame photometric value for [O] of $10^{15}$ cm$^{-3}$, and a reasonably estimated value of $10^{13}$ cm$^{-3}$ for [CH], a rate constant for reaction (5.1) of $1 \times 10^{-12}$ cm$^3$ sec$^{-1}$ can be estimated, in fair agreement with a value of $3 \times 10^{-12}$ cm$^3$ sec$^{-1}$ derived by considering the known rate of the reverse step of the isoelectronic reaction

$$N + O \rightarrow NO^+ + e^-$$

5) Other hydrocarbons and methyl alcohol fit the same general pattern, as far as these two positive ions are concenrned, and there is a broad agreement between their concentrations and the amounts of electronically excited CH obtained (as a relative quantity) by Padley[38] under similar conditions.[34]

Some general considerations of thermochemistry are also relevant. The heats of formation of CH, O, and $CHO^+$ [25] show reaction (5.1) to be about thermoneutral; it is unlikely that there will be significant energy of activation, and so one has an example of a chemi-ionization process of ground-state free radicals which should proceed with ease. An extended survey of other possible free radical reactions[22] shows that (5.1) is the only available one that is at all likely to fulfil the required conditions. Adoption of the accurate, revised value of the proton affinity of $H_2O$ obtained by Tal'rose and Frankevich[39] shows reaction (5.2) to be exothermic to the extent of 34 kcal., which again makes it a most probable process.

Although many of the ions found in hydrocarbon flames have been shown to be produced in the boundary layer, rather than in the flame itself (Sect. 2b), among the "true" flame ions are $H_3O^+$ and $CHO^+$, as the just-mentioned theory would require. Two other ions - $CH_3O^+$ and $C_3H_3^+$ - also fall into this class. Green and Sugden have suggested that these arise from the equilibria of reactions which are not far from thermoneutrality[22]

$$H_3O^+ + CH_2O \rightleftharpoons CH_3O^+ + H_2O \qquad (5.4)$$

$$CH_3O^+ + C_2H_2 \rightleftharpoons C_3H_3^+ + H_2O \qquad (5.5)$$

The reason for preferring these to reactions such as

$$H_3O^+ \quad + \quad C_3H_2 \quad \rightarrow \quad C_3H_3^+ \quad + \quad H_2O \qquad (5.6)$$

as postulated by Calcote[20] is that they require only major constituents or known relatively stable intermediates as neutral species, rather than unknown species such as $C_3H_2$. Their appear ance from a variety of hydrocarbons is in broad agreement with predictions from the equilibria (5.4), (5.5).

The Cambridge mass spectrometer can be used to study negative ions , but a brief investigation by Knewstubb and Sugden[40] shows them to be unimportant in these flames.

Brief mention also might be made here of ionization in the burnt gases of pure hydrocarbon flames which are fuel-rich enough to contain soot, and are therefore luminous. This is not the plac place to discuss the mechanism of soot formation, and it may be said simply that the soot particles are graphitic in structure, and although partially hydrogenated, will have work functions for removal of electrons not very different from that of pure graphite. The "vapor pressure" above an infinite surface of a solid of work function V is

$$p_e \quad = \quad \frac{(2\pi m)^{3/2}(kT)^{5/2}}{h^3} \quad \exp\,(-V/kT) \qquad (5.7)$$

where $p_e$ is the partial pressure of electron gas, and m is the electronic mass. For finite particles, the work function increases with successive removal of electrons, and expressions taking this into account, as well as the distribution of particle size, have been given (in order of increasing precision) by Thrush and Sugden[9], Einbinder,[41] and Smith.[42] The only experimental work would appear to be the few observations made with a microwave cavity,[9] anf this important field of ionization work largely remains to be explored.

## 6. "Transmitted" Chemi-ionization in Hydrocarbon Flames

Knewstubb and Sugden[43], using a radio-frequency method for electron concentration, observed that traces of certain metals - in particular those with ionization potentials in the range of 7-8 eV - gave results about two orders of magnitude greater than expected from the Saha equation in acetylene flames. This effect did not occur in hydrogen flames. The electron concentration was observed to decrease downstream from the reaction zone. The effect was most marked with lead, and was the same whether the addition was made as vapor of lead tetraethyl or as a spray of aqueous lead nitrate.

A more systematic investigation of this phenomenon was made by Padley and Sugden,[11] using the disk-shaped resonant cavity for conductivity. The electron concentration was found to decrease away from the reaction zone according to a second order law, but with a much smaller rate constant than has been found for reaction (5.3). Evidently one is dealing with an electron-ion recombination phenomenon, following chemi-ionization in the reaction zone. The following explanation was suggested: metal ions $X^+$ (e.g. $Pb^+$) are made by electron transfer reactions with ordinary flame ions such as $H_3O^+$

$$H_3O^+ + X \rightarrow H_2O + H + X^+ \qquad (6.1)$$

Now these recombine with electrons generated in the reaction zone by reaction (5.1), but since they are atomic ions they require a third body M for recombination to be effective, dissociative recombination as in reaction (5.3) being impossible

$$X^+ + e^- + M \rightarrow X + M \qquad (6.2)$$

This is of interest as being an example of transferred chemi-ionization, the primary mechanism of electron production being unchanged, but a different and much slower recombination mechanism being imposed by the additive.

## 7. Ionization of Alkali Metals and Related Substances.

Apart from the special work on gun-flash[1], in which rough agreement was obtained between the electron concentration measured by microwave attenuation, and that estimated on the basis of the equilibrium ionization of the known amount of potassium in the charge, a re-appraisal of some of the results and conclusions of the experiments on coal-gas and hydrogen flames with controlled traces of alkali metals (1 part in $10^5$ or less) can be attempted here in the light of subsequent advances of the knowledge of flame chemistry.

The work of Belcher and Sugden[7] established that alkali metals in coal-gas/air flames give electron concentrations which vary in the square law fashion with the amount of additive which would be expected for equilibrium ionization. Comparison of the various alkali metals showed that sodium and potassium gave the expected relative ionization from the Saha equation, but that the electron concentrations from caesium , and particularly from lithium, were low on this basis. Further observations following from this, by Smith and Sugden,[4 4] on alkali metals in hydrogen flames, confirmed that the effect arose from the formation of significant amounts of the stable gaseous hydroxides CsOH and LiOH. The checking of this by a flame photometric method[45] provided the starting point for the very fruitful field of the application of flame photometry of additives to flames and radical concentrations[46].

Belcher and Sugden[7] also introduced the idea that the electron concentration might be significantly reduced by formation of hydroxyl ions , and this received support from the subsequent work of Smith and Sugden[4,5] and from that of Page.[47] All this work, which was done by the direct attenuation method, pointed to a value for the electron affinity of the hydroxyl radical in the region of 60 kcal. mole$^{-1}$. Mass spectrometric observations, however, have failed to show anything approaching the amounts of $OH^-$ to be expected if this were true,[40] and photoionization work of Branscomb[48] comes out strongly in favour of a much lower value for the electron affinity of 40 kcal. mole$^{-1}$. It would seem that erroneous conclusions were drawn from the attenuation experiments, on account of neglect of possible negative ion formation in the cool boundaries of the flames and almost certainly because of neglect of variation of collsion frequency with flame gas composition in the comparison of the results for various flames, on the basis of which the electron affinity was estimated.

Belcher and Sugden[7] also considered the possible effects of halogen additives on the reduction of electron concentration because of salt formation and halogen ion formation, but today it is clear that their predictions lacked quantitative significance because they were made before the detailed nature of the excess free radicals in the burnt gases discussed in Sect.4 above was clear. They also report the first observations on the anomalous increases in ionization of alkali metals which can be obtained by addition of halogen under suitable conditions, but again sufficient information about flame reactions was not available for the interpretation of this phenomenon.

A new approach was made by Knewstubb and Sugden,[12] who used a radio-frequency apparatus to measure the ionization of alkali metals in hydrogen flames, and established that although potassium and caesium were ionized to an equilibrium extent effectively instantaneously, the approach to equilibrium with sodium was relatively slow, and could be followed by making measurements at increasing distances downstream in the gases. This allowed an assessment of the rate constant of the process

$$Na + M \rightarrow Na^+ + e^- + M \qquad (7.1)$$

where M is any other molecule in the flame gases. This work was then taken up in a more precise way, using the disk resonant cavity, by Padley and Sugden, who established the following points:[11] 1) sodium is ionized relatively slowly in flames of pure hydrogen, but eventually approaches the degree of ionization expected from the Saha equation; 2) the ionization process, as might be expected, is first order in the concentration of the alkali metal; and 3) the energy of activation associated with the ionization of sodium is some 50–60 kcal. less than the ionization potential. An explanation of this rather surprising result has been suggested by Smith[49] namely, that the rate-determining step is

$$Na + H_2O \rightarrow Na^+.H_2O + e^- \qquad (7.2)$$

followed by the relatively fast step

$$Na^+.H_2O + M \rightarrow Na^+ + H_2O + M \qquad (7.3)$$

The binding energy of the hydrate would be expected to be not far from 50 kcal. mole$^{-1}$. Further, hydrated alkali ions have been detected by the mass spectrometer,[50] although they may well have been boundary layer, rather than "true" flame ions.

Another result established by Padley and Sugden[11] was that if 1% of a hydrocarbon is added to the hydrogen, then equilibrium ionization is established effectively for sodium from the reaction zone onwards. It may be presumed that the rate-determining process (7.2) is replaced by a transfer process as discussed in Sect. 6 to provide an initial boost towards equilibri um electron concentration in the reaction zone.

Finally, a little may be said about the effects of halogens on the electron concentration provided by alkali metals. When amounts of halogen up to 1% are added to hydrogen flames, they are effectively shared between free halogen atoms B and the hydrogen halides HB by the balanced reaction[51]

$$HB + H \rightleftharpoons H_2 + B \qquad (7.4)$$

There also should be a tendency to form undissociated gaseous salt AB with alkali metal atoms A, and to form negative ions $B^-$. Both of these effects should result in a reduction of electron concentration, the first by removal of some of the alkali metal to form a compound less readily ionizable, and the second by direct removal of electrons. Nevertheless, the addition of the first small amounts of halogen (up to about 0.1%) often brings about an increase in electron concentration above the equilibrium value for no halogen. Further additions then produce the expected kind of decrease. This anomalous behavior has been found by Padley, Page and Sugden[51] to be more marked near the reaction zone of hydrogen flames (with 1% of acetylene to induce equilibrium ionization of sodium) and therefore has been linked by them with the excess concentrations of free radicals H and OH in the burnt gases.

They suggest that the introduction of the halogen allows two new balanced reactions to come into play

$$A + B \rightleftharpoons A^+ + B^- \qquad (7.5)$$

$$B^- + H \rightleftharpoons HB + e^- \qquad (7.6)$$

Algebraic addition (linear combination) of balanced reactions can be made, giving as the resultant of reactions (7.4) - (7.6)

$$A + H + H \rightleftharpoons A^+ + e^- + H_2 \qquad (7.7)$$

Thus the halogen can be thought of as introducing a new mechanism of ionization, which gives an electron concentration greater than the equilibrium one to the extent that the concentration of atomic hydrogen exceeds the equilibrium value. This mechanism is in competition with the overall mechanism

$$A + M \rightleftharpoons A^+ + e^- + M \qquad (7.8)$$

which maintains the electron concentration at its equilibrium (Saha) level in the absence of halogen. The ambiguous role of halogens in modifying electron concentrations is something which might have to be taken into account in applications directed either towards augmenting or diminishing them.

## 8. Conclusion

It will be clear from the foregoing sections that many of the basic problems of ionization in flames are well on the way to being understood. This has become possible because of the development of accurate, quantitative techniques for measuring ions and electrons. It is, perhaps, anomalous that present knowledge of ionization in flames is more complete than that of the reactions of combustion themselves. The reason for this, however, would seem to lie in the paucity of quantitative techniques for examining neutral species. Only in this way can a complete picture be obtained .

## References

[1] Andrew, R.E., Axford, D.W.E., and Sugden, T.M., "The measurement of ionization in a transient flame," Trans. Faraday Soc. 44 , 427-437 (1948).

[2] Bulewicz, E.M. and Padley, P.J., "Asymmetry of cyclotron resonance lines in the reaction zones of low pressure acetylene and cyanogen flames," J. Chem. Phys. 35, 1590-1593 (1961).

[3] Page, F.M. and Williams, H.E., private communication (1962).

[4] Smith, H. and Sugden, T.M., "Ionic equilibria in hydrogen/air flames containing alkali metal salts," Proc. Roy. Soc. (London) A 211, 31-58 (1952)

[5] Smith, H. and Sugden, T.M., "The stability of gaseous alkali hydroxides in flames," Proc. Roy. Soc. (London) A 211, 58-74 (1952).

[6] Belcher, H.E. and Sugden, T.M., "The determination of the collision frequency of electrons in coal-gas/air flames," Proc. Roy. Soc. (London) A 201, 480-488 (1950).

[7] Belcher, H.E. and Sugden, T.M., "Reactions governed by ionic equilibria in coal-gas/air flames containing alkali metal salts," Proc. Roy. Soc. (London) A, 202, 17-39 (1950).

[8] Page, F.M. and Sugden, T.M., "The mechanism of ionization of alkali metals in hydrogen flames," Trans. Faraday Soc. 53, 1092-1101 (1957).

[9] Sugden, T.M. and Thrush, B.A., "A cavity resonator method for electron concentrations in flames," Nature 168, 703-704 (1953).

[10] Sugden, T.M. and Wheeler, R.C., "The ions produced by traces of alkaline earths in flames," Discussions Faraday Soc. 19, 76-86 (1955).

[11] Padley, P.J. and Sugden, T.M., "The production and recombination of ions and electrons from metallic additives in hydrogen and hydrocarbon flames," 8th Symposium (International) on Combustion, Pasadena, 1960 (Williams and Wilkins Co., Baltimore, Md., 1962), p. 164 - 179.

[12] Knewstubb, P.F. and Sugden, T.M., "Observations on the kinetics of ionization of alkali metals in flame gases," Trans. Faraday Soc. 54, 372-380 (1958).

[13] Williams, H.E., "Further studies on the decay of free electrons in the mantle of an acetylene/air flame," 8th Symposium (International) on Combustion, Pasadena, 1960 (Williams and Wilkins

Co., Baltimore, Md., 1962), pp. 179-184.

[14] Margenau, H., "Conduction and dispersion of ionizes gases at high frequencies," Phys. Rev. 69, 508-513 (1945).

[15] Ingram, D.J.E., Discussions Faraday Soc. 19, 104 (1955).

[16] Bulewicz, E.M. and Padley, P.J., "A cyclotron resonance study of ionization in low pressure flames," 9th Symposium (International) on Combustion, Ithaca, 1962 (Academic Press, New York, to be published).

[17] Poncelet, J., Berendsen, R., and van Tiggelen, A., "Comparative study of ionization in acetylen-oxygen and acetylene-nitric oxide flames," 7th Symposium (international) on Combustion, Oxford 1958 (Butterworths Scientific Publications, London, 1959), pp. 256-262.

[18] Knewstubb, P.F. and Sugden, T.M., "Mass spectrometric studies of ionozation in flames," Proc. Roy. Soc. (London) A 255, 520-537 (1960).

[19] Deckers, J. and van Tiggelen, A., "Ion identification in flames," 7th Symposium (International) on Combustion, Oxford, 1958 (butterworths Scientific Publications, London, 1959), pp. 254-255.

[20] Calcots, H.F., "Ions in flames," AeroChem Research Laboratories, Inc., Princeton, N.J., Rep. AFBMD-TR-61-54, December, 1960 ; also Armed Services Technical Information Agency - AD 258 229; also AeroChem TP-24 (1960).

[21] Bascombe, K.N., Green, J.A., and Sugden, T.M., "The ionization produced by addition of acetylene to a hydrogen-oxygen-nitrogen flame," Joint Symposium on Mass Spectrometry, Am. Soc. Testing Mater. and Institute of Petroleum, Oxford, 1961 (Pergamon Press, Ltd., Oxford, 1962), pp. 66-86.

[22] Green, J.A. and Sugden, T.M., "Some observations on the mechanism of ionization in flames containing hydrocarbons," 9th Symposium (International) on Combustion, Ithaca, 1962 (Academic Press, New York, to be published).

[23] Saha, M.N., "Ionization in the solar chromosphere," Phil. Mag. 40, 472-488 (1920).

[24] Moelwyn-Hughes, E.A., Physical Chemistry (Pergamon Press Ltd., Oxford, 1957), p. 250.

[25] Field, F.M. and Franklin, J.L., Electron Impact Phenomena (Academic Press, Inc., New York, 1957), Appendix pp. 242-317.

[26] Padley, P.J. and Sugden, T.M., "Chemiluminescence and radical recombination in hydrogen flames," 7th Symposium (International) on Combustion, Oxford, 1958) Butterworths Scientific Publications, London, 1959), pp. 235-242.

[27] Gaydon, A.G., The spectroscopy of flames (Chapman and Hall, Ltd., London, 1957), Ch. VII.

[28] Gaydon, A.G. and Wolfhard, H.G., Flames, their structure, radiation and temperature (Chapman and Hall, Ltd., London, 1960), 2nd. ed., Ch. V - XV .

[29] Calcote, H.F., "Mechanisms for the formation of ions in flames," Combustion and Flame 1 ,385 (1957).

[30] Knewstubb, P.F. and Sugden, T.M., "Mass spectrometric observations of ions in hydrocarbon flames," Nature 181 , 1261 (1958).

[31] Knewstubb, P.F. and Sugden, T.M., "Mass spectrometry of the ions present in hydrocarbon flames," 7th Symposium (International) on Combustion, Oxford, 1958 (Butterworths Scientific Publications, London, 1959), pp. 247-253.

[32] De Jaegere, S., Deckers, J., and van Tiggelen, A., "Identity of the most abundant ions in some flames," 8th Symposium (International) on Combustion, Pasadena, 1960 (Williaims and Wilkins Co., Baltimore, Md., 1962), pp. 155-160.

[33] Deckers, J., and van Tiggelen, A., "Ion identification in flames," 7th Symposium (International) on Combustion, Oxford, 1958 (Butterworths Scientific Publications, London, 1959), pp. 254-255.

[34] Green, J.A., Larin, I.K., and Sugden, T.M., Trans. Faraday Soc., to be published (1963).

[35] Bulewicz, E.M., James, C.G., and Sugden, T.M., "The study of excess concentrations of hydrogen atoms in burnt gas mixtures," Proc. Roy. Soc. (London) A 235 , 89-106 (1956).

[36] James, C.G. and Sugden, T.M., "The source of the alkali metal continuum," Proc. Roy. Soc. (London) A 248 , 238-247 (1958).

[37] Bulewicz, E.M. and Sugden, T.M., "Studies of oxygen atoms at high temperatures," Chem. Soc. (London) Spec. Publ., No. 9 ,

81-96 (1958).

38 Padley, P.J., Ph.D. Thesis, University of Cambridge, Cambridge (1959).

39 Tal'rose, V.L. and Frankevich, E.L., "The ionic impact method for evaluating the affinity of molecules for the proton," Dokl. Akad. Nauk SSSR 111, 376-379 (1956).

40 Knewstubb, P.F. and Sugden, T.M., unpublished work (1960).

41 Einbinder, H., "Generalized equations for the ionization of solid particles," J. Chem. Phys. 26, 948-953 (1957).

42 Smith, F.T., "On the ionization of solid particles," J. Chem. Phys. 28, 746-747 (1958).

43 Knewstubb, P.F. and Sugden, T.M., "Ionization produced by compounds of lead in flames," Research Correspondence (London) 9, A1-6 (1956).

44 Smith, H. and Sugden, T.M., "The stability of gaseous lithium hydroxide at high temperatures," Proc. Roy. Soc. (London) A 219, 204-215 (1953).

45 James, C.G. and Sugden, T.M., "A general survey of the use of resonance radiation in the measurement of atomic concentration" Proc. Roy. Soc. (London) A 227, 312-329 (1955).

46 Sugden, T.M., "Excited species in flames," Ann. Rev. Phys. Chem. 13, 369-390 (1962).

47 Page, F.M., "Determination of the electron affinity of the hydroxyl radical by microwave measurements on flames," Discussions Faraday Soc. 19, 87-95 (1955).

48 Branscomb, L.M., Advances in Electronics and Electron Physics (Academic Press, New York, 1957), p. 66.

49 Smith, F.T., 8th Symposium (International) on Combustion, Pasadena, 1960 (Williams and Wilkins Co., Baltimore, Md., 1962), p. 178.

50 Knewstubb, P.F. and Sugden, T.M., "Mass spectrometric observations of ions in flames," Nature 181, 474-475 (1958).

51 Padley, P.J., Page, F.M., and Sugden, T.M., " Effects of halogens on the ionization in alkali-laden hydrogen and acetylene flames," Trans. Faraday Soc. 57, 1552-1562 (1961).

# IONIZATION PHENOMENA IN FLAMES

A. Van Tiggelen *
University of Louvain,
Louvain, Belgium

Abstract

This paper is a review of the work done in the author's laboratory during the past five years. It concerns three problems which rise in the study of ions in flames. First the elaboration of a method in order to measure the ion concentration in the flame, secondly the mass spectrometric identification of the ions present in the zone and thirdly a brief discussion of the possible processes of chemi-ionization.

## Introduction

A comprehensive study of ionization processes in flames comprises the solution of three main problems: 1) the development of an adequate method to measure the amount of ionization in the reaction zone of the flame; 2) the identification of the ions that are formed in the flame front and that disappear very rapidly (by recombination) in the burnt gases; and 3) the determination of the mechanism of ion formation, taking as an indisputed fact that ions are formed by chemi-ionization processes in unseeded flames.

---

Presented at the ARS Ions in Flames and Rocket Exhausts Conference, Palm Springs, Calif., October 10-12, 1962. The research in this document has been sponsored in part under contract No. AF61(052)-398, Aeronautical Research Laboratory, Office of Aerospace Research, U.S. Air Force, to which the author express thanks for the support given to their laboratory.

* Professor, Laboratoire de Chimie Inorganique et Analytique.

The research work summarized in this paper has been described more extensively in previously published papers and in a few Air Force Contract reports which will be referred to for more details concerning the equipment and the measurement techniques.[1-10]

## Measurement of Ion Concentration in Flames

Some difficulties are encountered in the experimental measurement of ion concentration in flames. No general method has been proposed which could be applied to any kind of flame.

The method of flame deflection in a external electric field [11-13] involves a rather complex interpretation of results. There may be uncertainties due to variation in burning velocity [14]. Microwave attenuation gives very valuable data concerning equilibrium and ionization processes in flame, and it has been successfully applied by Sugden and coworkers[15]. The method, however, requires a minimum volume (dimension at least equal to the wavelength) and therefore is not appropriate to the investigation of the flame front itself except, perhaps, in the case of very low-pressure flames.

The use of Langmuir probes [16,17] is able to give interesting data on the distribution of ions in the flame. It has shown that the chemi-ionization is located in the flame front. Nevertheless, the exploration of the flame front itself can be made only on very low-pressure flames where the flame front thickness exceeds the probe dimensions. Moreover, the probe method is restricted to the investigation of flames with a temperature low enough to make the thermo-emission of the probe (generally platinum wire) negligible as compared to the ion concentration of the flame.

It has been possible to maintain the principle of the probe method but to modify profoundly its application in determing the conductivity of the reaction zone in the flame. Two different ways have been followed successively.

### A. Method of the Circular Slit Burner +

In order to avoid the limitation and difficulties

+ This section was written with the collaboration of J. Poncelet.[4,5]

encountered with conventional probes, a technique has been evolved which fulfills the following requirements: 1) ability to measure specifically the ionization in the flame front (excluding the burnt gases); 2) applicability even to very hot flames; and 3) absence of flame front perturbation by quenching effects.

All of these conditions seem to be achieved in the method of the circular burner, for which a schematic representation is given in Figs. 1 and 2.

The electric current intensity in the flame is the quantity observed when a potential difference is applied between the inner and outer burner tube. The necessary equations are summarized briefly here:

$$i = \frac{E}{R_a + R_f + R_c} \tag{1}$$

where $i$ = current intensity
$E$ = potential applied between the inner and outer tube
$R_a, R_c$ = contact resistance at anode and cathode
$R_f$ = flame resistance

$$R_f = \frac{\rho\, l}{L\, x_o} \tag{2}$$

where $\rho$ = specific resistance of flame
$l$ = length of flame (see Fig. 2)
$L$ = perimeter of annular flame = $2\,\pi r$
$x_o$ = thickness of flame front

$$V_o = \frac{D}{S} = \frac{D}{2\,\pi r l} \tag{3}$$

where $D$ = gas flow
$S$ = flame surface
$V_o$ = burning velocity

By substitution of (2) and (3) in (1), one obtains

$$\frac{1}{i} = \frac{R_a + R_c}{E} + \frac{\rho\, D}{(2\pi r)^2 V_o x_o E} \tag{4}$$

From theoretical considerations on burning velocities and from experimental facts, it has been shown that [18]

$$V_o x_o = 0.96 \times 10^{-2} \sqrt{T_m} \quad (5)$$

where $T_m$ = mean flame temperature [19] = $T_i + 0.74\ (T_f - T_i)$

$T_f$ = flame temperature (final)
$T_i$ = initial temperature of fresh gases

Equation (4) becomes now

$$\frac{1}{i} = \frac{R_a + R_c}{E} + \frac{\rho D}{10^{-2} 0.96\ (2\pi r)^2 E \sqrt{T_m}} \quad (6)$$

If it is supposed that $R_a$ and $R_c$ remain constant when the gas flow increases, $\rho$ can be derived from the slope of a plot of 1/i against gas flow D. Since

$$\frac{1}{\rho} = n\ e\ (k+ + k-)$$

where e = charge of electron
k+, k- = mobility of positive ion and electron
then k- = 2500 cm/sec >> k+

Finally one obtains n = number of ions per cubic centimeter.

The method of measuring the concentration of the ions present in a flame, as described here, has been applied to a few different mixtures (containing a given oxidant and a given fuel). It definitely has been shown that for a given degree of dilution the maximum ion concentration coincides very nearly with the maximum burning velocity [4,5]. Furthermore, a very rough estimation of the activation energy for the process of ion formation shows that it varies between very broad limits around 50 to 100 kcal for widely different fuels mixed with oxygen. It does not exceed 150 kcal [4,5,8].

This apparent activation energy is calculated graphically from a plot of $\log(nT_m/Y)$ vs $1/T_m$ according to Ref. 4 for mixtures with constant ratio of fuel to oxidant but increasing dilution with nitrogen:

$$\log \frac{n\ T_m}{Y} = K - \frac{E \log e}{R\ T_m} \quad (8)$$

where Y = 100% inert gas/100 (taking the dilution into account).

The values of the activation energies so obtained prove definitely that ionization other than thermal is involved because all reasonable values of ionization potential for expected species are around 10 eV = 230.000 cal. Lastly, it will be noted that the order of magnitude of ion concentration in stoichiometric hydrocarbon-air mixtures has been found by this method to be equal to about $10^{10}/cm^3$.

## B. The Rectangular Slit Burner Method ≠

The theory of the conductivity of ionized flame gases has been elaborated by Wilson.[20] He considered the potential distribution between two plane and parallel electrodes in a hot gas where a steady concentration of ions and electrons is maintained by a process of ion formation which compensates the losses by recombination. The total potential applied to the electrodes V can be considered as a sum of three terms :

$$V = V_a + V_f + V_c$$

where $V_a$ and $V_c$ represent the anodic and cathodic potential drops, whereas $V_f$ corresponds to the potential drop in the flame itself where a constant electric field is assumed.

It is proposed by Wilson that the current intensity obeys the following equation :

$$V = Ai^2 + Bid$$

where d is the distance between the two electrode plates. A and B are constants for a given flame and given electrodes.

The term Bid then should be equal to $V_f$. This means that Ohm's law is obeyed in the flame, at least as long as the applied potential does not reach prohibitively high values. The term $Ai^2$ corresponds to the potential drop near the electrodes as a result of the phenomenon of space charge.

Evidently, the specific resistivity of the flame should be obtained by the ratio of B to the area

≠ This section was written with the collaboration of H. Philippaerts.[10]

of the transverse section of the flame. The relation between V and i allows an evaluation of B by plotting V/i vs i, since the intercept of the straight line on the ordinate axis would be equal to Bd. Indeed

$$V/i = Bd + Ai$$

The author has tried repeatedly to use this equation and the corresponding plots in order to measure the flame resistance but without any success because of inconsistency in the experimental results. It is his belief that the equation proposed by Wilson is only approximate.

The new method that is presented in this section is based on the equation $V = V_a + V_c + V_f$ presented in the foregoing. It differs essentially from the conductivity method proposed earlier in this laboratory and described in the preceding section by the important improvement that it is not based on the assumption used previously that the dead space near the electrodes is independent of the gas flow feeding the "circular slit burner". It has been observed, indeed, that this assumption is not correct and that the method does not allow the measurement of the conductivity of the flame itself but that, rather, it consists in the measurement of the cathodic potential drop. As a consequence, all ion concentrations previously obtained by the circular slit burner technique are not correct. It will be shown that the true values are higher.

1. Description of experimental equipment. The burner is described schematically in Fig. 3. The laminar premixed flame burns on a slit about 0.1 cm wide and 24.5 cm long. The water jacket provides efficient cooling of the upper part of the burner. The two electrodes (Fig. 3) are made of massive bridge-shaped copper pieces. They are cooled by sucking a strong air current through a cylindrical boring, which passes through the metal. A small hole pierced just above the burner slit allows the elimination by suction of the flame gases that otherwise would cause a deformation of the flame against the electrode. This device allows the gas profile to be uniform not only between the electrodes but even up to the region of contact between the flame and the electrodes. Variable distances between the latter can be obtained by means of a

fine screw adjustment system.

2. Theoretical considerations. The potential V applied to two electrode plates is given by

$$V = V_a + V_c + \rho \frac{i\ d}{S}$$

where S = transverse area of the flame (parallel to electrodes)
d = distance between electrodes.

Because the ion concentration in the burned gases is much smaller than in the flame front itself, it may be assumed that (Fig. 3)

$$S = x_o l$$

where $x_o$ = flame front thickness
l = length developed by the flame front above the slit.

Furthermore, the burning velocity $V_o$ is given by

$$V_o = D/Ll$$

where L = total length of the slit
D = total gas flow of flammable mixture feeding the burner.

By combination of the three last equations, one obtains

$$V = V_a + V_c + \rho \frac{i\ L\ V_o d}{D\ x_o} \qquad (9)$$

Two procedures now are possible:

1) If the distance between the two electrodes is changed, the applied potential V can be ajusted in such a manner as to keep the intensity i constant. If the necessary potential V then is plotted vs the electrode distance d, a straight line is obtained with a slope

$$\text{tg } a = \rho \frac{i\ L\ V_o\ d}{D\ x_o} \qquad (10)$$

combining with $1/\rho = n\ e\ k$, where n = ion concentration = electron concentration, e = charge of electron = $1.6\ 10^{-19}$ Cb, k = mobility of electron = 2500 cm/sec/V.[20] One obtains

$$n = \frac{i\ L\ V_o}{k\ e\ D\ x_o\ tga} \qquad (11)$$

Since it has been shown [18] that

$$x_o = 0.96\ 10^{-2}\ T_m/V_o$$

$$n = \frac{i\ L\ V_o^2\ 10^2}{0.96\ k\ e\ D\ tga\ \ T_m} \qquad (12)$$

When L = 24.5 cm and all constants are combined,

$$n = 6.4\ 10^{18}\ \frac{i\ V_o^2}{D\ tga\ \ T_m} \qquad (13)$$

where i is in amps, $V_o$ in cm/sec, and D is in $cm^3$. $T_m$ is the mean temperature of flame front and is given by [19]

$$T_i + 0.74\ (T_f - T_i)$$

2) By using two identical air-cooled probes placed between the two electrodes at a given distance d' from each other, the potential drop between the probes divided by d' gives directly the value of tga from one single measurement. This simpler method presents the advantage that it does not need more than one measurement. It is strictly valid only if the two probes are perfectly identical, since the perturbations due to the two probes are exactly compensated only under this condition.

The concentration obtained by the two methods are always in the same ratio: the ion concentration obtained by the first method is always about twice the concentration obtained by the second method. Because of its simplicity, the latter method has been used to obtain the experimental results given below because only the relative values of the ion concentration when the composition of the mixture is changed were of interest.

3. <u>Experimental testing of the method</u>. In order to verify the relation

$$V = V_a + V_c + \frac{p\ i\ d}{S}$$

different plots were made of V vs d for different values of the total potential $V_t$ applied to the electrodes. The observed linearity proves the constancy of $V_a$ and $V_c$ as long as the total potential $V_t$ remaints constant. Furthermore, it can be seen that the total electrode potential drop in the dead spaces at the two electrodes is much larger than in the flame.

The linearity of $V_f$ vs the current intensity i is illustrated in Fig. 4, at least as long as the current intensity remains sufficiently low. It is necessary in the application of the method to work under the conditions where the linearity holds.

### C. Brief Summary of Experimental Observations [†]

The rectangular slit burner method, described in the preceding pages, seems to give more reproducible and reliable results than the circular slit burner method. It is to be noted that the results obtained for the concentration of ions in hydrocarbon-air flames are about 100 times larger. As an illustration, some data are presented in Fig.5. The abscissa R represents the ratio $CH_4/(CH_4+O_2)$.

Furthermore, a constant value is obtained now for the activation energies of the formation of the ions when Eq.(8) is applied to the data of Fig.5. The obtained activation energy is about equal to 50 kcal and it remains constant all over the range of the ratio R (Fig.6). The temperatures $T_m$ that have been used are based on measured flame temperatures. The value of 50 kcal is very near to the activation energy of 40 kcal found for the branching process in this flame.[21]

Related measurements are being continued presently in this laboratory on many different kinds of flames. It is premature to draw conclusions yet, but some interesting observations have been made on the influence of additives. Some indications of the observed facts are reported elsewhere.[10]

The author would like to mention also a few preliminary results obtained in a systematic study of the correlation between the intensities of the different band spectra emitted in a flame and the

[†] This section was written with the collaboration of H. Philippaerts and J. D'Olieslager.

of radicals. Although such a correlation exists, especially between OH emission and ion concentration in hydrocarbon flames, it is difficult to propose yet a relation between the sources of the excited radicals and the ions. The latter might be considered as the extreme case of excited states.

Finally, the experimental proof that the ionization of alkali metals in flame is purely thermal will be mentioned. This has been shown by the fact that, although hydrogen-oxygen flames contain a negligible amount of ions, the addition of some Na salt is accompanied by both the formation of $Na^+$ ions and by the emission of the Na doublet. It has been found that

$$\log \frac{n}{(Na)} = K - \frac{66{,}000}{RT_m}$$

where (Na) represents the intensity of the resonance line.

The value 66,000 cal is very near the energy difference 69,000 cal between the ionization and resonance energy of the Na atom.[6]

Ion extraction and identification in a flame.

In order to determine the nature of the ions that are formed in a flame, they may be extracted from the flame and brought into a high vacuum system, where, after separation from the burnt gases, they can be accelerated and projected on to a narrow slit that serves as the entrance to a conventional mass spectrometer.

The possibility of the direct identification of the flame ions by means of mass spectrometry has been shown in three short notes [2] and in a later paper.[3] Independently and almost at the same period, Knewstubb and Sugden published a related work using a notably different technique.[22,23] Two different versions have been elaborated in this laboratory.

## A. Low-Pressure Burning Flames [3,5] */

Because a flame cannot be maintained at the very

*/ This section was written with the collaboration of J. Deckers and J. De Jaegere.

low working pressure ($10^{-5}$mm Hg) of a mass spectrometer, a sample of the flame gases has to be expanded through a hole. Behind the latter, the burnt gases have to be pumped away while the ions are kept in a straight beam by means of an electrostatic field maintained by a system of electrodes. If the hole is too narrow, most of the ions may lose their charge by collision with the walls. If the hole is large enough, no pumping installation would be sufficient to obtain a satisfactory vacuum. The equipment that has been built makes it possible to bring the ions from a flame burning at pressures from 3 to 10 mm Hg into an intermediate region at $10^{-2}$mm Hg and from there into a vacuum of $10^{-4}$mm Hg. The passage in the intermediate region was made as short as possible in order to avoid a randomized redistribution of the initially nearly homogeneous ion velocities.

Figure 7 gives a schematic description of the apparatus. It consists of four chambers: the combustion chamber I, the separation chamber II, the acceleration chamber III, and the mass analyzer IV.

The Pyrex combustion chamber I (65 mm diam) contains the burner. It is connected to a pump and has a side arm for the ignition system of the flame. The burner is a copper tube (30 mm diam) moving in the chamber through a "simmer ring." The polished "lava" bottom of chamber I is pressed against the rotating disk by means of a flexible membrane. The separation chamber II is made of a steel cylinder (300-mm diam) closed by two steel plates with a rubber seal. A system of "simmer rings" allows rotation of the disk while a vacuum of $10^{-2}$mm Hg is maintained in this chamber. Single holes in the lava bottom and in the rotating disk are 2 mm in diameter and at a distance of 70 mm from the central axis. The communication between I and II lasts 1/220, since the revolution speed is 6 rps. The communication between chambers II and III consists of a hole (2-mm diam) in a copper plate, supported for insulation by Plexiglas and fitted with a rubber seal. A vacuum of $10^{-4}$mm Hg is obtained in the acceleration chamber by means of an oil diffusion pump.

The separation between chambers III and IV consists of the narrow (1-mm broad and 15-mm high) entrance slit of the mass spectrometer. Both electrostatic and magnetic fields can be modified in order to register the different masses caught on the collector that is placed at the end of the analyzer tube (chamber IV). The total accelerating field is divided between the different electrodes, which are represented separately in Fig.9. Burner, flame, and disk are at the same positive potential, and only about 10 V drop is used between the latter and the first accelerating lens $L_1$. The resolving power of the system was only 50 m/$\Delta$m.

B. Atmospheric Pressure Burning Flames[9,10] *

The principle of ion extraction from the flame is the same. However, since the flame is burning at higher pressure, the diameter of the periodically opened hole has been reduced to 0.5 mm. The intermittent opening is obtained by the rotation of a stainless steel disk tightly pressed against a nylon surface (Figs.8 and 9). Other materials have been tried with much less succes (e.g. lava, teflon and aluminium). The bearing surface is a circle of only about 1 $cm^2$. No undue heating occurs even when experiments last more than 1/4 to 1/2 hr.

Four perforations (0.5 mm) in the rotating disk are drilled in a circular grove at about a distance of 12 mm from its edge; the grove was made in order to reduce the thickness of the disk so that the depth of the holes in the stainless steel material was only 0.5 mm. The gliding part of the disk was polished very carefully with a precision of 1/100 mm. In order to obtain a coincidence of the extraction hole, and the holes in the disk (with a precision of 1/10 mm); six centering screws allowed changes in the orientation of the disk axis, which was in a well anchored, hollow, heavy-walled, metal cylinder.

It will be noticed on Fig.9 that the first part of the equipment consists of two chambers where the ions first are extracted and thereafter accelerated. Separating both chambers is a conical-shaped wall favoring the capture of the ions and the evacuation of the gas molecules in the extraction chamber. With a rotating oil pump, a vacuum

* This section was written with the collaboration of J.N. Bertrand.

of about $10^{-2}$mm Hg can be maintained which is reduced only in a negligible manner when the disk rotates at the normal speed of 108 rpm. This means that the duration of one single opening is about 1/900 sec.

The whole system of chambers is made of stainless steel, and all insulating material is made of Teflon in order to prevent all corrosion.

The pumping system connected to the acceleration chamber includes a large oil diffusion pump with a capacity of 1000 liters/sec. It is backed by a mechanical pump with a capacity of 15 $m^3$/hr. This system can maintain in the acceleration chamber a vacuum of about $10^{-5}$mm Hg. A third pumping assembly (oil diffusion pump 100 liters/sec and mechanical pump 6 $m^3$/hr) maintains a similar vacuum in the $90^{\circ}$ deflection tube that is placed between the polar pieces of the electromagnet. The latter allows a continuous variation of the magnetic field from 150 to 3500 gauss with a precision better than 1/50,000 when the fluctuations of the power supply are less than 5%.

The total acceleration potential also can be modified up to a maximum value of 2000 V by means of a bleeder connected to the different electrostatic lenses or electrodes, which are numbered from $L_0$ to $L_5$ on Fig.10.

Very soon it was observed that the flame needs to be brought much nearer to the disk than was thought at first.

Sufficient ions are extracted only if the distance between the burner and the disk is smaller than 8 mm when the flame length (visible cone height) is about 6 mm.

The best disposition and separation of electrodes as presented here (Fig.10) are the result of a long and tedious investigation that very often led to results contrary to theoretical expectations. It seems that the best separation and position are very sensitive to a large number of factors for which an explanation is completely impossible in the actual state of affairs. One must not forget indeed that at each opening of the extraction hole some unknown gas pressure gradient in the extraction chamber is established. The best conditions for the maintenance of a sufficiently focused ion beam are for the moment purely empirical. They depend probably also on the geometry of the chamber.

By placing a collector in turn after each lens (or electrode or slit), a progressive step study of the best potential distribution has been made. The same kind of difficulties were encountered in choosing the best size and shape of each accelerating electrode. These difficulties also were solved by testing with a collector placed after each electrode in turn. All the flames studied were burning on a tube with a conical-shaped opening of 2 mm.

Since no ion source other than the flame itself could be inserted in the equipment, some internal calibration had to be made with some known flame ions.

Fortunately, the earlier work, where deuterium compounds were used,[3] enabled the investigators to make use of the fact that $H_3O^+$ forms one of the most abundant ions in many hydrocarbon flames. Furthermore, it already was known that some mixtures lead to the formation of a noticeable amount of $NH_4^+$ ion. Therefore, the distance between the two peaks 19 and 18 gives immediately the order of magnitude of the one-mass-difference peak distance over all the range.

For higher masses, the salted flames could be used: with Na salts, mass 23 ($Na^+$); with K salts, mass 39 and 41 ($K^+$ isotopes); and with Cu salts, mass 63 and 65 ($Cu^+$ isotopes).

The presence of nitrogen compounds in a flame provokes the appearance of the predominant $NO^+$ ion with mass 30.

In sum, the complete calibration was made possible by means of the following well defined and abundant ions: $NH_4^+ = 18$; $H_3O^+ = 19$; $Na^+ = 23$; $NO^+ = 30$; $K^+ = 39$ and 41; $Cu^+ = 63$ and 65.

It was observed that the resolving power of the mass analyzer was about 150 m/$\Delta$m.

## C. Summary of Experimental Results **

A first remark is that there is, at least for

** This section was written with the collaboration of J. Deckers, S. De Jaegere, and J.N. Bertrand.

the time being, no essential difference between the results obtained with flames burning at low and at atmospheric pressures.

Some typical results are presented in Figs. 11, 12 for the three hydrocarbon flames. Of course the relative intensity of the peaks varies with the richness of the mixture and also with the distance between the flame and the extracting hole. Because of this, it would be impossible to present here a detailed report of the large amount of data that has been obtained in this laboratory.

However, there are a few observations that have been made and from which impprtant conclusions can be derived. The observations first simply will be enumerated successively. They will be discussed in the following section.

1) As soon as a nitrogen compound is present in the flammable mixture either as fuel ($NH_3$ for example) or as oxidizer ($N_2O$ for example), the most abundant ion becomes the $NO^+$.[7,8]

2) Abundant ions are often the addition products of a proton and a polar molecule. This occurs, for instance, when $NH_3$ is added to the mixture with the appearance of $NH_4^+$. This explains probably the occurence of $H_2CN^+$ in nitrogen compound flames, of $H_2Cl^+$ in chlorinated flammable mixtures, etc.[7,8]

3) The transient formation of some compounds in the flame is proved nicely when these compounds are polar molecules. In that case they appear as the product of addition to a proton. Some examples are $(CH_3OH)H^+$ (mass 33) (coinciding, however, with $HO_2^+$), $(CH_2O)H^+$ (mass 31), $(CH_2CO)H^+$ (mass 43), $(CH_3COOH)H^+$ (mass 49), etc.

4) Some flames are so poor in ions that they are detected neither by the technique of ion concentration measurement nor by mass spectrometry. Such flames are $H_2/O_2$, $H_2 + CO/O_2$, $CH_2O/O_2$, $H_2S/O_2$, $CS_2/O_2$.

5) The addition of very small amounts of an ion-producing fuel to one of the preceding ion-poor flames provokes the abundant production of ions and the occurence of conductivity. This observation constitutes an easy way to

test qualitatively different fuels in order to find out if they are able to form ions when they are burned with oxygen.[9]

6) The $(CN)_2/O_2$ flame contains the following ions: $NO^+$, $N^+$, $CO^+$, $C^+$ and $CN^+$ ($NO^+$ being the most abundant).

7) Addition of NO (6%) to $CO/O_2$ flames does not lead to formation of ions. On the other hand, only 3% $(CN)_2$ provokes the appearence of abundant $NO^+$ ions.

8) In $NH_3/O_2$ and $N_2O/H_2$ flames, the predominant ion is $NO^+$.

### Mechanism of Chemi-Ionization [7,9] ++

A very thorough discussion of the possible processes of ion formation by a chemical reaction in a flame has been presented by Eyring et al.[24] and also by Calcote.[25]

Most interesting is the observation that, in $CH_4/O_2$ and $CH_3OH/O_2$ flames, ions are produced, whereas in $CH_2O/O_2$, $HCOOH/O_2$, and $CO + H_2/O_2$ flames, no ions are found. This fact proves that the process of chemi-ionization involves the reaction of a small hydrocarbon fragment containing one carbon atom that is not double-bonded to an oxygen atom. It is presumably the strong exothermicity of the formation of this first CO double bond which liberates a large part of the necessary ionization energy. The fragment might be the radical CH, and a possible chemi-ionization process then would be

$$\text{unexcited } CH + O_2 \longrightarrow CO_2H^+ + e^-$$

which is nearly thermoneutral or only slightly endothermic.

Many authors are definitely in favour of the following process which should be responsible for the primary ion production:

$$CH + O \longrightarrow COH^+ + e^-$$

Evidently a sufficient amount of enrgy is available but some problem rises concerning observed

++ This section was written with the collaboration of J.N. Bertrand, H. Philippaerts and J. D'Olieslager.

apparent activation energy of about 50 kcal which has been found for the overall process of ion formation. The rate of the last process can be written as follows:

$$\frac{dn}{dt} = k\,(CH)(O) = k_o\,(CH)(O)e^{-E/RT}$$

Accepting that E = 0, because of the exothermicity of the proposed process, there is still a relation between the concentration of radical and the temperature:

$$(CH) = K_1\,e^{-E_1/RT} \quad \text{and} \quad (O) = K_2\,e^{-E_2/RT}$$

and therefore

$$\frac{dn}{dt} = K\,e^{-(E_1 + E_2)/RT}$$

at a given mixture composition.

Now from data [10] $E_1$ = 50 kcal and, presumably, $E_2$ should be of the same order of magnitude so that:

$$\frac{dn}{dt} = K\,e^{-100000/RT}$$

which is in complete disagreement with the observations reported here.

In flames with nitrogen-containing compounds, an analogous process might be

$$\text{unexcited } NH + O_2 \longrightarrow NO^+ + OH^-$$

$$\text{excited } NH^* + O_2 \longrightarrow NO^+ + OH^- + 10\text{ kcal}$$

The latter is more probable.

The proton transfer to the polar molecules occurs very easily in a later stage; for instance

$$CO_2H^+ + H_2O \longrightarrow CO_2 + H_3O^+$$

or, generally speaking

$$IH^+ + P \longrightarrow I + PH^+$$

where $IH^+$ is any ion and P is a polar molecule.

The following equilibrium also can be established :

$$PH^+ + P' \rightleftharpoons P'H^+ + P$$

Proton exchanges are distinctly noticeable when the intensities of the mass peaks are investigated.

at different distances between the burner and the extracting hole, as illustrated in Fig.13. All ions are disappearing in favor of the $H_3O^+$, which, afterwards,decays very rapidly.The existence of a small maximum in the $C_3H_3^+$ion peak intensity is an argument against the opinion that it would be a primary ion. $C_3H_3^+$ becomes more important in richer flames (curve 5).

Most surprising is the observation (Fig.13) that the decay of the ion concentration is much less rapid when a small amount of methyl-bromide is added to a stoichiometric mixture of $CH_4$-$O_2$ ($N_2$=50%). King has suggested [26] that the recombination of negative ions occurs less easily when their mobility decreases. This, of course, would be the case when the heavy $Br^-$ takes the place of the lighter $OH^-$. The decay of thermal ions ($K^+$) is very much slower than the decay of chemi-ions. It parallels the cooling of the burned gases (Fig.15).

The addition of $N_2O$ leads to the formation of $NO^+$, which is definitely a chemi-ion.

As soon as NO molecules are present in the burned gases, charge transfer occurs because of the low ionization energy of NO:

$$IH^+ + NO \longrightarrow IH + NO^+$$

There is still a problem concerning the formation of the ions in the cyanogen flames. In this case, of course, no CH or NH radicals are formed in the reaction zone. However, excited $C_2^*$ radicals are produced which are able to react according to

$$C_2^* + O_2 \longrightarrow CO^4 + CO + e^- + 14 \text{ kcal}$$

There is also another possibility:

$$CN^* + O_2 \longrightarrow NO^+ + CO + e^- - 39 \text{ kcal}$$

This last reaction would mean that the formation process of the ions occurs with an activation energy that is the sum of 39 plus the activation energy of the formation process of the excited $CN^*$ radical:

$$C_2 + NO \longrightarrow CN^* + CO$$

However, the presence of ions in $CO/O_2$ flames with additives such as $CCl_4$ or $C_2Cl_4$ would indicate that the process involving $C_2^*$ is more likely.

Indeed, the addition of these halogenated compounds is accompanied by a marked $C_2^*$ emission of the flame.

The short discussion set forth in the foregoing shows that some light has been thrown on the problem of chemi-ionization, although many points still remain obscure. However, as compared to the situation existing only a few years ago, it now seems quite likely that the final solution is to be found by continuing the study of the ion identity in flames that differ widely in the nature of the constituting atoms.

## References

[1] Deckers, J. and Van Tiggelen, A., "Extraction of ions from a flame," Combustion and Flame 1, 281-286 (1957; also "Une méthode d'extraction des ions dans la flamme," Rev. Inst. Franç. Pétr. 13, 359-365 (1958)

[2] Deckers, J. and Van Tiggelen, A., "Note on ion identification in flames," Bull. Soc. Chim. Belg. 66, 664-665 (1957); also "Ion identification in flames by mass spectrometry," Nature 181, 1460 (1958) and Nature 182, 863 (1958).

[3] Deckers, J. and Van Tiggelen, A., "Ion indentification in flames," Seventh Symposium (International) on Combustion, (Butterworths Sci. Pub. London 1959), p. 254-255.

[4] Poncelet, J., Berendsen, R. and Van Tiggelen A., "Comparative study of ionization in acetylen-oxygen and acetylen-nitrous oxide flames," Same as ref.[3], p. 256-262.

[5] Van Tiggelen, A., "Identification of ions in flames and measurement of their concentration," T.R. 1, Armed Service Tech. Info. Agency Doc. 204474, Contract AF61(514)-1099, (May 1958).

[6] Berendsen, R., Taelemans, G.,and Van Tiggelen, A., "Induced ionization in hydrogen-oxygen flames," Bull. Soc. Chim. Belg. 69, 32-44 (1960).

[7] De Jaegere, S., Deckers, J.,and Van Tiggelen, A., "Identity of the most abundant ions in some flames," Eigth Symposium (International) on Combustion (The Williams and Wilkins Cy., Baltimore, Maryland, 1962),p. 155-160.

[8] Van Tiggelen, A., "Ion identification and ionization processes in flames," T.R. (Final), Armed Service Techn. Info. Agency Doc. 242292, Contract AF61(514)-1099 (May 1960).

[9] Van Tiggelen, A, "Experimental investigation of ionization processes in flames," T.R. 1, Armed Service Techn. Info. Agency Doc. 258050, Contract AF61(052)-398 (May 1961).

[10] Van Tiggelen, A., "Ion production and concentration in flames," T.R. 2, Contract AF61(052)-398 (May 1962).

[11] Lewis, B., "The effect of an electrical field on flames and their propagation," J. Am. Chem. Soc. 53, 1304-1313 (1931).

[12] Calcote, H.F., "Electrical properties of flames," Third Symposium (International) on Combustion (The Williams and Wilkins Cy., Baltimore, 1949), p. 245-253.

[13] Calcote, H.F. and Pease, R.M., "Electrical Properties of flames," Ind. Eng. Chem. 43, 2726-2731 (1951).

[14] Gaydon, A.G., and Wolfhard, H.G., "Flames " (Chapman and Hall, London, 1953), p. 281.

[15] Sugden, T.M., "Microwave studies in the ionization of alkali-metals in flame gases and related phenomena," Fifth Symposium (International) on Combustion (Reinholds Pub.Cy., N.Y. 1955),p.406-411.

[16] Loeb, L.B., "Basic Processes of gaseous electronics " (University of California Press, Berkeley, Calif., 1955), p. 332.

[17] Calcote, H.F. and King, I.R., "Effect of probe size on ion concentration measurements in flame," J. Chem. Phys. 23, 2203-2204 (1955); also "Studies

of ionization in flames by means of Langmuir probes," Fifth Symposium (Intenational) on Combustion (see ref.15), p. 423-434.

18 Van Wonterghem, J. and Van Tiggelen, A., "L'épaisseur et la vitesse de propagation du front de flamme," Bull. Soc. Chim. Belg. 63, 235-260 (1954).

19 Van Tiggelen, A., "Contribution à l'étude cinétique des réactions de combustion et d'inflammation," Mém. Acad. Roy. Belg. Cl. Sci. 27, 1-27 (1952).

20 Wilson, H., "Electrical Conductivity in flames" Rev. Mod. Phys. 3, 156-189 (1931).

21 Vandenabeele, H., Corbeels, R., and Van Tiggelen, A., "Activation energy and reaction order in methane-oxygen flames," Combustion and Flame 4, 253-260 (1960).

22 Knewstubb, P.F. and Sugden, T.M., "Letter to Editor," Nature 181, 474-475 (1958)

23 Knewstubb, P.F. and Sugden, T.M., "Mass spectrometry of the ions present in hydrocarbon flames," Seventh Symposium (International) on Combustion (Butterworths Sci. Pub., London 1959), p. 247-253.

24 Mukheyee, N.R., Fueno, T., Eyring, H., and Ree, T., "Mechanism of ion formation in high temperature flames," Eighth Symposium (International) on Combustion (The Williams and Wilkins Cy., Baltimore, Maryland, 1962), p. 222-229.

25 Calcote, H.F., "Ion production and recombination in flames," Eighth Symposium (International) on Combustion (see ref.[24]), p. 184-199.

26 King, J.R., "Recombination of ions in flames," T.R. 174A, Contract AF49(638)-650 (December 1961).

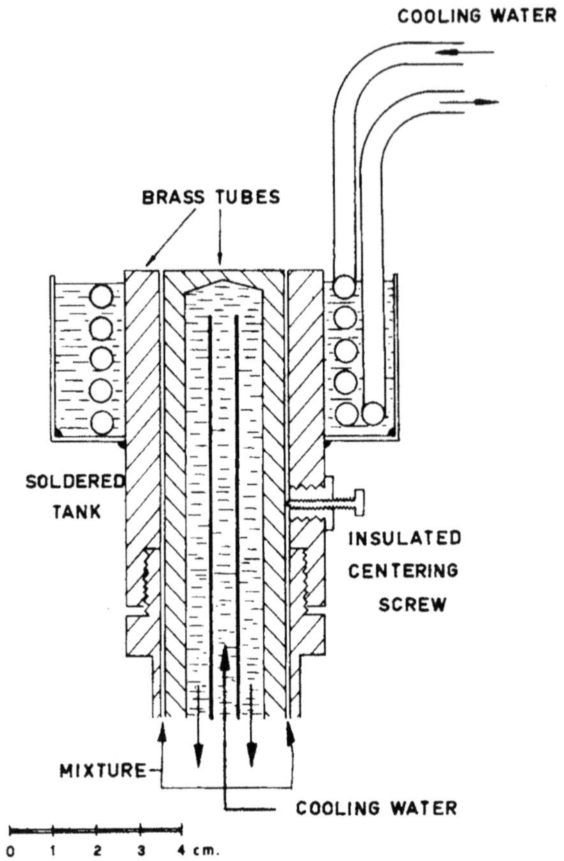


Fig. 1 Circular slit burner cooling system

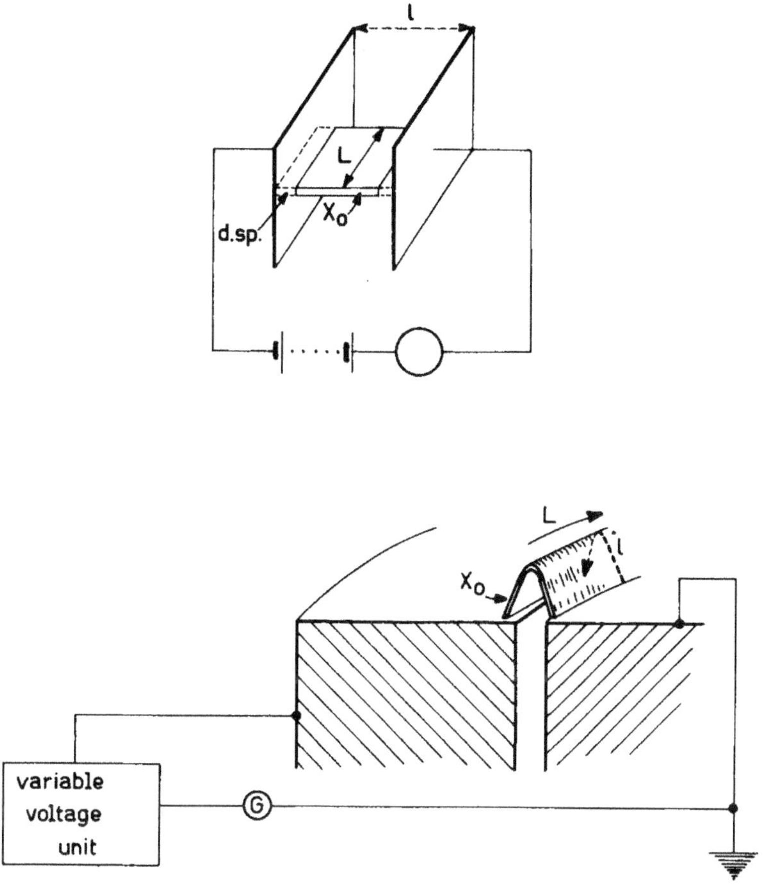


Fig. 2 Schematic drawing of conductivity measurement in the circular slit burner flame

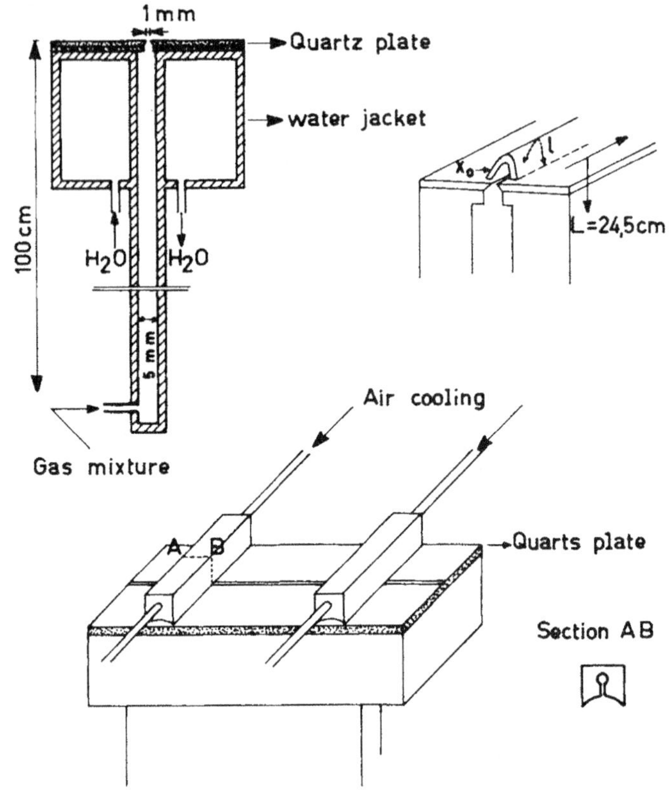


Fig. 3 Rectangular slit burner; shape of flame; measuring electrodes

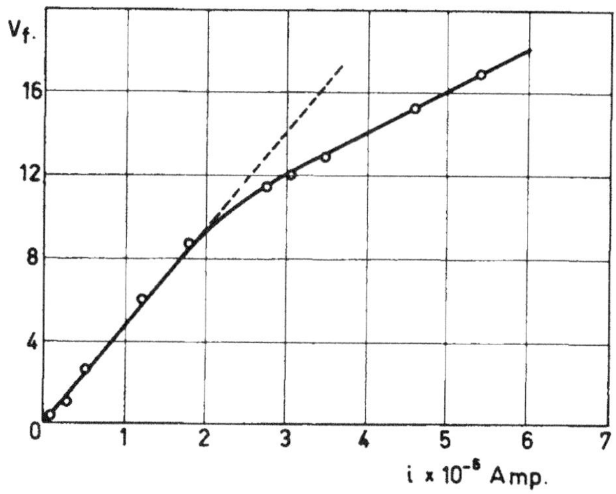


Fig. 4 Linearity of current intensity vs potential drop in the flame (up to a critical value), $CH_4$-$O_2$-$N_2$ stoichiometric mixtures, 65% $N_2$

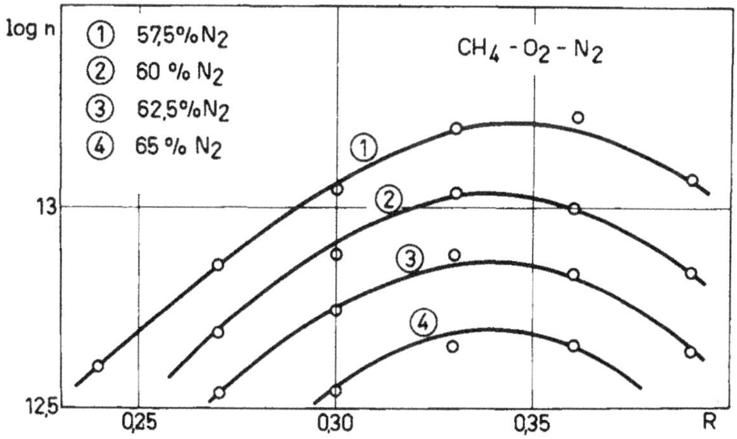


Fig. 5 Logarithmic plot of ion concentration vs ratio $CH_4/CH_4 + O_2 = R$. Mixtures are stoichiometric at $R = 0.33$

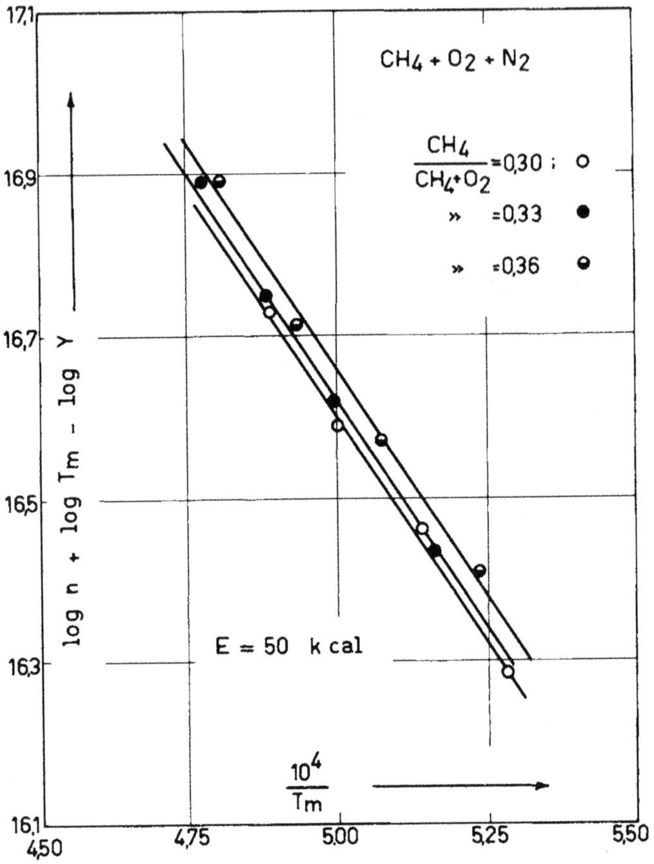


Fig. 6 Apparent activation energy of the formation process of the ions

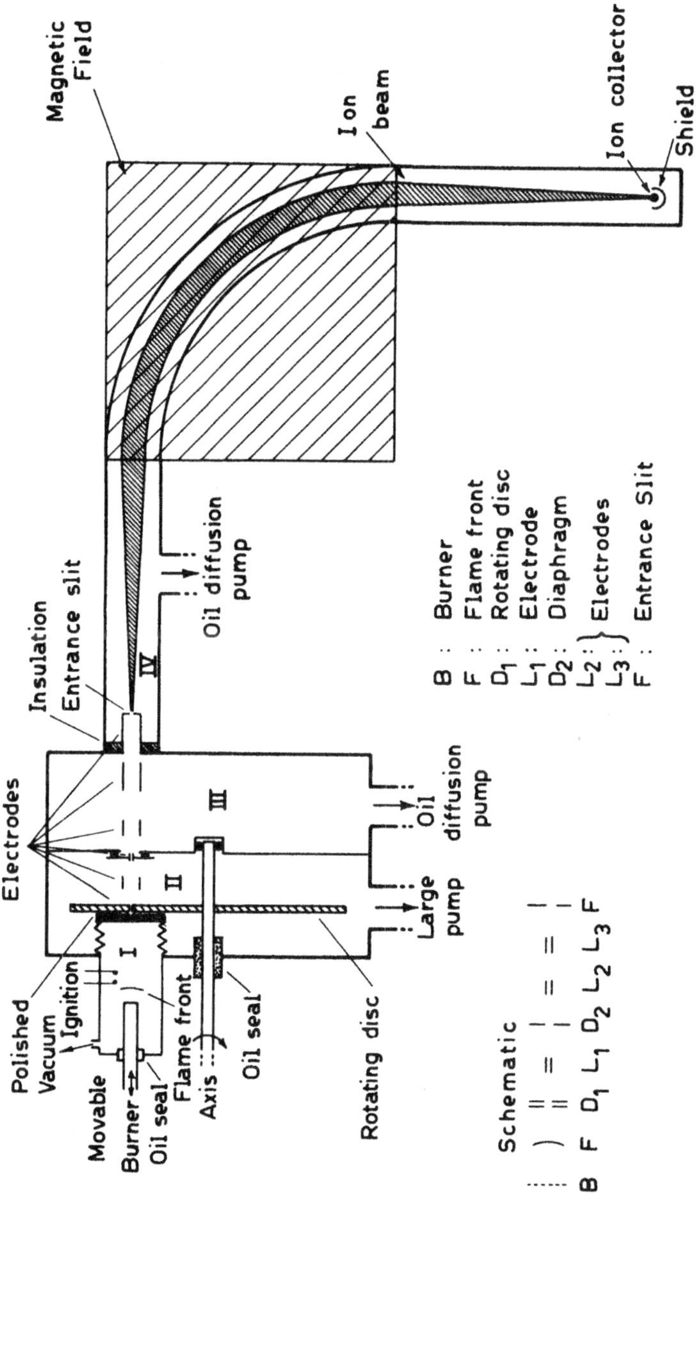


Fig. 7 Ion extraction and identification equipment for low pressure flames

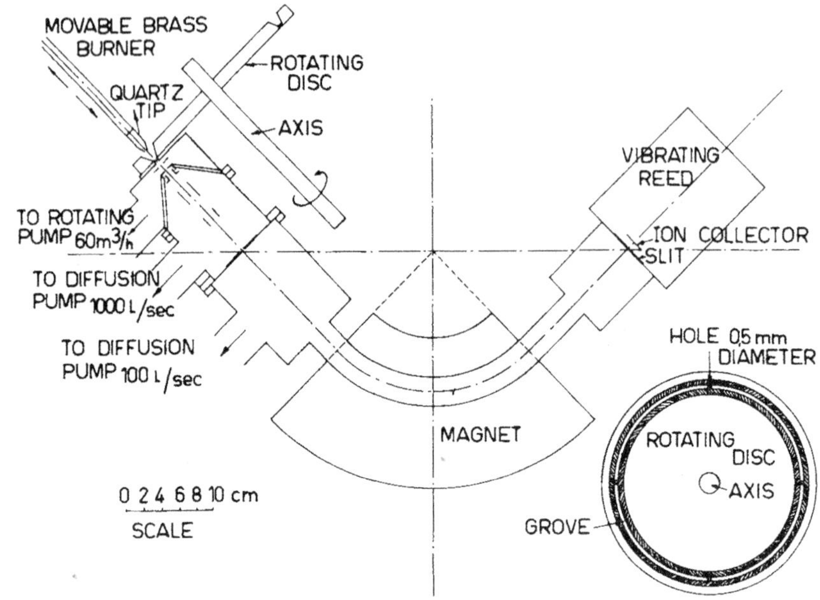


Fig. 8 Ion extraction and identification equipment for atmospheric pressure flames

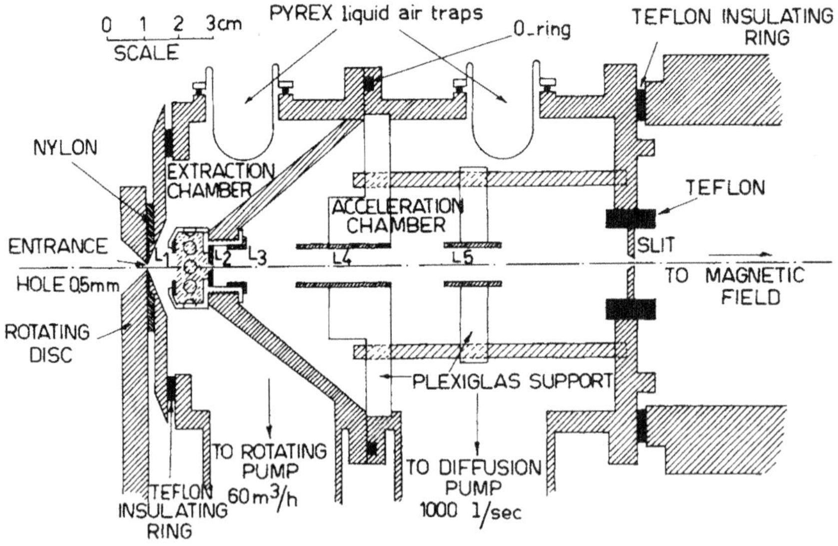


Fig. 9 Detailed schematic drawing of the ion extraction device

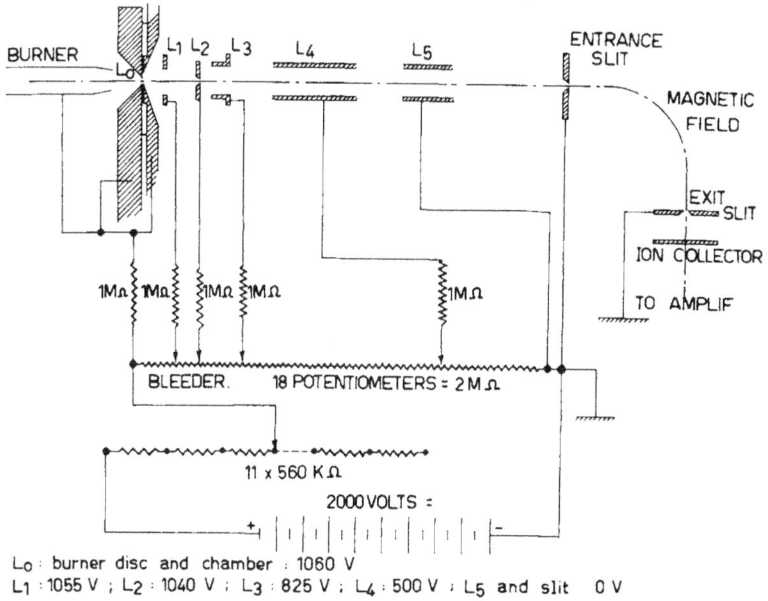


Fig. 10 Potential distribution in the ion extraction and acceleration chambers

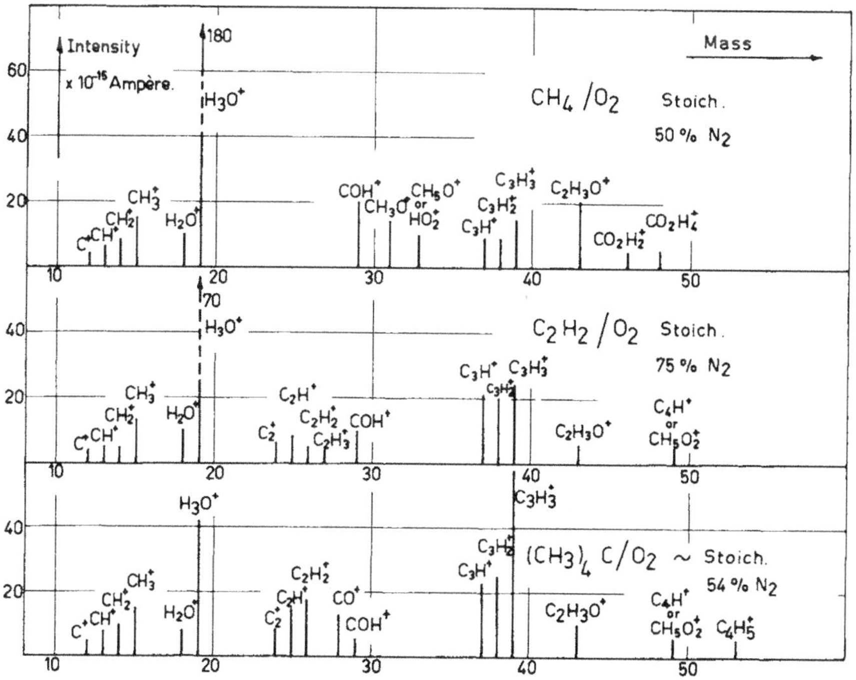


Fig. 11 Relative peak intensities for different masses in three hydrocarbon flames

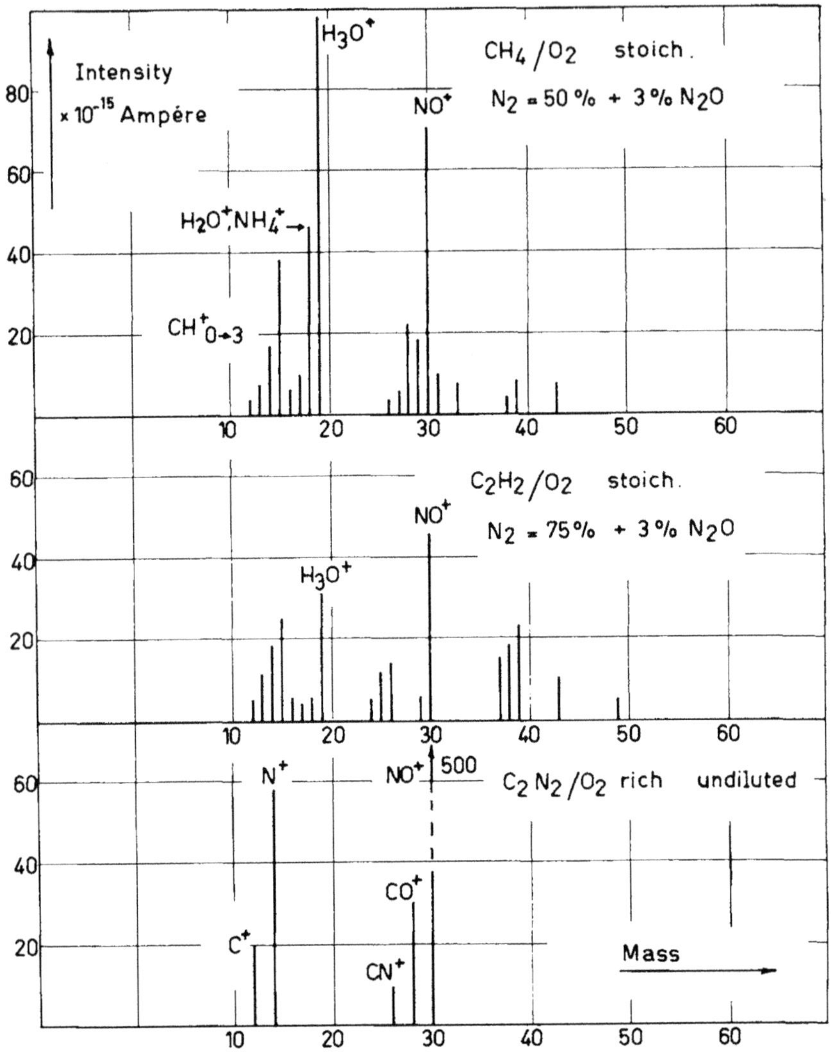


Fig. 12 Relative peak intensities for different masses when the nature of the fuel is profoundly changed

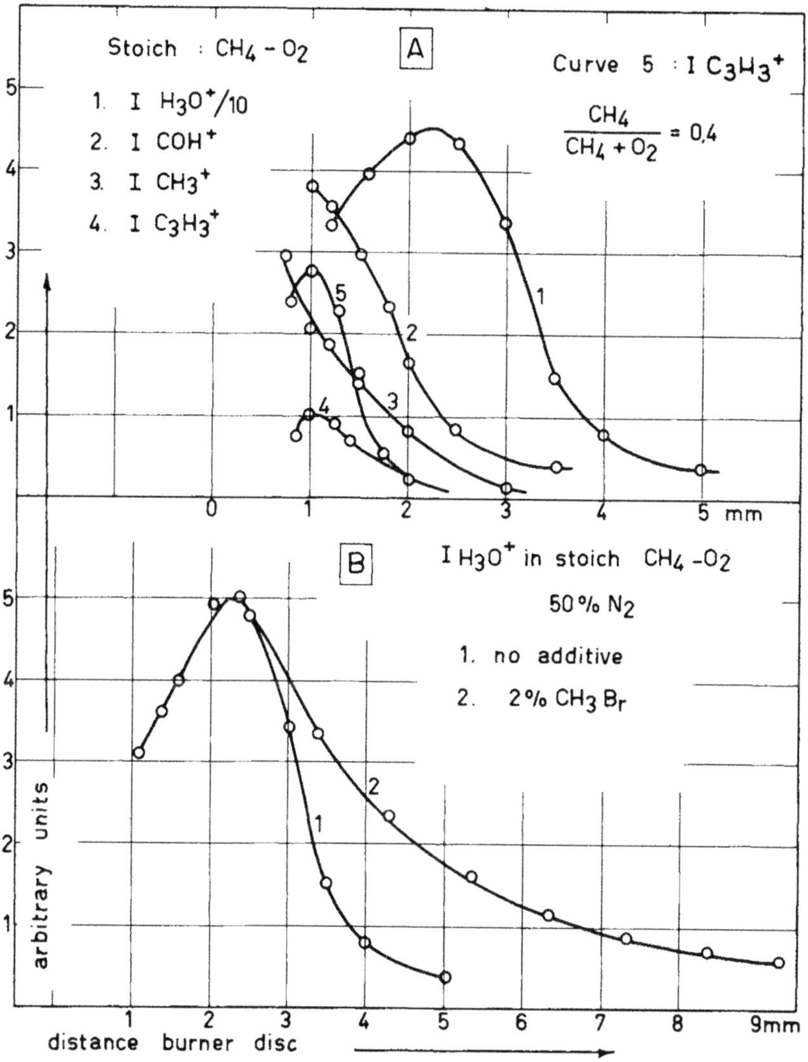


Fig. 13 Influence of flame distance from extraction hole on the peak intensities

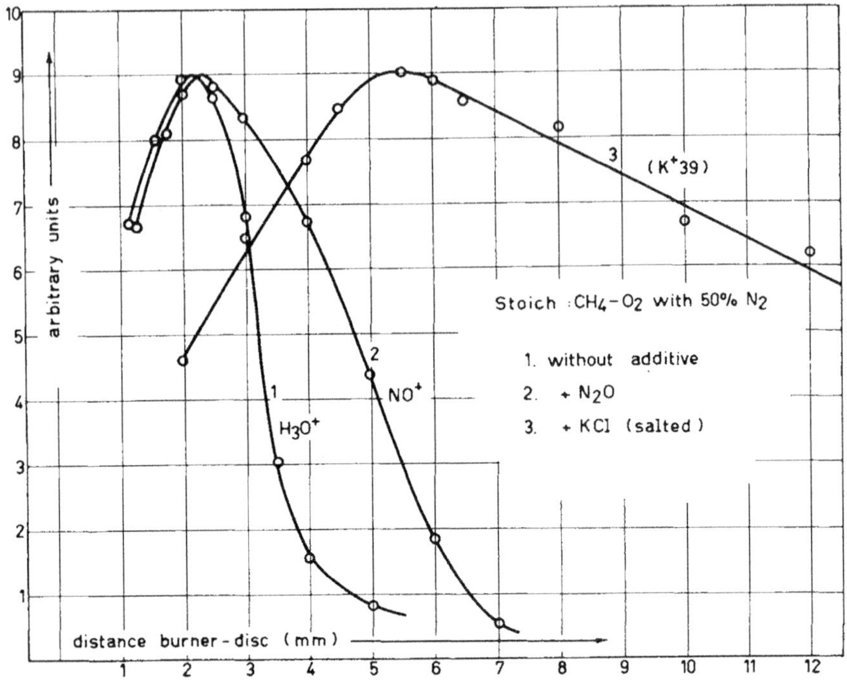


Fig. 14 Illustration of the difference between the decay of thermal and chemi-ions

# RECOMBINATION OF IONS IN FLAMES

I.R. King*

Texaco Experiment Incorporated, Richmond, Va.

## Abstract

The recombination of ions in flames has been studied by the Langmuir-probe technique. The probe technique is discussed, and some results showing the effect of temperature, pressure, and the presence of electronegative gases on recombination are presented. Both normal hydrocarbon-air flames and hydrocarbon-air flames seeded with alkali metals have been studied. Recombination of natural flame ions has been found to be much faster than recombination of alkali-metal ions. Although electronegative additives are quite effective in reducing free-electron concentrations in flames, these same additives show only a small effect on recombination rates.

## Introduction

The abnormally high concentration of ions found in the combustion zone of hydrocarbon-air flames has puzzled scientists for years. Neither the source of this ionization nor the mechanism by which these ions recombine once they have been formed is understood fully. Although considerable effort has been devoted to studying the mechanism of ion formation in flames,[1-11] little attention has been paid to the mechanism of ion recombination. Yet such information would be of considerable value to those concerned with the attenuation of electromagnetic radiation by rocket exhaust as well as to those interested in ion propulsion. The present paper summarizes some of the results of ion-recombination studies conducted in this laboratory in both normal hydrocarbon-air flames and hydrocarbon-air flames seeded with alkali metals.

---

Presented at the ARS Ions in Flames and Rocket Exhausts Conference, Palm Springs, Calif., October 10-12, 1962. This research was sponsored principally by the Air Force Office of Scientific Research of the Office of Aerospace Research under contract AF 49(638)-650.

*Senior Physicist, Chemistry and Physics Group.

## Theory

If a single species of each charge sign is present in a gas where no ions are being formed, the second-order rate law for recombination is

$$dn^+/dt = - \alpha n^+ n^- \qquad (1)$$

where $n^+$ and $n^-$ represent the number densities of positive and negative species, respectively, t is the time, and $\alpha$ is called the recombination coefficient. If $n^+$ and $n^-$ are equal, Eq. (1) integrates to

$$1/n^+ = \alpha t + \text{const} \qquad (2)$$

If, then $1/n^+$ is a linear function of time, the recombination is called second order. Similiarly, the first-order rate law

$$dn^+/dt = - \alpha' n^+ \qquad (3)$$

leads to

$$\ln n^+ = - \alpha' t + \text{const} \qquad (4)$$

In a flame, if one looks at an element of gas moving downstream, there is first a period of time in which ions are produced by chemical reaction and then a period of recombination. If the gas velocity is constant during the latter period and the recombination is second order, the reciprocal of the positive-ion concentration should be a linear function of distance beyond the reaction zone. Or, if recombination is first order, the logarithm of the concentration should be linear in distance.

The concentration of position ions can be measured[2,12,13] by means of Langmuir probes as a function of position in a flame. The requirement of constant velocity beyond the reaction zone can be met by a flat flame about 10 cm in diam, since observation that the temperature is uniform[14,15] rules out the possibility that the flame gases mix with the surrounding air within the first few centimeters downstream from the reaction zone. Thus, it is possible to determine recombination rates of the ions found in flames.

When alkali-metal salts are introduced into a hydrocarbon flame, they are ionized to an abnormal extent--greater than predicted from purely thermal considerations--as they pass through the reaction zone of the flame. This abnormal ionization then decays to that of thermal equilibrium as the gas moves downstream. It is this decay of abnormal ionization that allows one to measure the recombination rate. Unlike natural flame ions, alkali-metal ions are formed not only in the combustion zone but are also present in the burned-gas region of the flame to an appreciable extent as a result of thermal ionization. Because of this thermal ionization ocurring downstream of the reaction zone, the second-order rate law must be modified to

$$dn^+/dt = q - \alpha n^+ n^- \qquad (5)$$

where q is the volume production rate, $\alpha$ the recombination rate, and $n^+$ and $n^-$ the concentration of positive and negative species, respectively. Assuming charge balance and after integration, this equation becomes

$$\text{arc coth } (n^+/n_\infty^+) = n_\infty^+ \alpha t + \text{const} \qquad (6)$$

where $n_\infty^+$, the equilibrium ion concentration, is given by

$$n_\infty^+ = (q/\alpha)^{1/2} \qquad (7)$$

The recombination coefficient then may be determined from a plot of arc coth $(n^+/n_\infty^+)$ against time. Since it is not possible to carry out the measurements far enough downstream to obtain a reliable value of the equilibrium concentration, $n_\infty^+$ is treated as an adjustable parameter and chosen to give the best straight line. For this reason the precision of the measurements is not high enough to determine the order of recombination, which has been assumed to be second order.

## Experimental

The experimental technique, described in detail elsewhere,[2] involves inserting a fine wire probe into the flame under investigation and measuring the current to the wire as a function of applied voltage with respect to the burner. An analysis of the current-voltage curve thus obtained allows many properties of the plasma, or flame (including positive-ion concentration), to be determined. Experimental equipment consists of a burner system capable of producing flames over a range of pressures and an electrical system for obtaining current-voltage probe curves.

The low-pressure burner system has been described previously.[16] Measurements at atmospheric pressure were conducted on a Meker-type burner 5.5 cm in diam. The burner was sufficiently large so that the temperature of the central section of the flame gases remained practically constant for a distance of some 3 cm downstream of the reaction zone. Alkali-metal salts, when used, were introduced into the unburned gas stream from an aqueous solution by an atomizer. The atomizer is unique in the fact that it requires very little air and produces a very fine spray or mist of almost uniform drop size and will operate at pressures as low as 2 cm Hg. Furthermore, the salt solution is introduced from outside the system, thus allowing solutions to be changed easily.

The probing technique involves the use of two similar probes mounted on a common shaft at right angles to each other so that only one probe is in the flame at any one time. The probe-support shaft may be rotated through 90° so that either probe may be immersed in the flame. The two probes differ in length by a fixed amount, so by measuring the current to each probe separately, at a given distance above the flame zone, and subtracting the two it is possible to obtain the current density at the center of the flame. This technique eliminates to a large extent the effect of variations of concentration across the diameter of the flame. From current-density measurements obtained in this manner the positive-ion concentration at any height in the flame may be calculated.

Since recombination coefficients are determined by observing the rate of decay of ionization after passage through the flame front, it is necessary to measure the ion concentration at various distances above the flame zone. In accomplishing this, the probe mechanism is adjusted by a micrometer assembly. Such an arrangement allows the position of the probe to be determined accurately at any time. As a further check of probe position and to be sure the two probes are at the same height and location in the flame, a cathetometer or traveling telescope is used in conjunction with the micrometer assembly. Distance measurements are then converted to time by measuring the total flow to the burner at atmospheric pressure and room temperature and making the necessary corrections for increase in volume due to pressure and temperature changes. From the cross-sectional area of the burner and the total flow of gas through the burner, the linear flow velocity may be calculated. In the present experiments the velocity was not corrected for the difference in molecular weights of the burned and unburned gases. However, this difference is always small when the oxidizer is air.

The electrical circuit is described in detail elsewhere.[2] The positive-ion current is measured and divided by the surface area of the probe to obtain the positive-ion current density, $J^+$. Then Langmuir-probe theory permits the calculation of the positive-ion number density:

$$n^+ = e^{-1} J^+ (2\pi m^+/kT^+)^{1/2} \tag{8}$$

Here e is the charge of an electron (the positive ions are assumed to be singly charged), $m^+$ is the mass of a positive ion, k is Boltzmann's constant, and $T^+$ is the translational temperature of the positive ions. For a normal hydrocarbon-air flame it is assumed that the molecular weight of the positive ions is 20. The temperature of the ions is taken to be the adiabatic flame temperature. Previous experiments[17] have shown that the positive-ion concentration obtained from Eq. (8) is accurate only at extremely low pressures where the length of the mean free path of the ion is considerably greater than the diameter of the measuring probe. This was one of Langmuir's original assumptions. However, it has been shown[13] that good agreement between true ion concentration and measured ion concentrations may be obtained by multiplying the measured value by the ratio of the probe radius to the mean free path of the ions. Values used throughout this paper have been "corrected" in this manner. A more refined correction, based on the work of Bohm, Burlop, and Massey, has recently been published by Calcote.[7]

## Results

### Natural Flames

#### Comparison of several fuels

In some earlier studies[2,6] it was observed that ion concentration varied considerably from fuel to fuel, depending upon fuel type. Hydrogen-air, carbon disulfide-air, and hydrogen sulfide-air flames gave ion concentrations too low to measure accurately (below $10^7$ ions/cm$^3$). Hydrocarbon-air flames on the other hand, gave quite high ion concentrations--in the neighborhood of $10^{12}$ ions/cm$^3$--and showed very little variation from one fuel to the next. This would indicate that in hydrocarbon flames the predominant ionization is most probably due to the same species. This view is further supported by recombination data that indicate that these same hydrocarbon-air flames exhibit practically the same recombination rate. Table 1 summarizes the results of some earlier measurements of recombination coefficients in different hydrocarbon flames. The data were obtained at reduced pressures on a flat-flame burner 3 cm in diam. Impressively straight lines were obtained in all cases when the data were treated according to the second-order law.

Effect of pressure

The effect of pressure on recombination has been investigated over a pressure range of 0.087 to 1.0 atm. In all cases, impressively straight lines were obtained when the data were treated according to the second-order law [Eq. (2)]. A typical curve is shown in Fig. 1. The recombination coefficient is determined from the slope of the curve and Eq. (2). Results of the study are summarized in Fig. 2, where the recombination coefficient in a propane-air flame burning at two different equivalence ratios is plotted as a function of pressure. The logarithm of the recombination coefficient varies inversely as the logarithm of the pressure, giving a negative pressure index of -0.64.

Effect of temperature

These measurements were conducted in a propane-air flame burning on a Meker-type burner 5.5 cm in diam with the alternate-probe technique. All measurements were at 1 atm. Flame temperatures were varied by four different methods: 1) by varying the fuel-air ratio, 2) by diluting the flame with nitrogen, 3) by using a water-cooled burner grid to extract heat from the flame, and 4) by replacing the nitrogen in the air with argon or helium. Good agreement was obtained between the various methods with the exception of method 4. Here, mass-spectrometric data[9] have shown that different types of ions are formed when nitrogen is replaced with argon, and so one would suspect perhaps that temperature may have a different effect on the recombination rate. Results of the study are shown in Fig. 3. Although the temperature range covered is fairly small (400°K) and a certain amount of scatter in the results is unavoidable, the results show a definite trend: the recombination coefficient increases slightly with increasing temperature. The order of magnitude of the recombination coefficients ($10^{-7}$ $cm^3/sec$) indicates recombination is most probably occurring by either a third-body[18,19] or a dissociative[20] process.

Effect of electronegative gases

When an electronegative gas is introduced into a hydrocarbon-air flame, many of the free electrons normally created in the combustion process will become attached to these electronegative atoms to form negative ions. If the attachment process is sufficiently rapid, electron attachment will occur before the electrons have had a chance to recombine with a positive ion. Under these conditions recombination will occur primarily between positive and negative ions. On the other hand, if attachment is a relatively slow process, electrons will disappear primarily by recombination with positive ions. Since recombination rates are dependent upon the type of species involved,[18] it should be possible to influence the recombination rate by introducing electronegative gases into the flame. Results shown in Fig. 4 confirm this prediction. Here $\alpha$ is plotted as a function of the percent chlorine in the total flow of a propane-air flame burning at 1 atm. It is obvious from the graph that electronegative gases cause a reduction in $\alpha$ and that the first traces of additive have by far the greatest effect. It is rather surprising that the introduction of such an additive causes a reduction in the recombination rate, since it is generally agreed that ion-ion recombination is considerably faster than ion-electron recombination.[18] Introduction of an electronegative gas should cause an increase in the number of negative ions and so emphasize ion-ion recombination. However, results show a decrease in $\alpha$ with the addition of small amounts of halogens. Furthermore, the positive-ion concentration showed an appreciable increase when small amounts of chlorine were added. Similar results have been observed by Van Tiggelen[4,5] using a mass spectrometer to extract and identify ions formed in the combustion process. He finds a marked increase in the $H_3O_2^+$ ions in flames containing halogen compounds.

## Seeded Flames

### Comparison of several salts

Since alkali-metal atoms are ionized to an abnormal extent as they pass through the reaction zone of a hydrocarbon-air flame, such a flame offers a convenient environment in which to study the recombination of these ions. Once through the reaction zone, the ions recombine until thermal equilibrium is attained. Recombination coefficients have been determined for a number of alkali-metal ions and the results summarized in Table 2. All measurements were conducted at atmospheric pressure in a lean propane-air flame burning at 1970°K.

Lithium, sodium, potassium, rubidium, and cesium salts (usually carbonates or chlorides) were used. Generally speaking, the chloride compounds gave somewhat lower values than the carbonates. The chloride compounds also gave more positive ions than the carbonates. This is in agreement with results obtained when chlorine gas was added to a hydrocarbon-air flame. That is, the chlorine caused an increase in positive-ion concentration and a decrease in electron concentration at a fixed point in the flame. Of considerable interest is the fact that the recombination coefficients not only vary from salt to salt but they also correlate roughly with the ionization potential for the salts. For example, both ionization potentials and recombination coefficients vary in the other

$$Cs < Rb < K < Na < Li$$

In all cases the recombination coefficients in flames containing alkali-metal salts were some two orders of magnitude smaller than those for natural ions, which indicate that recombination is most likely occurring between an ion and an electron rather than between two ions.

Effect of pressure

The effect of pressure on the recombination of alkali-metal ions was investigated over a pressure range of 0.15 to 1 atm. Results of the study are shown in Fig. 5. Because of the difficulty encountered in introducing alkali-metal salts into a flame burning at reduced pressures, only a limited number of points have been obtained to date. Although considerable scatter is evidenced in the results, indications are that pressure has little if any effect on the recombination of alkali-metal ions. These results differ considerably from those obtained in natural flames and indicate that two different mechanisms of recombination are involved. The data reported here were obtained with a 0.1 N NaCl solution in a propane-air flame with an adiabatic flame temperature of 1970°K.

Effect of temperature

As in the case of natural flame ions, these measurements were carried out at a pressure of 1 atm in a propane-air flame. Flame temperatures were varied either by varying the equivalence ratio or by diluting the flame with nitrogen, although the latter technique was used in most cases. The method of varying flame temperature had little effect on the results. Results of the study are shown in Fig. 6. A 0.1 N NaCl solution was used in all cases. Although the temperature range investigated was fairly small, a number of interesting observations are possible. The recombination coefficient at first decreases quite rapidly with increasing temperature, reaching a minimum at about 1950°K, and then increases rather slowly with increasing temperature. The peculiar behavior suggests two different mechanisms probably are operative here, although additional information is required before a complete explanation is possible. It is interesting to compare values obtained in this investigation with those obtained by Knewstubb and Sugden[21] using an electromagnetic attenuation technique. They obtained an approximate value of $3.5 \times 10^{-9}$ $cm^3/sec$ for the recombination coefficient of $Na^+$ ions at 2500°K. If the curve reported here is extended to 2500°K, the value obtained by Knewstubb and Sugden would fall almost exactly on the curve.

Effect of electronegative additives

The effect of halogens and halogenated compounds on the recombination of alkali-metal ions was also investigated in some detail. The purpose of the investigations was to determine how much recombination rates could be influenced by the introduction of additives. Results showed that although the introduction of minute traces of electronegative additives caused an appreciable reduction in the concentration of free electrons, this same additive had relatively little effect on the recombination rate. Chlorine, used in most of these studies, was obtained by introducing carbon tetrachloride vapor into the flame. The effectiveness of carbon tetrachloride in reducing free-electron concentrations in flames is shown in Table 3. To eliminate the effect of ions other than those of the alkali metals, these studies were conducted in a carbon monoxide-air flame. This is one of the few flames which has no ions of its own and does not cause abnormal excitation of the alkali metals. Electron concentrations were measured with an electromagnetic attenuation technique described previously.[16] The ratio of free-electron concentrations for a given salt with and without halogen present has been used as an indication of the effectiveness of the halogen in removing free electrons. Although limited in scope, the data do allow several tentative conclusions to be drawn. Although the normality of the solutions vary from salt to salt, it is apparent that the greatest effect in electron reduction occurs in the case of cesium, followed by potassium, sodium, and lithium in that order. That is, the effectiveness decreases as the ionization potential increases. Even when the number of ions was held constant by adjusting the normality of the salt solutions, the effectiveness of the halogen still varied from salt to salt in the order indicated. For example, a 0.0001 N $K_2CO_3$ solution gave $2.8 \times 10^{10}$ electrons/$cm^3$, and a 0.01 N $Na_2CO_3$ solution gave $2.68 \times 10^{10}$ electrons/$cm^3$. Yet the same amount of carbon tetrachloride caused a nineteenfold reduction in $K_2CO_3$ electron concentration and only a fivefold reduction in the $Na_2CO_3$ electron concentration. With the exception of potassium, the effect of halogens on the recombination rates follows the same general trend. The recombination studies were conducted in a seeded propane-air flame burning at a pressure of 1 atm.

Results are shown in Table 4 where the ratio of the recombination coefficient in seeded flames without additive to the recombination coefficient in similar flames with additive is taken as an indication of the effectiveness of the additive. As in the case of natural flame ions, no great change in $\alpha$ was observed when halogen was introduced. The greatest effect occurred in the case of potassium where $\alpha$ was decreased by a factor of eight.

## Conclusions

Recombination coefficients in hydrocarbon-air flames have been found to be of the order of $10^{-7}$ cm$^3$/sec and to vary little from one fuel to another. The order of magnitude of the recombination coefficient suggests either a dissociative process or an ion-ion process. The effects of temperature, pressure, and the introduction of electronegative gases favor the ion-ion process. Recombination between an $H_3O^+$ ion, already identified as the most prominent ion downstream of the combustion zone in hydrocarbon-air flames, and an $OH^-$ ion has been suggested.[16] Although the introduction of electronegative additives causes some reduction in the recombination rates in these flames, it appears unlikely that any great change can be effected in this manner.

In the case of alkali-metal ions, recombination coefficients have been found to be of the order of $10^{-9}$ cm$^3$/sec and to vary appreciably from one alkali metal to another. This variation follows the same order as the ionization potential of these elements. For sodium, the recombination coefficient was found to be independent of pressure but to vary considerably with temperature, reaching a minimum at a temperature of about 1950°K and increasing in value both above and below this. As in the case of natural flame ions, introduction of electronegative gases caused a considerable reduction in the concentration of free electrons but caused only a small reduction in recombination rates. Again, indications are that no great change in recombination rates can be effected by the introduction of additives to the flame.

## References

1 Calcote, H.F., "Electrical properties of flames--burner flames in transverse electric fields," Third Symposium on Combustion, Flame and Explosion Phenomena (The Williams and Wilkins Co., Baltimore, Md., 1949), p. 245.

[2] Calcote, H.F. and King, I.R. "Studies of ionization in flames by means of Langmuir probes," Fifth Symposium (International) on Combustion (Reinhold Publishing Corp., New York, 1955), p. 423.

[3] King, I.R., "Effect of pressure on ion formation in propane-air flames," J. Chem. Phys. 29, 681-82 (1958).

[4] Van Tiggelen, A., "Ion identification and ionization processes in flames," Laboratory for Inorganic Chemistry, Universite Catholique de Louvain, Louvain, Belgium (May 1960).

[5] Deckers, J. and Van Tiggelen A., "Comparative study of ionization in acetylene-oxygen and acetylene-nitrous oxide flames," Seventh Symposium (International) on Combustion (Butterworths Scientific Publications Ltd., London, 1959), p. 256.

[6] King, I.R., "Comparison of ionization and electronic excitation in flames," J. Chem. Phys. 31, 855 (1959).

[7] Calcote, H.F., "Ion production and recombination in flames," Eight Symposium (International) on Combustion (The Williams and Wilkins Co., Baltimore, Md., 1962), p. 184.

[8] King, I.R., "Recombination of ions in flames--effect of temperature," J. Chem. Phys. 35, 380-81 (1961).

[9] Knewstubb, P.E. and Sugden, T.M., "Mass spectrometry of the ions present in hydrocarbon flames," Seventh Symposium (International) on Combustion (Butterworths Scientific Publications Ltd., London, 1959), p. 247.

[10] Shuler, K.E. and Weber, J., "A microwave investigation of the ionization of hydrogen-oxygen and acetylene-oxygen flames," J. Chem. Phys. 22, 491-502 (1954).

[11] Calcote, H.F., "Mechanisms for the formation of ions in flames," Combust. Flame 1, 385-403 (1957).

[12] King, I.R., "Ion recombination rates in methane-air flames," J. Chem. Phys. 27, 817-18 (1957).

[13] King, I.R. and Calcote, H.F., "Effect of probe size on ion concentration measurements in flames," J. Chem. Phys. 23, 2203-04 (1955).

[14] Friedman, R. and Burke, J., "Measurement of temperature distribution in a low-pressure flat flame," J. Chem. Phys. 22, 824-30 (1954).

[15] Fristrom, R.M. and Westenberg, A.A., "Flame zone studies, IV. Microstructure and material transport in a laminar propane-air flame front," Combust. Flame 1, 217-18 (1957).

[16] King, I.R., "Recombination of ions in flames," J. Chem. Phys. 37, 74-80 (1962).

[17] Bohm, P., Burhop, E.H.S., and Massey, H.S.W., The Characteristics of Electrical Discharges in Gases, edited by A. Guthrie and R. Wakerling (McGraw-Hill Book Co. Inc., New York, 1949), Chap. 2.

[18] Loeb, L.B., Basic Processes of Gaseous Electronics (University of California Press, Berkeley, Calif., 1955), Chap. 6.

[19] Von Engel, A., Ionized Gases (Clarendon Press, Oxford, 1955), Chap. 6.

[20] Bates, D.R., "Dissociative recombination," Phys. Rev. 78, 492-93 (1950).

[21] Knewstubb, P.E. and Sugden, T.M., "Observations on the kinetics of the ionization of alkali metals in flame gases," Trans. Faraday Soc. 54, 372-80 (1958).

Table 1 Recombination in flames of several fuels[a]

| Fuel | Equivalence ratio[b] | Ambient pressure, atm | Recombination coefficient, $cm^3/sec$ |
|---|---|---|---|
| Methane | 1.08 | 0.087 | $2.5 \times 10^{-7}$ |
| Propane | 1.18 | 0.072 | $2.9 \times 10^{-7}$ |
| Acetylene | 1.15 | 0.026 | $2.8 \times 10^{-7}$ |

[a] Oxidizer air. Platinum probe; radius, 0.007 cm.

[b] Equivalence ratio $\phi$ is the stoichiometric air-fuel ratio divided by the actual air-fuel ratio.

Table 2 Recombination of alkali-metal ions in a propane-air flame at 1 atm and a flame temperature of 1970°K

| Salts | Recombination coefficients, $cm^3/sec$ 0.1 N | 0.01 N | Ionization potential of metal, ev |
|---|---|---|---|
| $Cs_2CO_3$ | $0.53 \times 10^{-9}$ | $0.32 \times 10^{-9}$ | 3.89 |
| CsCl | 0.17 | 0.25 | |
| $Rb_2CO_3$ | 0.19 | 0.36 | 4.18 |
| RbCl | 0.40 | 0.36 | |
| $K_2CO_3$ | 0.25 | 0.70 | 4.34 |
| KCl | 0.18 | 0.57 | |
| $Na_2CO_3$ [a] | 3.1 | 4.0 | 5.14 |
| NaCl | 1.9 | 2.8 | |
| $Li_2CO_3$ [a] | 9.2 | 7.8 | 5.39 |
| LiCl | 8.5 | 9.0 | |

[a] Using an electromagnetic attenuation technique to follow electron disappearance, Knewstubb and Sugden[21] obtained values of $3.5 \times 10^{-9}$ and $5.5 \times 10^{-9}$ $cm^3/sec$ for the recombination coefficient of sodium and lithium ions, respectively. The agreement between the two sets of data is surprisingly good considering that two entirely different techniques were used. Knewstubb and Sugden's values were obtained from electron decay, while values in the present article were obtained from positive-ion measurements.

Table 3 Effect of carbon tetrachloride in reducing free-electron concentrations in a carbon monoxide-air flame at 1 atm and a flame temperature of 2000°K

| Salt | $Ne_1$ (without additive) | $Ne_2$ (with additive) | $Ne_1/Ne_2$ |
|---|---|---|---|
| 0.001 N $Cs_2CO_3$ | $34.6 \times 10^{10}$ | $5.5 \times 10^9$ | 62 |
| 0.0001 N $Cs_2CO_3$ | 10.9 | 2.7 | 40 |
| 0.001 N $K_2CO_3$ | 8.9 | 3.3 | 27 |
| 0.0001 N $K_2CO_3$ | 2.8 | 1.5 | 19 |
| 0.01 N $Na_2CO_3$ | 2.68 | 5.2 | 5 |
| 0.001 N $Na_2CO_3$ | 0.81 | 2.5 | 3.2 |
| 0.0001 N $Na_2CO_3$ | 0.25 | 1.95 | 1.3 |
| 0.1 N $Li_2CO_3$ | 4.1 | 5.0 | 8 |
| 0.01 N $Li_2CO_3$ | 1.3 | 3.3 | 4 |

Table 4 Effect of carbon tetrachloride on recombination rates of alkali-metal ions in a propane-air flame at 1 atm and a flame temperature of 1840°K

| Alkali metal | $\alpha$(salt)/$\alpha$(salt + $CCl_4$)[a] | Ionization potential, ev |
|---|---|---|
| Li | 0.75 | 5.36 |
| Na | 1.3 | 5.12 |
| K | 8.0 | 4.318 |
| Rb | 2.0 | 4.159 |
| Cs | 5.5 | 3.87 |

[a] Values given are average values; 0.02% by volume of carbon tetrachloride vapor was added to the total flow in all cases.

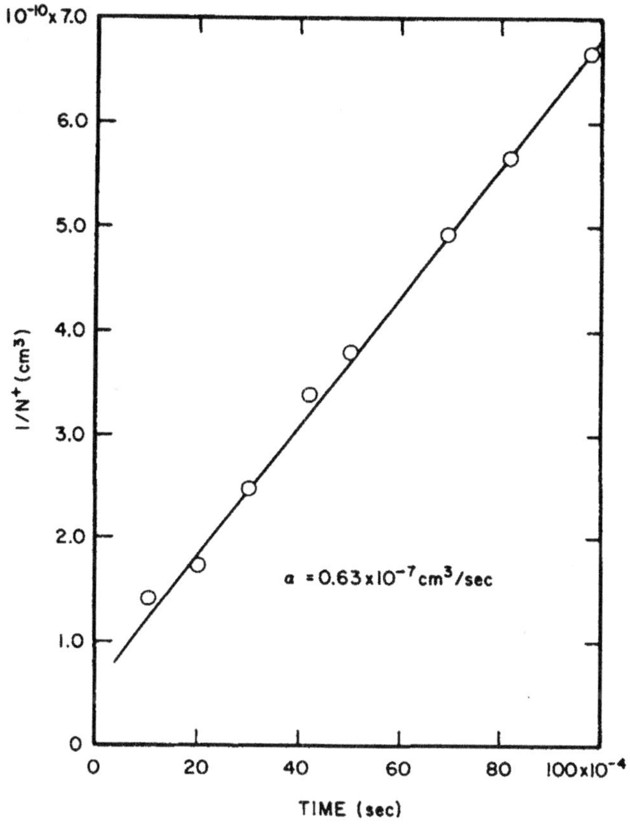


Fig. 1 Typical second-order recombination plot; propane-air flame at 1 atm; equivalence ratio, 0.79; flame temperature, 2020°K

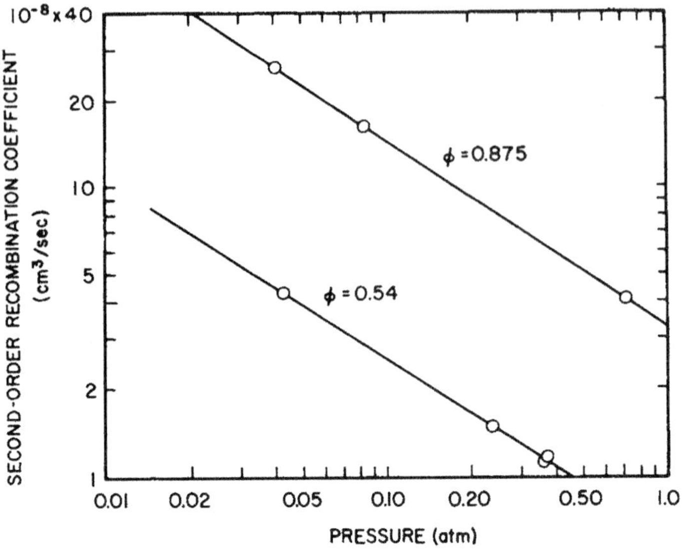


Fig. 2 Effect of pressure on recombination in a propane-air flame. Recently, Calcote, working at pressures below 30 mm Hg with a small burner, reported that under these conditions ambipolar diffusion was the predominant decay process. Correcting for this diffusion, he reported a recombination coefficient which was independent of pressure[7]

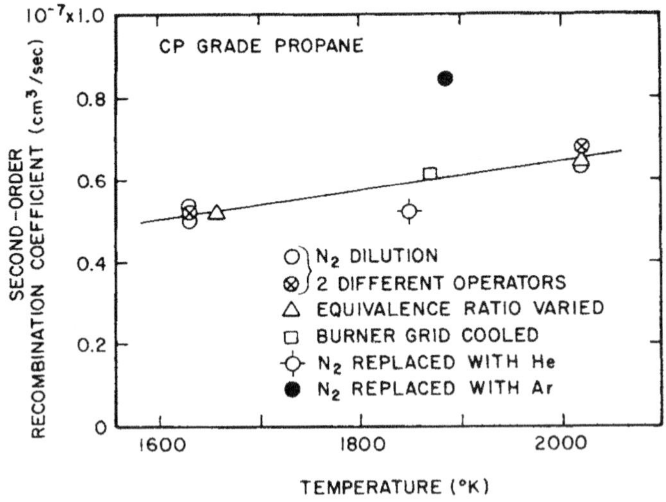


Fig. 3 Effect of temperature on recombination in a propane-air flame at 1 atm

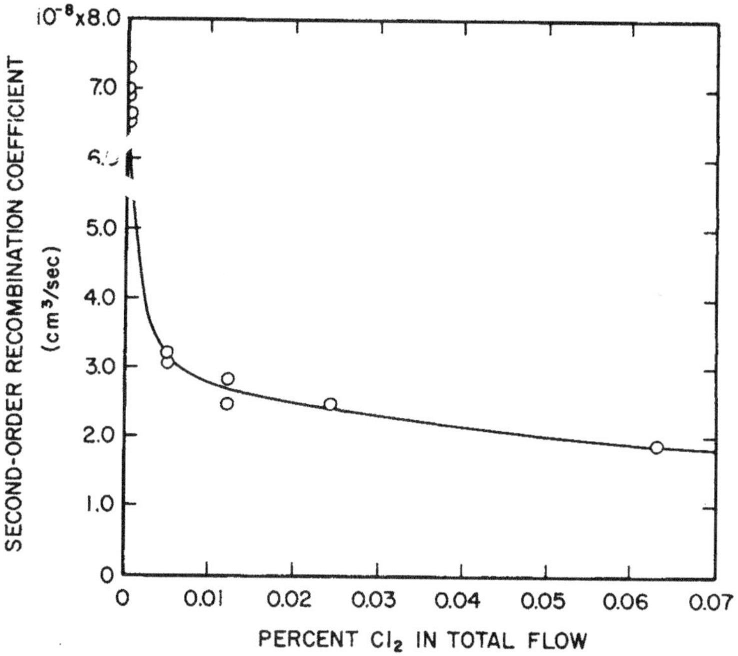


Fig. 4 Effect of chlorine on recombination in a propane-air flame at 1 atm

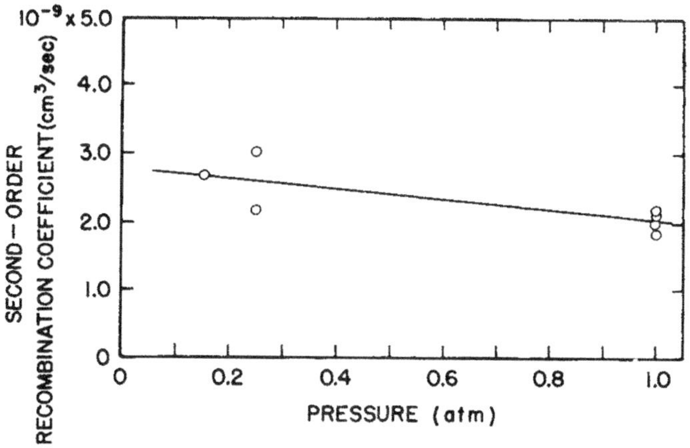


Fig. 5 Effect of pressure on recombination of alkali-metal ions; NaCl in a propane-air flame; flame temperature, 1970°K

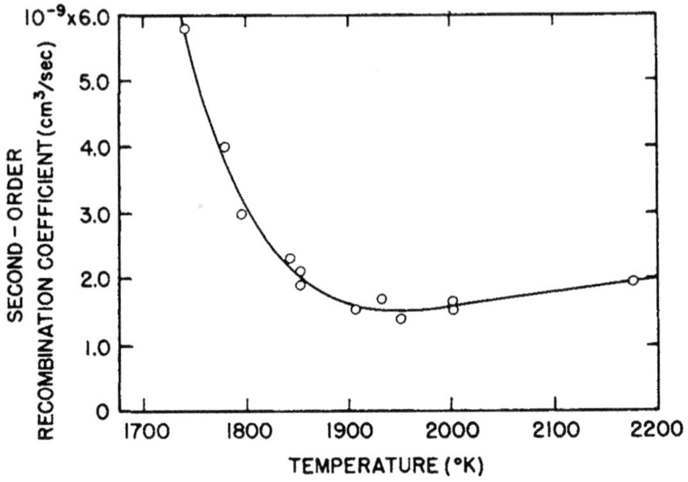


Fig. 6 Effect of temperature on recombination of alkali-metal ions; NaCl in a propane-air flame; pressure, 1 atm

# III

# IONIZATION IN SHOCK AND DETONATION WAVES

# IONIZATION BEHIND SHOCK WAVES

J. Derek Teare*

Avco-Everett Research Laboratory,
Everett, Massachusetts

Abstract

The paper provides an introduction to the phenomena associated with ionization occurring behind strong shock waves, including both equilibrium and nonequilibrium effects. The treatment of equilibrium phenomena covers calculations of equilibrium conditions behind shock waves, and discusses transport processes, including electrical conductivity, radiation, and convective heat transfer. The nonequilibrium material is specialized toward the ionization processes occurring in argon and in air. Details of the mechanisms which produce the initial ionization behind a shock wave in argon are still unknown, but ionization by electron-atom collisions rapidly becomes the dominant process. The rate of ionization is then determined by the rate of energy transfer to electrons by elastic collisions with atoms and ions, and is relatively independent of the inelastic ionization cross section. A theoretical model for the chemical and ionization processes occurring in air is compared with experiment. For shock speeds less than 7 km/sec, the dominant mechanism is shown to be the reaction $N + O \rightarrow NO^+ + e$. Ionization processes by other neutral-neutral collisions, and by electron-neutral collisions, are also considered, together with the effects of photoionization and of charge exchange reactions. Extension of the model to higher shock speeds is discussed. Experiments which produce shock velocities of approximately 400 km/sec in hydrogen are briefly reviewed.

---

Submitted for publication to AIAA October 14, 1963.

The author is pleased to acknowledge many helpful discussions with R. A. Allen, R. L. Taylor, B. Kivel and E. T. Gerry, and wishes to thank B. W. Bova for editorial assistance. He is especially grateful to J. C. Keck, S. C. Lin and H. E. Petschek for their advice and constructive criticism during the preparation of the paper.

Figures 10-21 are reproduced from Ref. 54 by courtesy of the American Institute of Physics.

*Principal Research Scientist.

## Introduction

Ionization phenomena occurring behind shock waves have been of considerable interest to physicists for several years, and much research into the fundamental processes has been accomplished. The shock waves were produced initially by various techniques, but the bulk of the experimental work has been carried out in conventional shock tubes, which have the outstanding merit of producing a homogeneous gas sample[1] in which the thermodynamic state of the gas is readily calculable. Most of the early experiments were carried out with noble gases,[2, 3] in which large degrees of ionization are achievable at relatively moderate shock speeds. Apart from yielding data on transport phenomena, such as heat transfer and electrical conductivity, many interesting experiments have been carried out which involve interaction of the ionized gas with a magnetic field.

More recently, particular attention has been focused on the ionization phenomena associated with shock heated air under the stimulus of the familiar "re-entry" problem. The presence of charged particles in the hot gas surrounding a hypersonic vehicle raises problems of communication and can affect the heat transfer to the vehicle surface. The radar characteristics of the flow field associated with all types of hypersonic objects are of paramount importance to the problems of detection of ballistic missiles. The radar signatures can be affected by the rate of ionization behind the bow shock wave, by the degree of completion of the ionization processes in the gas cap, and by the diffusion, convection and recombination processes in the cooler wake. These considerations have motivated considerable research into the properties of ionized air, including shock tube, shock tunnel, ballistic range, and in-flight experiments. Many of these same phenomena are also of interest in the study of meteor trails.

The advent of cis-lunar vehicles and interplanetary probes requires consideration of the phenomena associated with atmospheric re-entry at flight speeds up to 55,000 ft/sec (~17 km/sec). In this speed range a very large fraction of the internal energy of the gas behind a normal shock wave is invested in ionization, and this can influence both the radiative and convective contributions to the heat transfer to a vehicle surface. Recent experimental data obtained in shock tubes and ballistic ranges are of importance in determination of both of these aspects of the heating problem.

At the far end of the velocity scale are the experiments in which electromagnetic means are used to accelerate plasmas to velocities as high as 500 km/sec. Some of these experiments are directed toward production of plasma samples in which the mean free path is so large that the gas is virtually "collision-free," and some of the experimental configurations do result in "clean" shock waves behind which the equilibrium conditions may be calculated by application of the appropriate conservation equations. The ionization processes occurring within such a shock wave are of great interest, and play a major role in determining the details of the shock structure.

This paper is restricted to consideration of ionization phenomena behind shock waves, with only brief discussions of the interaction of the subject material with other aspects mentioned in the previous paragraphs. Most of the experimental data discussed were obtained in shock tube experiments, and the gases considered are either argon or air. The treatment covers both equilibrium phenomena, including electrical conductivity, heat transfer and radiation, as well as the nonequilibrium effects, in which attention is directed mainly to the ionization mechanisms themselves. The paper, intended as an introduction to the subject, is written from the viewpoint of the aerophysicist. While the references are intended to give as broad a coverage as possible, they are somewhat heavily weighted in view of the author's own involvement in the high temperature gas dynamics research carried out at the Avco-Everett Research Laboratory over the past several years.

## Equilibrium Ionization Behind a Shock Wave

The conditions behind a normal shock wave with one-dimensional flow are determined by the equations of conservation of mass, momentum and energy, in the form

$$p = p_1 + \rho_1 U_s^2 (1 - \rho_1/\rho) \tag{1}$$

$$H = H_1 + \frac{1}{2} U_s^2 \left\{ 1 - (\rho_1/\rho)^2 \right\} \tag{2}$$

where p, $\rho$, and H denote the pressure, density and enthalpy of the gas, respectively, $U_s$ is the shock velocity, and the suffix 1 denotes conditions in the undisturbed gas ahead of the shock wave. Immediately behind the shock, $p = R\rho T$, where R is the gas constant in units of energy per unit volume per $^{\circ}$K. Also, the enthalpy is linearly related to the temperature, T, with $H = \frac{5}{2} RT$ for a monatomic gas, and $H = \frac{7}{2} RT$ for a diatomic gas with the translational and rotational degrees of freedom fully excited. For either of these cases, Eqs. (1) and (2) are equivalent to the usual Rankine-Hugoniot conditions, for which p, $\rho$ and T may be expressed directly in terms of the shock Mach number $M_s = U_s/a_1$, with $a_1^2 = \gamma RT$. Equations (1) and (2) thus yield

$$\frac{\rho}{\rho_1} = \frac{\gamma + 1}{\gamma - 1 + 2/M_s^2} \tag{3}$$

so that for large values of $M_s$ the ratio $\rho/\rho_1$ approaches limiting values of 4 and 6 for monatomic and diatomic gases, respectively. The corresponding values of the specific heat ratio, $\gamma$, are 5/3 and 7/5, respectively.

When other degrees of freedom (including vibration, dissociation, and ionization) are excited, this simple relationship for the density ratio no longer holds. Since the pressure and enthalpy given by Eqs. (1) and (2) remain roughly constant, the excitation of these additional degrees of freedom requires a decrease in translational temperature, T, and a corresponding increase in density ratio, $\rho/\rho_1$, with the equation of state given by $p = RZ\rho T$. Here Z denotes the number of particles per original gas particle, when dissociation or ionization has occurred.

Tabulations of enthalpy and Z as functions of T and $\rho$ under equilibrium thermodynamic conditions are available for most gases of interest, so that Eqs. (1) and (2) can always be solved, usually by iterative means, to obtain the equilibrium conditions behind a shock wave in any given gas. Calculations of equilibrium conditions behind shock waves in argon were published by Resler, et al.[2] and by Bond,[4] for example, while similar calculations have been published by many authors for shock conditions in air. Figure 1 shows the equilibrium ionization achievable behind normal shock waves in air as a function of shock velocity. It is based on calculations similar to those reported by Hilsenrath and Beckett.[5] A recent report by Hilsenrath and Klein[6] presents revised calculations of the thermodynamic properties of air, with inclusion of corrections for the Debye-Hückel limiting-law and the second virial coefficients. This report also includes a survey of many of the earlier sources of thermodynamic equilibria and shock calculations for air. Other recent calculations of equilibrium thermodynamic properties include those of Lick and Emmons[7] for helium, and those of Drellishak, et al.[8] for argon.

In general, the equilibrium calculations are on a very sound footing, being based on statistical mechanical considerations which use very accurate spectroscopic data. The major uncertainties are associated with determining the appropriate cut-off for the partition functions of atoms and ions, since the classical expressions for these partition functions are divergent if all electronic energy levels are included. In general, the thermodynamic equilibria are not sensitive to the cut-off used, but under conditions of high temperature and very high electron density ($n_e \gtrsim 10^{18}$ cm$^{-3}$) greater care must be taken in the choice of appropriate cut-off. Under these conditions, too, the high electron density reduces the effective ionization potential of the neutral particles. This effect must also be included in the calculation of equilibrium thermodynamic properties.

Measurements of equilibrium thermodynamic properties behind shock waves have served mainly to establish the validity of experimental techniques, and to verify the behavior of the instrumentation and of the shock tube itself. Thus, it is seen that there is little merit in further equilibrium measurements of this type, except in circumstances where such a measurement would help clarify an uncertainty in one of the fundamental properties which could not be resolved spectroscopically. Much greater interest lies in determination of the transport properties in the shock-heated gas, and these will be briefly surveyed.

## Electrical Conductivity

From the derivations of Chapman and Cowling,[9] the electrical conductivity of a partially ionized gas may be expressed in the form

$$\sigma_{dc} = \eta n_e e^2/m_e \nu_e \qquad (4)$$

where e, $m_e$, $n_e$ and $\nu_e$ are the electron charge, mass, concentration and collision frequency, respectively, and $\eta$ is a nondimensional coefficient of order unity. For a multicomponent gas, the collision frequency is given by

$$\nu_e = \overline{c}_e \Sigma n_i Q_i \qquad (5)$$

where $Q_i$ is the averaged elastic momentum transfer cross section for collision of electrons with the $i^{th}$ species, of concentration $n_i$. The mean electron velocity in Eq. (5) is given by $\overline{c}_e = (8\,kT/\pi m_e)^{1/2}$, where k is Boltzmann's constant and T is the electron temperature. For a gas in which the neutral particles are heavy rigid spheres of cross section Q and of concentration $n_o$, the conductivity becomes $\sigma = \eta\, n_e e^2/m_e \overline{c}_e n_o Q$, and for compatibility with the calculated value of Ref. 9 the numerical coefficient is required to be $\eta = 0.85$.

For large degrees of ionization, the conductivity becomes dominated by distant encounters between electrons and ions, and Eqs. (4) and (5) then yield

$$\sigma = 0.85\ e^2/m_e \bar{c}_e Q_{ion} \tag{6}$$

This expression may be compared directly with the theory of Spitzer and Härm,[10] in which the conductivity of a gas which is completely singly ionized is given in the form

$$\sigma = \frac{0.591\ (kT)^{3/2}}{m_e^{1/2}\, e^2 \ell n\,(h/b_o)}$$

which is equivalent to

$$\sigma = \frac{0.943\ e^2\ (kT/e^2)^2}{m_e \bar{c}_e\ \ell n\ (h/b_o)} \tag{7}$$

In this expression, h is the Debye shielding distance is the plasma, given by

$$h^2 = kT/8\pi n_e e^2 \tag{8}$$

and $b_o \equiv e^2/3kT$ is the impact parameter at which a positive ion deflects a mean-energy electron by $90^\circ$. From comparison of Eqs. (6) and (7), an effective cross section for ion-electron collisions may be deduced in the form

$$Q_{ion} = 8.1\ b_o^2\ \ell n\ (h/b_o)\ cm^2 \tag{9}$$

The above summarizes the approach adopted by Lin, et al.[11] in comparing their measurements of electrical conductivity in shock-heated argon with theory. Some of their data points are reproduced in Fig. 2. The dashed curves show the theoretical conductivity according to Eq. (7), which is affected by initial pressure only through the variation of Debye distance, h, caused by variation of $n_e$. The solid curves are computed by means of Eq. (4), with inclusion of both the electron-ion interactions and the "close encounter" electron-neutral effects in Eq. (5), using an argon momentum transfer cross section discussed below.

Lamb and Lin[12] obtained measurements of shock-heated air with the same technique used in argon. Their data points are shown on Fig. 3, together with a theoretical curve based on Eq. (4). In order to compute $\nu_e$, it is necessary to assume values for the averaged elastic momentum transfer cross sections for collisions of electrons with $N_2$, $O_2$, NO, O, and N.

Shkarofsky, et al.[13] published a review of the various cross sections available in the literature. Figure 4 shows their preferred values for argon, $O_2$, $N_2$ and NO; their cross sections were presented as functions of electron energy, but they are reproduced in Fig. 4 as functions of electron temperature, after averaging over a Maxwellian velocity distribution. The argon curve of Fig. 4 was used in the conductivity calculations of Fig. 2.

The appropriate choices for O and N cross sections are less clear-cut. Available measurements (for example, Neynaber, et al.[14, 15]) are at too high an electron energy to be of direct application to the range of conductivity data. However, Lin and Kivel[16] have published values of electron collision frequency behind a shock wave in a mixture of 95% $O_2$ - 5% $N_2$. For this mixture, the major contributions to $\nu_e$ are from collisions between electrons and oxygen atoms or molecules, and the relative role of these two species changes by a factor of 2.4 over the range of the measurements ($T = 3800 \pm 200^\circ K$). Agreement within 30% between calculated and measured values of $\nu_e$ is obtained by using cross sections of $10^{-15}$ $cm^2$ for $N_2$ and NO (as in Fig. 4), $3 \times 10^{-15}$ $cm^2$ for $O_2$ and $1.5 \times 10^{-16}$ $cm^2$ for oxygen atoms. This value for the $O_2$ cross section is appreciably lower than the value given by the solid curve of Fig. 4, but the discrepancy is of the same order as the uncertainties in measurements of cross sections at such low electron energies.

The theoretical curve of Fig. 3 uses the solid curves of Fig. 4, together with the dashed lines shown for the O and N cross sections. These dashed lines are rather arbitrarily drawn, representing compromises between the calculations and measurements reported in Refs. 13 through 16. It is of interest to assess the changes which would be produced in the theoretical conductivity of Fig. 3 by changes in the choice of momentum transfer cross sections. For $U_s = 3.5$ km/sec, the $N_2$ cross section is responsible for 87% of the summation in Eq. (5). The other contributors of significance are NO (5%), O (4%), and $O_2$ (4%). Over the whole shock speed range of the data, the O contribution to $\nu_e$ is less than 10%, so that large changes in O cross section would have relatively small effect on theoretical conductivity. The N contribution becomes appreciable above $U_s = 5$ km/sec, and for $U_s = 6$ km/sec the major contributions are from $N_2$ (49%), N (31%), O (6%) and from electron-ion collisions (13%). The dashed curves of Fig. 3 show the effect of substantial changes ($\pm$ 50%) in the atomic nitrogen momentum transfer cross section from the value of Fig. 4.

The high values of conductivity achievable in shock-heated

argon and air have made possible many interesting shock tube experiments involving interactions between conducting fluids and magnetic fields. Among these may be cited the argon experiments of Patrick and Brogan,[17] Pain and Smy,[18] Dolder and Hide[19] and Fishman, et al.[20] Experimental investigations of the magnetohydrodynamically-supported shock layer in air have also been reported.[21]

## Radiation

Spectroscopic studies of the visible radiation from partially ionized argon behind a shock wave were reported by Petschek, et al.[22] The radiation consisted of a continuum on which were superimposed argon lines which were broadened and shifted. Interpretation of the line shift permitted ion concentrations to be inferred which were in agreement with the calculated equilibrium ionization behind the shock wave. The measurements of the continuum radiation compared well with the Kramers-Unsöld theory,[23] and provided a determination of the factor $Z_{eff}^2$ in the theoretical intensity.

The equilibrium radiation from shock-heated air has received extensive treatment over the past few years, and recent summary papers have been presented by Treanor[24] and by Keck, et al.[25] At moderate shock velocities the molecular bands provide the major contributions to the radiation, but at sufficiently high shock velocities the degree of ionization behind the shock becomes large and the radiation is dominated by the continuum associated with the deflection or capture of electrons. Various authors[26-29] have estimated the Kramers' radiation by positive ions ($N^+$ and $O^+$) which is expected to be the largest contributor at temperatures above about 10,000°K. Recent unpublished shock tube data at 5000 A obtained by Allen show a higher level of equilibrium radiation than predicted by the most-favored Kramers' calculation made by Biberman and Norman.[29] An additional contribution to the radiation in the range of Allen's measurements could arise from scattering or capture of electrons by neutral atoms.

For either type of continuum, the Unsöld expression[23] gives the radiation intensity per cubic centimeter per unit frequency interval in the form $\xi \propto n_e n_i Z_{eff}^2$, where $n_e$ is the electron concentration, $n_i$ the concentration of the particles producing the radiation, and $Z_{eff}$ the effective charge of these particles. For fully dissociated, partially ionized air, $n_e \propto \rho^{0.5}$. Hence, for the continuum produced by positive ions, since $n_{ion} \propto \rho^{0.5}$, the intensity is $\xi \propto \rho^{1.0}$. On the other hand, for the neutral continuum, since $n_{atom} \propto \rho^{1.0}$, the intensity is $\xi \propto \rho^{1.5}$. Allen has pointed out that this difference in density dependence could provide a means of distinguishing between the two types of continua.

Taylor[30] has measured $Z_{eff}^2 \approx 0.01$ for the $N_{ff}$ (free-free, or Bremsstrahlung) radiation in the infrared at wavelengths beyond $2\mu$, indicating that free-free scattering of electrons by nitrogen atoms should not be important in Allen's measurements at 5000 A. However, in interpretation of arc measurements made in nitrogen at atmospheric pressure, Boldt[31] suggests the presence of $N^-$ continuum radiation produced by the capture of electrons by nitrogen atoms in the first excited state, with a binding energy of the electron to this state estimated to be 1.1 ev. It should be possible to verify or disprove the presence of such an $N^-$ continuum in the radiation from shock-heated air by means of measurements over a wide range of density, as indicated above.

## Heat Transfer to a Wall

The calculation of the heat transfer from a partially ionized gas to a cold wall requires a knowledge of both the transport properties of the gas and of the chemical behavior of the gas as it is cooled by the wall. Several experimental investigations of heat transfer in argon have been reported in the literature. Van der Noordaa[32] measured the heat transfer rates on the side wall of a shock tube behind the incident shock. Bershader and Rutowski[33] measured the heat transfer at the stagnation point of a hemispherical body, while Smiley[34] measured heat transfer at the end wall behind a reflected shock. Camac, et al.[35] have also reported measurements on the end wall, covering a temperature range up to 65,000°K for un-ionized argon. The latter temperature is obtained behind the reflected shock when the incident shock velocity is 5.5 km/sec, and for an initial pressure $p_1$ = 1 mm Hg, this temperature persists for about one microsecond before the ionization process sets in. Heat transfer measurements during this first microsecond are made for a fully frozen boundary layer, and provide excellent experimental confirmation of an assumed temperature dependence, $\kappa \propto T^{3/4}$, for the argon thermal conductivity, $\kappa$, over the above temperature range. At slightly later times, the ionization process sets in very rapidly, drastically reducing the translational temperature. The experiments showed no significant change in heat transfer rate during and after the ionization, whereas calculations for an equilibrium boundary layer would predict a reduction of 20-30% in heat transfer rate. It is probable that a more complicated boundary layer theory[36] which takes account of possible differences between electron temperature and atom temperature would be necessary to account for the discrepancy.

Boundary layer theories for high temperature air usually take account of the dissociation processes, particularly for stagnation point heat transfer to a hypersonic vehicle. At flight speeds above 35,000 ft/sec (~ 11 km/sec), the energy content and thermal transport properties of the ionized components of the shock-heated air are likely to influence the boundary layer behavior. Most of the interparticle cross sections are sufficiently well known to permit reasonably accurate calculation of transport properties of air for temperatures well into the region where ionization is important.[37] Equilibrium boundary layers are calculable[38-40] with the aid of such transport properties. However, frozen boundary layer calculations require the solution of diffusion equations for all of the important species in the gas outside the boundary layer. Fay and Kemp[41] have recently introduced a simplified "binary diffusion" model for a partially ionized diatomic gas, in which the charge exchange cross section for atom-ion collisions is taken to be infinite, thus preventing relative diffusion of the atoms and the ion-electron pairs. Diffusion can then be treated as if there were only two components, molecules and atom-ion-electron particles, though the thermodynamic and transport properties are evaluated using all four components. Calculations of stagnation point heat transfer made with this model for both frozen and equilibrium boundary layers in nitrogen are shown in Fig. 5, together with equilibrium calculations from other sources. The fairly large differences between equilibrium and frozen heat transfer make it a matter of some importance that criteria be developed for making the correct choice in any given design situation. Shock tube data are available for shock velocities up to 40,000 ft/sec (see, for example, Ref. 42), simulating flight velocities up to 55,000 ft/sec. The data are, in general, in good agreement with the spread of theory in Fig. 5, but do not permit any clear distinction between frozen or equilibrium situations. In general, for a particular body size it is possible to assess the density at which the nitrogen dissociation-recombination process would be frozen in the boundary layer, but more accurate rate constant information for the various electron recombination processes is necessary before the problem can be completely determined. This particular uncertainty is one of the many motivations for study of the specific rate constants for the various ionization processes.

## The Approach to Equilibrium Ionization in Argon

Petschek and Byron[43] investigated the various possible ionization mechanisms in shock-heated argon, and obtained comparisons between theory and experiment. They considered the possibilities of ionization by electron-atom collisions, ionization by atom-atom collisions, and photoionization. Except for a region very close to the shock front where the electron concentration would be vanishingly small, it was expected that the first of these mechanisms would dominate the process. However, it was recognized that the electron gas would lose energy rapidly in producing such ionization, and that this energy must be replaced by elastic collisions of hot atoms and ions with colder electrons. The concept of different temperatures for atoms and electrons in the same gas is quite reasonable, since an electron requires of the order of $10^5$ collisions with atoms to change its energy significantly, but only one collision with another electron to approach a Maxwellian distribution. Thus an electron temperature and corresponding Maxwellian distribution exist as long as an electron has less than $10^5$ collisions with atoms for every collision with another electron. Because of the difference in cross section for these collisions, the necessary condition is satisfied if the degree of ionization is greater than $10^{-8}$.

## Inelastic Collisions

For this ionization process, the important inelastic cross section is that for excitation to the first excited state at 11.5 ev. It is reasonable to assume that essentially all of the atoms which get excited from the ground state will be further excited, and eventually ionized, by collision with electrons. Thus the various excited state populations will be in local equilibrium with the electrons, and the initial excitation will be the rate-determining step. The required cross section is not very well determined, but it should have a threshold at 11.5 ev. Based on an interpretation of the data of Maier-Leibnitz,[44] Petschek and Byron use

$$Q_{in} = 7 \times 10^{-18}\ (w - 11.5)\ \ \mathrm{cm}^2 \tag{10}$$

where w is the electron energy in electron volts. After integrating over a Maxwellian distribution for the electrons, the rate of ionization† is given by

$$\frac{dn_e}{dt} = 8.8 \times 10^{-6}\, n_A\, n_e \left[\frac{2(kT_e)^3}{\pi m_e}\right]^{1/2} \cdot \left(\frac{T_{exc}}{T_e} + 2\right) \exp\left(-T_{exc}/T_e\right)\ \ \mathrm{cm}^{-3}\,\mathrm{sec}^{-1} \tag{11}$$

† Equation (11) includes a factor of 2 omitted from the corresponding equation in Ref. 43.

where $n_e$ and $n_A$ are the concentrations of electrons and atoms, respectively, $T_e$ is the electron temperature, and $T_{exc} = 134,000^\circ K$ is the energy of the first excited state. A competing process would be the direct ionization by electron-atom collision, but the relevant cross section for this would be similar to Eq. (10) with the threshold approximately 4 ev higher. Thus for an electron temperature of 1 ev, the ionization rate would be smaller than that given in Eq. (11) by a factor $e^4 \approx 50$.

The total energy gained by each new electron after it has become equilibrated with the other electrons is $k\,(T_{ion} + \frac{5}{2} T_e)$, where $T_{ion} = 183,000^\circ K$ is the ionization potential. Thus the rate of energy loss by the electrons through inelastic collisions is

$$\dot{E}_{ion} = k\,(T_{ion} + \frac{5}{2} T_e)\,\frac{dn_e}{dt} \quad \text{ergs/cm}^3\text{ sec} \tag{12}$$

and this must be replaced by elastic collisions.

<u>Elastic Collisions</u>

For elastic collisions between electrons and the heavy particles, the rate of energy transfer can be written

$$\dot{E}_{el} = (m_e{}^2/m_A)\,(T_A/T_e - 1)\,n_A n_e\,\overline{Q(c_e)\,c_e{}^3} \quad \text{ergs/cm}^3\text{ sec} \tag{13}$$

where $m_A$ is the mass of the heavy particles whose temperature is $T_A$, $c_e$ is the electron velocity, and $Q\,(c_e)$ is the appropriate cross section for the collision involved. For atom-electron collisions the rate $(\dot{E}_{el})_{atom}$ is obtained by taking Q to be the diffusion cross section as used above in the section on electrical conductivity. For electron-ion collisions, the coulomb cross section is given by Eq. (9), and, after integration over the electron distribution, the corresponding rate of energy transfer to the electrons is given by

$$\left(\dot{E}_{el}\right)_{ion} = \frac{8.1\,e^4\,n_e^2}{m_A}\left(\frac{2m_e}{\pi k T_e}\right)^{1/2} \cdot \left(\frac{T_A}{T_e} - 1\right) \ell n\,(h/b_o) \quad \text{ergs/cm}^3\text{ sec} \tag{14}$$

The expressions for $(\dot{E}_{el})_{atom}$ and $(\dot{E}_{el})_{ion}$ give roughly equal rates of energy transfer when the degree of ionization is $5 \times 10^{-4}$ and $T_e = 10^4$ °K. The total energy transfer is given by their sum, but at higher degrees of ionization the electron-ion process will dominate.

## Energy Balance

To determine the ionization produced behind a shock wave by the mechanisms discussed, it is necessary to follow a fluid element from the shock front, integrating the rate equations as functions of time measured from the shock front. Immediately behind the shock, the temperature and density are given by Eqs. (1), (2), and (3) for $\gamma = 5/3$. As the ionization process proceeds, the translational temperature $T_A$ will decrease and the density will increase. The behavior of the electron temperature $T_e$ and the rate of ionization $dn_e/dt$ are determined from the electron energy balance

$$\dot{E}_{el} - \dot{E}_{ion} = \frac{3}{2} n_e k \frac{dT_e}{dt} \tag{15}$$

In practice the right-hand side of Eq. (15) can be neglected in comparison with either term on the left, so that the electron temperature becomes determined explicitly as the ionization proceeds. Figure 6 shows the electron temperature as a function of atom temperature for a range of shock velocities. Each of the heavy solid lines traces the behavior of electron temperature and degree of ionization for a given shock velocity characterized by the initial value of $T_A$ immediately behind the shock. It is seen that even for shock velocities as high as 7 km/sec, with corresponding initial temperatures of 45,000°K, the electron temperature does not rise appreciably above 16,000°K. Figure 6 is computed on the basis of neglecting electron-atom contributions to the elastic energy transfer. If Eq. (13) is used with the same argon diffusion cross section as in Fig. 4, negligible changes in Fig. 6 are obtained. However, the more serious uncertainty is in the cross section of Eq. (10), but Petschek and Byron estimate that a factor-of-two change in this cross section affects the overall rate of ionization by rather less than 20%. The detailed shapes of the lines in Fig. 6 would be affected by a change in Eq. (10), of course, since a much smaller inelastic cross section would permit $T_e$ to approach $T_A$ more rapidly, and to exhibit a greater overshoot relative to the final equilibrium temperature. On the other hand, a much larger cross section could require such large values of $(T_A - T_e)$ to provide the elastic collisional heating that $T_e$ would approach the equilibrium from below. These effects are illustrated in Fig. 6 for a single shock speed ($T_A = 25,000^{\circ}K$ initially) by the dashed curves, computed for factor-of-two changes in inelastic cross section.

The result obtained for the above mechanism gives good agreement with the experimental rate of ionization deduced from time-resolved continuum radiation measurements for conditions in the neighborhood of the maximum value of $dn_e/dt$. However, integration of the expression for $dn_e/dt$ back toward zero electron concentration at the shock front does not yield a finite value for the total ionization time, and cannot give agreement with the experimentally observed time for the onset of ionization. Thus, the dashed curve in Fig. 7 shows the profile of $n_e$ obtained by integrating toward the shock front from the point A, with the zero of the time scale located on the basis of experimental data. Inclusion of the electron-neutral contribution to the elastic heating, $(\dot{E}_{el})_{atom}$, makes only small improvement. Even though the electron-atom mechanism is in agreement with experiment over the major part of the ionization process, it is clear that some other process must account for production of approximately 10% of the equilibrium electron density during the early stages of ionization.

Initial Production of Electrons

Petschek and Byron carried out careful experiments to determine the possible source of these electrons, without completely resolving the question. The possibilities that photoionization or diffusion of electrons could produce sufficient ionization ahead of the shock wave were eliminated. An apparent correlation with impurity level was established, even when extreme care was used to reduce impurities to a minimum. For a fractional impurity level $X_{imp}$ in the test gas, an empirical rate law of the form

$$\frac{dn_e}{dt} = AX_{imp}\, n_A^2 \exp\left(-T_{act}/T_A\right) \quad cm^{-3}\, sec^{-1} \tag{16}$$

was found to give correlation with all the data, with the constants adjusted to values $A = 8 \times 10^{-8}\ cm^3/sec$ and $T_{act} = 134,000^{\circ}K$. Equation (16) was used in conjunction with the electron-atom impact mechanism. The ionization rate obtained by the addition of Eqs. (11) and (16) could be integrated from zero time and zero electron concentration at the shock front to give excellent agreement with experiment over a wide range of shock velocity, impurity level and initial shock tube pressure. Such a calculation is shown as the solid line in Fig. 7. $\tau$ is defined rather arbitrarily as the time at which $n_e^2 = 0.5\,(n_e)^2_{eq}$, corresponding to a continuum radiation intensity of one-half of the equilibrium level. The variation of the experimental ionization time $\tau$ is shown in Fig. 8, along with the calculations made for different levels of the nondimensional impurity concentration $X_{imp}$. The ordinate $p_1 \tau$ permits data obtained at different shock tube pressures to be compared directly, since the relaxation times for a given value of $X_{imp}$ will vary inversely with $\rho_1$, as all processes considered involve binary collisions.

In general, it can be seen that the mechanism for the initial production of electrons in argon will be a difficult one to resolve, but it is clear that the electron-atom collision process determines the major portion of the ionization profile, and that it is controlled principally by the rate of energy transfer by elastic collisions between electrons and ions. Since the appropriate coulomb cross sections are fairly well determined, the rate of ionization can be calculated with more precision than is available in the measurements of inelastic cross sections.

## Recombination

Knowledge of the recombination processes in high temperature argon is necessary for applications such as nozzle flows, or boundary layer studies as discussed in the heat transfer section above. For the overall reaction path

$$e + Ar \underset{k_R}{\overset{k_F}{\rightleftarrows}} 2e + Ar^+$$

the rate constant is obtained from Eq. (11) in the form

$$k_F = 3.75 \times 10^{-16} T_e^{0.5} (T_{exc} + 2T_e) \exp(-T_{exc}/T_e) \quad cm^3/sec$$

Using the principle of detailed balancing, with

$$K = 2.9 \times 10^{16} T_e^{1.5} \exp(-T_{ion}/T_e) \quad cm^{-3}$$

from the Saha equation, the three-body recombination rate coefficient may be written

$$k_R \equiv k_F/K = (1.3 \times 10^{-32}/T_e)(T_{exc} + 2T_e) \exp(49000/T_e) \quad cm^6/sec \tag{17}$$

This form of the rate coefficient was used by Bray[45] in theoretical studies of argon flow in a hypersonic nozzle. It is applicable as long as no alternative mechanisms provide comparable or faster rates of recombination; Bray examined several alternative possibilities, including radiative recombination.

Defining $\alpha = n_e^{-2}(dn_e/dt)$, and using Eqs. (12) and (15) with $\dot{E}_{el} \approx \dot{E}_{ion}$ one obtains

$$\alpha \approx \frac{|\dot{E}_{el}| / n_e^2}{k(T_{ion} + \frac{5}{2} T_e)} \quad cm^3/sec$$

For elastic collisions between electrons and ions, $\dot{E}_{el}$ is given by Eq. (14), yielding a value for $\alpha$ which is virtually independent of $n_e$, and which shows only weak dependence on $T_e$, for $T_e \gg T_A$. It is applicable as long as $T_e$ is controlled by the elastic collisions with heavy particles. Thus when $T_e$ is varying slowly, $\alpha$ will be almost constant and the rate of disappearance of electrons will be proportional to $n_e^2$, giving the appearance of a binary recombination process, as pointed out by Byron, et al.[46]

However, the above value of $k_R$ can only be valid as long as de-activation from the first excited state provides the rate limiting step. Calculations have been made by Bates, et al.[47] and by Byron et al.,[46] using collisional de-excitation cross sections of Gryzinski,[48] which show that as $T_e$ is reduced this rate-limiting step occurs at some higher electronic state. Above the level of the "bottleneck" the excited state populations remain in equilibrium with the electrons, but below the critical level the populations are able to approach equilibrium with the ground state. Makin and Keck[49] have obtained a similar result using the variational theory of reaction rates to calculate the three-body recombination rate coefficient. They obtained a relatively simple expression,

$$k_R = 2.3 \times 10^{-8} \, T_e^{-9/2} \quad cm^6/sec \qquad (18)$$

and found that the rate-limiting step is de-activation at an energy level 2.5 kT below the dissociation limit. This value for $k_R$ is in close agreement with the calculations of Refs. 46 and 47. All three calculations assume a hydrogen-like atom.

The behavior of Eqs. (17) and (18) is shown in Fig. 9, together with the numerical results of Ref. 47. It is seen that Eq. (17) predicts much too high a value of $k_R$ at low values of

$T_e$, since the "bottleneck" then occurs at a level higher than 11.5 ev. On the other hand, Eq. (18) predicts too high a value of $k_R$ at high values of $T_e$, since the calculations then require the bottleneck to be lower than the first excited state of the argon atom. In the absence of radiative recombination, Camac[50] has suggested that the solid line of Fig. 9 will give a reasonable indication of the variation of $k_R$ with electron temperature.

## Approach to Equilibrium Ionization in Air

In shock-heated argon,ionization is the only major process which contributes to the reduction in translational temperature from the value given by the Rankine-Hugoniot relations. In air, or any other diatomic gas, the situation is entirely different, since the initial translational energy imparted by the shock wave is fed into vibrational and electronic excitation and dissociation, as well as into ionization. Furthermore, there are many processes which can contribute appreciably to the ionization, so that it is no longer clear that the electron-neutral mechanisms should be dominant, nor that the transfer of energy to the electrons should control the ionization rate.

Lamb and Lin,[12] in carrying out their study of electrical conductivity, observed that the ionization process in shock-heated air progressed several orders of magnitude faster than in argon at corresponding shock strength.‡ The rate of rise of ionization was found to be beyond the resolution of available instrumentation in conventional shock tubes. In an attempt to circumvent this difficulty, Lin made measurements in $O_2$ - $N_2$ mixtures. By using 1/4% $O_2$ in $N_2$, the rate of production of electrons was reduced to a measurable value, which established very strong indications that the dominant process was the reaction $N + O \rightleftarrows NO^+ + e$. Values of a cross section for this reaction were estimated on a basis of quasi-steady state concentrations of O and N atoms during the ionization process, and these were subsequently revised in the light of calculations of the O and N concentrations based on finite rate chemistry. The revised cross sections are shown in Fig. 10. These data preceded a much more extensive study of nonequilibrium ionization in air, reported in Ref. 51, and motivated a detailed series of attempts to formulate and improve a theoretical model for the description of the chemistry occurring behind shock waves in air.

‡Equivalent shock strengths for such a comparison would maintain the kinetic energy per particle imparted by the shock wave to be proportional to the ionization potential; i.e., $U_s$ would be chosen such that the ratio $U_s^2$ (Mol. Wt.)/(I.P.) has the same value for both argon and air.

The experiments were carried out for initial pressures in the range $0.02 \leq p_1 \leq 0.2$ mm Hg, over a shock speed range $5 \leq U_s \leq 7$ km/sec. As in the case of argon, the time required for ionization can be scaled inversely with the initial pressure, bringing the characteristic ionization rise distance into the range 2 ~ 10 cm for $p_1 = 0.02$ mm Hg. This is well in excess of the spatial resolution of the microwave and magnetic induction probes used in the experiment. Since conventional shock tubes (diameter ≈ 1.5 in) cannot be operated satisfactorily at pressures below ~0.5 mm, the experiments were carried out in a 24-inch diameter shock tube specially designed for low pressure operation. The experimental measurements showed a rapid (but resolved) rise of electron density behind the shock front after a brief incubation time, and a transient peak value considerably higher than the corresponding final equilibrium value. Rise distances varied between 10 and 50 times the viscosity mean free path in the undisturbed gas ahead of the shock. Some typical time histories of the electron concentration, and summary information showing rise distances and maximum density gradients, are shown in later figures in this section.

## Air Chemistry Model

Attempts to correlate Lin's measurements in the 24-inch diameter shock tube with theoretical calculations of ionization depend on the availability of information on the behavior of T, $\rho$, and particle concentrations throughout the region in which the ionization processes are occurring. The model used to calculate this behavior has been built up over the past few years, and continually modified to incorporate the best available rate constant information. It is described in Refs. 52 and 53, as well as in Ref. 54, from which much of the following material is reproduced.

For the degree of ionization occurring behind air shocks at velocities less than 7 km/sec, the ionization is thermally insignificant, and may be considered as decoupled from the processes which produce the decrease in T. The reactions included in the chemical model are the dissociation processes

$$O_2 + X + 5.1 \text{ ev} \rightleftarrows O + O + X$$

$$N_2 + X + 9.8 \text{ ev} \rightleftarrows N + N + X$$

$$NO + X + 6.5 \text{ ev} \rightleftarrows N + O + X$$

in which X may be any of the species $O_2$, $N_2$, NO, O or N, and the "shuffle" or "exchange" processes

$$O + N_2 + 3.3\ \text{ev} \rightleftarrows NO + N$$

$$N + O_2 - 1.4\ \text{ev} \rightleftarrows NO + O$$

$$N_2 + O_2 + 1.9\ \text{ev} \rightleftarrows NO + NO$$

The rate equations for these reactions, together with vibrational excitation expressions in the form

$$\frac{d\epsilon_v}{dt} = \frac{(\epsilon_v)_{eq} - \epsilon_v}{\tau_v}$$

can be combined to provide a set of differential equations governing the various species concentrations. These equations can be integrated from initial conditions at the shock front, subject to constraints applied by the conservation relationships (Eq. (1) and Eq. (2)). The rate constants and values of $\tau_v$ as functions of T and $\rho$ are discussed in Ref. 53.

Typical results of such calculations are shown in Figs. 11 and 12. The particle concentration histories show that oxygen and nitrogen atoms are initially produced rather rapidly behind the shock front, in contrast to the overall time taken to reach final adiabatic equilibrium. For example, at a distance of 10 mean free paths behind the translational-rotational shock front, dissociation of $O_2$ and of $N_2$ have reached 75% and 50% of the final equilibrium level, respectively, and the translational temperature has dropped by a factor of two from its initial value. On the other hand, for $p_1 = 0.02$ mm Hg, a distance of several hundred mean free paths is required for $N_2$ dissociation and the translational temperature to come within 10% of final equilibrium. The rapid initial rates of change in temperature and composition imply that the various competing ionization processes may have differing importance at different distances behind the shock front. The long chemical relaxation time toward the equilibrium level implies that the degree of ionization may also be expected to depart from its equilibrium level for a large distance behind the shock front.

The justification for use of this model lies in its satisfactory description of shock wave chemistry in the region of moderate shock velocities, characterized by initial temperatures around 10,000°K. There are many dangers inherent in extrapolating the model to the temperature range of interest in the ionization experiments, since the rate constant information is not available at temperatures near 25,000°K, nor is there any certainty that the coupling between vibration and dissociation incorporated in the model will give a satisfactory description of the processes at such high temperatures. Another uncertainty is in the extent to which the initial translational-rotational relaxation can be considered to occur independently of the subsequent chemistry and excitation, as assumed in the model. However, sensitivity of the calculated ionization to the chemical model was explored, and it is felt that quite substantial changes in the rate constants used could be tolerated without necessity for any major revision of the conclusions reached about the ionization processes.

Once the temperature, density and neutral particle concentration histories have been obtained, it is a straightforward procedure to integrate rate equations for ionizing reactions, with initial conditions specified at the translational-rotational shock front.

A large number of ionizing reactions was considered, but many were discarded as being unimportant under the conditions of the experiment. For the reactions which were included in the calculations, a set of "basic" rate constants was selected, on the basis of available data or by drawing on knowledge of similar reactions for other gases. Calculations were made with these basic rate constants, in order that the relative role of different mechanisms at different times within the relaxation zone could be assessed. It was then possible to investigate the effect of moderate (factor of 3) changes in the rate constants for the dominant processes, and to assess the sensitivity to large changes in the rate constants for the less significant reactions. The choice of the set of reactions considered, and of the "basic" rate constants, is discussed briefly below. Greater detail is given in Ref. 54.

Ionization by Atomic and Molecular Collisions

The rate of ionization due to inelastic collisions between neutral particles in a shock-heated gas is in general negligible compared with other ionizing processes which can occur, as long as the mean kinetic energy per particle is small compared with the ionization potential of the particle being ionized. However, when chemical energy is available in the colliding system, as in the reaction

$$N + O + 2.8\ ev \underset{\kappa_{R_1}}{\overset{\kappa_{F_1}}{\rightleftarrows}} NO^+ + e \qquad (R1)$$

the amount of energy required to produce the ion-electron pair can be much less than the ionization potential of the molecule formed in the collision. The relatively large cross section for this reaction at 5000°K deduced from the data of Fig. 10 results from the crossing, or near-crossing, of the potential energy curves for the N + O system near the potential energy minimum of the ground state of $NO^+$. The theory of Bates and Massey[55] permits extremes of temperature dependence for this cross section to be deduced in the form

$$Q_a = 1.7 \times 10^{-12} T^{-1} \exp(-55,700/T) \quad cm^2$$

and

$$Q_b = 10^{-14} T^{-1} \exp(-32,500/T) \quad cm^2$$

These two expressions yield the lines shown in Fig. 10, each passing through the center of the data points. $Q_a$ gives the steepest possible temperature dependence, and involves an activation energy of 4.8 ev. $Q_b$ uses the minimum activation energy of 2.8 ev for reaction (R1). Multiplication by the mean thermal speed based on the reduced mass of the colliding O and N atoms, and use of the equilibrium constant for reaction (R1), yields two extreme values of the "dissociative recombination" rate constant for the reaction at low temperature. These are

$$\kappa_{R_{1a}} \cong 6 \times 10^{-1} T^{-3/2} \exp(-23,200/T) \ cm^3/sec$$

and

$$\kappa_{R_{1b}} \cong 3 \times 10^{-3} T^{-3/2} \quad cm^3/sec$$

The steep temperature dependence of $\kappa_{R_{1a}}$ would rule out any experimental observation of dissociative recombination for $NO^+$ at or near room temperature, whereas $\kappa_{R_{1b}}$ gives a room temperature rate constant quite compatible with experimental values cited in Ref. 54. Accordingly, in setting up a "basic" set of rate constants for ionization processes to be used in comparison with the experimental data, $\kappa_{R_1}$ for reaction (R1) may be assigned the value $3 \times 10^{-3} T^{-3/2}$ $cm^3/sec$. The corresponding forward rate constant may be written in the form

$$\kappa_{F_1} \cong 5 \times 10^{-11} T^{-1/2} \exp(-32,500/T) \quad cm^3/sec$$

Similar considerations apply to the reactions

$$N + N + 5.8 \text{ ev} \underset{\kappa_{R_2}}{\rightleftarrows} N_2^+ + e \qquad (R2)$$

and

$$O + O + 6.9\ ev \underset{\kappa_{R_3}}{\rightleftarrows} O_2^+ + e \qquad (R3)$$

for which the measurements of Kasner, et al.[56] suggest that the dissociative recombination does not require appreciable activation energy. Thus in the selection of "basic" rate constants, values of $\kappa_{R_2} = 3 \times 10^{-3}\ T^{-3/2}$ and $\kappa_{R_3} = 2 \times 10^{-3}\ T^{-3/2}$ $cm^3/sec$ are assigned for reactions (R2) and (R3), respectively.

For the molecule-molecule collision processes, cross sections for ionization of $N_2$ and $O_2$ as determined by molecular beam experiments have been published by Utterback and Miller[57] and Utterback.[58] By extrapolation of these measured cross sections toward the ionization threshold, $w_i$, it is possible to obtain an approximate fit in the form $A\left(\frac{w}{w_i} - 1\right)^n\ cm^2$, where w is the kinetic energy of the colliding system, relative to appropriate center of mass coordinates. Integrating over a Maxwellian distribution of molecular velocity, forward rate constants may be deduced for

$$O_2 + O_2 + 12.1\ ev \overset{\kappa_{F_4}}{\rightleftarrows} O_2 + O_2^+ + e \qquad (R4)$$

using $A = 1.0 \times 10^{-18}\ cm^2$ and $n = 5$, and for

$$N_2 + N_2 + 15.6\ ev \overset{\kappa_{F_5}}{\rightleftarrows} N_2 + N_2^+ + e \qquad (R5)$$

using $A = 1.8 \times 10^{-17}\ cm^2$ and $n = 4$. The results obtained are

$$\kappa_{F_4} = 10^{-33}\ T^{9/2}\ (1 + 4 \times 10^{-5}\ T)\ \exp(-141,000/T)\ \ cm^3/sec$$

and

$$\kappa_{F_5} = 3 \times 10^{-28}\ T^{7/2}\ (1 + 3 \times 10^{-5} T)\ \exp(-181,000/T)\ cm^3/sec$$

It may be noted that the pre-exponential factors in the expressions for $\kappa_{F_4}$ and $\kappa_{F_5}$ become comparable with the pre-exponential factor in $\kappa_{F_1}$ for $T \gtrsim 20,000^\circ K$. There are many other reactions similar to the five considered above which could produce ionization, but for which there is no direct experimental information available in the literature. Several such reactions were considered in the computational model of Ref. 54, with forward rate constants expressed in the simple Arrhenius form

$$\kappa_{F_j} = A_j\,(T)\,\exp\,(-\Delta E_j^{\,o}/kT)$$

with $A_j(T)$ chosen similar to the pre-exponential factor in $\kappa_{F1}$ or $\kappa_{F4}$, and with $\Delta E_j^{\,o}$ chosen as the lowest possible activation energy for the reaction in question. As none of the reactions considered proved of importance at any stage of the computed ionization histories, further discussion of these reactions is omitted from the present paper.

## Electron-Impact Ionization

A rigorous treatment of the effect of reactions of the type

$$X + e \rightarrow X^{+} + e + e \qquad \text{(R6)}$$

would follow the lines of the Petschek and Byron theory of ionization in argon given above. In terms of the local electron energy distribution function $f(w_e)$, and the total ionization cross section $Q_X\,(w_e)$ for collisions between electrons of energy $w_e$ and molecules or atoms of type X, the instantaneous rate of production of ion pairs per unit volume following a fluid element behind the shock front is given by[59]

$$\left(\frac{dn_e}{dt}\right)_{R6} = n_e \sum_X n_X \int_{w_X}^{\infty} c_e\, Q_X\,(w_e)\, f\,(w_e)\, dw_e \;\text{cm}^{-3}\text{sec}^{-1}$$

where $w_X$ is the threshold energy for the ionization cross section $Q_X$, and other quantities are as previously defined. For $kT_e \ll w_X$, this expression may be approximated in the form

$$\left(\frac{dn_e}{dt}\right)_{R6} = \bar{c}_e n_e \sum_X n_X w_X Q'_X\,(w_X)\,\exp\,(-w_X/kT_e)\;\text{cm}^{-3}\;\text{sec}^{-1} \qquad (19)$$

where $Q'_X\,(w_X)$ is the slope of the $Q_X$ vs $w_e$ curve at (or near) the threshold energy $w_X$.

Evaluation of the contribution of the electron-impact process thus requires knowledge of the variation of electron temperature behind the shock front, and of the appropriate values of the quantities $w_X$ and $Q'_X$ for the different neutral molecules and atoms within the relaxation zone. As in the case of argon ionization, it may be expected that the appropriate threshold energy $w_X$ will be appreciably less than the ionization potential for species X, corresponding to multi-step ionization via an excited state whose population is maintained in local equilibrium with the electrons. Again, as in the case of argon ionization, the electron temperature may be expected to be lower than the translational temperature, since energy lost by the electron gas through inelastic collisions must be replaced through elastic collisions with heavy particles. However, over the shock speed range of the measurements of Ref. 51, the degree of ionization is too low for coulomb interactions to dominate the transfer of energy to the electrons through elastic collisions, so that a reliable estimate of the rate of transfer of energy to the electrons becomes much more difficult than in the case of argon ionization. To circumvent the difficulty of computing an electron temperature history, a compromise approach was adopted in Ref. 54. The electron temperature was assumed at all times to equal the translational temperature, but the value of $w_X$ was set equal to the ionization potential for the particle X. The effect of these assumptions will be re-examined in the discussion of the computed ionization histories later in this section.

Except for the specification of the products $w_X Q'_X$, Eq. (19) is then available for use, in conjunction with rate equations for the corresponding backward processes, in the determination of the overall electron production behind the shock front. For $N_2$, $O_2$ and NO the apparent ionization cross section by electron impact has been deduced[59] from room temperature measurements. In each case the product $w_X Q'_X$ is approximately $\pi a_0^2 = 8.8 \times 10^{-17}$ cm$^2$ where $a_0 = 5.29 \times 10^{-9}$ cm is the first Bohr radius. In the absence of specific information about the corresponding cross sections for atomic oxygen and atomic nitrogen, the value $w_X Q'_X = \pi a_0^2$ is also used for these species. It may be noted from Eq. (10) that the corresponding product for argon is $11.5 \times 7 \times 10^{-18}$ cm$^2$ which is also comparable to $\pi a_0^2$.

Photoionization

For the reaction

$$X + h\nu \rightarrow X^+ + e \qquad \text{(R7)}$$

the rate of production of ion pairs per unit volume following a fluid element can be expressed as

$$\left(\frac{dn_e}{dt}\right)_{R7} = c \sum_X n_X \int_{(w_X)/h}^{\infty} Q_{fX}(\nu)\, n(\nu)\, d\nu \ \text{cm}^{-3}\,\text{sec}^{-1} \qquad (20)$$

where c is the velocity of light, h is Planck's constant, $Q_{fX}(\nu)$ is the photoionization cross section for the molecule X at light frequency $\nu$, and n ($\nu$) d$\nu$ is the local number density of photons in the frequency interval between $\nu$ and $\nu$ + d$\nu$.

The importance of photoionization as a mechanism for producing electrons in shock-heated air will thus depend both on the intensity of ultraviolet radiation at wavelengths for which h$\nu$ exceeds the ionization threshold $w_X$, and on the photoionization cross section $Q_{fX}$ of the particle X exposed to such radiation. A limited amount of information, obtained from photoelectric measurements,[60] is available on the far ultraviolet radiation from the relaxation zone behind strong shock waves in air ($U_s \approx 7$ km/sec). Hammerling[61] has tentatively concluded that most of this uv radiation may be attributed to the b' $^1\Sigma_u^+ \rightarrow X\ ^1\Sigma_g^+$ transition of $N_2$, so that most of the photons would have energies in the neighborhood of 12.9 ev corresponding to the energy difference between these two electronic states (measured from their respective ground vibrational levels). If the excitation processes were fast compared with the chemical processes which are responsible for the decrease in translational temperature, the total photon emission from this transition could approach a limiting value $N_\nu = (n_{N_2}/\tau_r) \exp(-E^*/kT)$ photons/cm$^3$, where $\tau_r = 4 \times 10^{-9}$ sec is the radiative lifetime for the upper state, and $n_{N_2}$ is the local number density of $N_2$ molecules, and where $E^* = 12.9$ ev. Fortunately the measurements are well below this upper limit, but the radiation intensity is too high to permit the possible photoionization effect to be neglected in interpretation of the data of Ref. 51.

From Ref. 62 it may be seen that $Q_{fX} \approx 2 \times 10^{-17}$ cm$^2$ for $O_2$, $N_2$ and NO molecules at photon energies slightly higher than the ionization threshold. Thus the photon mean free path, $\ell_{\nu 1}$, in the undisturbed gas ahead of the shock may be related to the upstream viscosity mean free path, $\ell_1$, by the expression

$$\ell_{\nu_1}/\ell_1 = 2^{1/2}\, Q_{kinetic}/Q_{fX} \approx 300$$

Behind the shock front, where $\rho/\rho_1 \approx 10$, the photon mean free path, $\ell_{\nu 2}$, is thus approximately 30 $\ell_1$ which is comparable with the thickness of the relaxation zone. This implies that the local photon density n ($\nu$) d$\nu$ in the far ultraviolet can only be determined from a detailed consideration of the excitation, emission and radiative transport processes within the relaxation zone. This is not attempted in Ref. 54, but a rough estimate of the photoionization rate is made in the following manner.

It is assumed that the uv radiation intensity exhibits a transient maximum, similar to maxima observed in time-resolved measurements of the radiation from other nitrogen band systems (e.g. $N_2^+$ first negative system[60]) in air. The gross characteristics of the radiation from this overshoot region can then be expressed, following Hammerling,[61] in the form

$$F_\nu = \frac{L\,\bar{n}_{N_2}\,\exp(-E^*/kT^*)}{2\,(\tau_r + \tau_c)} \quad \text{photons/cm}^2 \qquad (21)$$

where L is the effective thickness of the overshoot region, $\bar{n}_{N_2}$ is the averaged number density of $N_2$ molecules within the region, $\tau_c$ is the collisional de-excitation time for the upper electronic state, and $T^*$ is a temperature which characterizes the population of the upper state at the time of maximum radiation intensity. This expression involves a large number of highly empirical factors, but these can be chosen to give agreement with the measured radiative flux for $U_s = 6.8$ km/sec. With the aid of this effective normalization, Eq. (21) probably provides a reasonable method of estimating the variation of $F_\nu$ with shock speed. The numerical values assigned to L, $T^*$ and $\tau_c$ for use in the computed ionization histories are discussed in Ref. 54.

Thus the value of $F_\nu$ may be specified as a function of $U_s$ and $p_1$, while the photoionization cross section may be assumed at the value $Q_{fX} = 2 \times 10^{-17} \approx \pi a_o^2/4$. Consequently, the rate of production of ion pairs is given by

$$\left(\frac{dn_e}{dt}\right)_{R7} = (n_{O_2} + n_{NO})\,\frac{\pi a_o^2}{4}\,F_\nu \quad \text{cm}^{-3}\ \text{sec}^{-1} \qquad (22)$$

The summation $\sum_X n_X$ in Eq. (20) has been replaced by $(n_{O_2} + n_{NO})$ since photons of 12.9 ev energy could ionize $O_2$ and NO molecules, but not other chemical species behind the shock front. Equation (22) is used to describe the local rate of photoionization at a distance x behind the shock front, for $0 < x < L$. Behind the overshoot region, for $x > L$, the rate given by Eq. (22) should be multiplied by a factor $\exp\left\{-(x - L)/\ell_{\nu 2}\right\}$, thus providing a crude correction for the effect of absorption in decreasing the photon flux at large distances behind the shock front.

## Charge Exchange

There are many charge exchange reactions which involve

a transfer of unit charge from one molecule (or atom) to another during a collision, and which do not contribute to the net electron production rate during the early part of the ionization history. However, at later times when the electron-ion recombination rates become comparable to the ionization rates, such charge exchange reactions could influence the electron density profile through a continuous reshuffling of the relative populations among the different positive ions. This effect would be especially significant if there exist large differences in recombination rates for the different kinds of positive ions, as would be expected from comparing dissociative recombination for molecular ions with radiative recombination for atomic ions.

For the oxygen-nitrogen system under consideration, comprising the neutral particles $N_2$, $O_2$, NO, O and N, and the five corresponding positive ions, there are ten reactions of the type

$$A^+ + B \rightarrow A + B^+ \qquad \text{(R8)}$$

The required energy of reaction ranges from less than 1 ev (in transferring a charge from $O^+$ to N) to 6.3 ev (in transferring a charge from $NO^+$ to $N_2$). Quantitative information on the cross sections for these reactions is very meager, especially at the relatively low particle energy under consideration. Existing data,[63] however, seem to indicate that the cross section for a typical exothermic charge transfer of this type at about 1-ev particle energy is of the order of a few times $\pi a_0{}^2$ (i.e., about $3 \times 10^{-16}$ cm$^2$), which is not much smaller than the gas kinetic (momentum transfer) cross section.

Other types of charge exchange can occur, through reactions which involve simultaneous charge transfer and explicit rearrangement of atoms during a collision, and which range from

$$O_2{}^+ + N + 0.2 \text{ ev} \rightleftarrows NO + O^+ \qquad \text{(R9)}$$

to

$$NO^+ + O + 6.7 \text{ ev} \rightleftarrows O_2 + N^+ \qquad \text{(R10)}$$

These reactions are very similar to the "shuffle" reactions listed for the chemical processes considered in obtaining time histories of neutral concentrations behind the shock front. Eight other charge exchange reactions similar to (R9) and (R10) can be immediately listed. There appears to be no rate constant information for these reactions, but their similarity to the neutral processes suggests that corresponding exothermic rate constants should be of comparable magnitude, which would require cross sections for exothermic charge transfer of this type to be about $\pi a_0^2$.

All twenty of these charge transfer reactions are included in the computational model, with "basic" rate constants corresponding to exothermic cross sections of magnitude $3\pi a_0^2$. Since many of the charge exchange processes tend to achieve situations of local equilibrium throughout a large portion of the relaxation zone, they cannot be included as simple rate equations, since the integration would then encounter serious numerical instability. In practice, the calculation of the ionization history involves only one differential equation,

$$\frac{dn_e}{dt} = \text{overall net rate of production of electron-ion pairs} \qquad (23)$$

At the end of each integration step, the positive ions are redistributed to satisfy five equations of the type

$$\frac{dn_{ion}}{dt} = (\text{net rate of production by ionization}) + (\text{net rate of production by charge transfer}) \qquad (24)$$

Since the overall rate of production of ions by charge transfer is zero, the sum of the five equations similar to Eq. (24) is exactly equivalent to Eq. (23). This ensures that $n_e$ is equal to the sum of the positive ions, as required by the condition of electrical neutrality in the gas.

The various contributions to Eq. (23) arise from the atomic and molecular processes typified by reactions (R1) through (R5), from the electron impact mechanisms similar to reaction (R6), and from the photoionization process given by Eq. (22). The magnitude of each of the contributions at a given distance behind the translational-rotational shock front depends on the local translational temperature and on the local values of neutral particle concentrations. In addition, as soon as backward reactions become important, the instantaneous recombination rates will also depend on local positive ion concentrations. Calculations using the above description of the ionization processes have been made for various shock conditions.

## Calculated Ionization Histories

The results of such calculations for a shock speed $U_s = 6.9$ km/sec are shown in Figs. 13 through 18. These calculations used the T, $\rho$ and neutral particle concentration histories of Figs. 11 and 12. Figure 13 shows the normalized electron and ion densities as functions of distance behind the translational-rotational shock front. The abscissa is made nondimensional in terms of the upstream viscosity mean free path, $\ell_1$. (For $p_1 = 0.02$ mm Hg, $\ell_1 \approx 0.25$ cm.) Figure 14 shows the variation of specific ionization rates for the various groups of ionization processes, as functions of nondimensional distance behind the shock front. The ordinate of this figure is also nondimensional, defined as the absolute value of the time rate of change of normalized electron density (following the mass motion of the fluid) multiplied by the time, $\ell_1/U_s$, during which the shock traverses a distance of one upstream mean free path. $\ell_1/U_s$ is approximately equal to the mean binary collision time behind the shock. The curve labelled "atom-atom" refers to the sum of the contributions from reactions (R1), (R2), and (R3), while the curve labelled "molecule-molecule and atom-molecule" refers to the sum of all other reactions in this category, typified by reactions (R4) and (R5).

All processes except the "atom-atom" process continue to provide a positive rate of electron production, even at large values of x. However, the electron concentration overshoots the equilibrium level, so that the net rate of production of electrons becomes negative for $x/\ell_1 \gtrsim 20$. The "atom-atom" processes reverse direction and remove the excess electrons by dissociative recombination.

Figures 13 and 14 also include results of ionization calculations for $p_1 = 2$ mm Hg. In the nondimensional plots of Figs. 11 and 12, the histories for the two pressures shown are indistinguishable for $x/\ell_1 < 100$, since all the neutral processes are binary over this distance. In Fig. 14, the photoionization rate is seen to exhibit a dependence on $p_1$, but the overall effect on the nondimensional ionization history remains insignificant. Slight differences are observed for $x/\ell_1 > 100$ in Figs. 13 and 14 owing to the difference in final equilibrium conditions for the two values of $p_1$.

For $p_1 = 0.02$ mm Hg, the respective contributions of reactions (R1), (R2), and (R3) are shown in Fig. 15. Reaction (R1) stands out as the dominant ionization process, though calculations made over a range of shock velocities show that reaction (R2) gains in importance as $U_s$ is increased. The sensitivity of the calculations to changes in the rate constant for reaction (R1) will be exhibited in later figures.

First, however, it is necessary to explore the effect of changes in the assumed charge exchange cross sections on the calculated ionization histories. Results are shown in Fig. 16 for the "basic" cross section of $3\pi a_o^2$, and for cross sections of 300 $\pi a_o^2$ and 0.03 $\pi a_o^2$. The solid lines in this figure are identical with those of Fig. 13. The drastic reduction in cross section delays the production of atomic ions, reduces their overshoot in concentration, and slows down the rate of removal of the overshoot. Correspondingly, the molecular ion concentrations have higher peak values. The "atom-atom" reactions thus reverse direction at a slightly smaller value of x, producing a 10% reduction in peak electron concentration. A substantial increase in charge exchange cross section produces almost negligible effects on the concentration histories, showing that the "basic" value is already large enough to maintain local quasi-statistical equilibrium between the various ions thoughout most of the relaxation zone.

Finally, it is of interest to examine the effects of upstream photoionization and electron diffusion on the resultant ionization profile behind the shock. From the previous discussion, it is clear that sufficiently energetic photons could escape from the radiation overshoot region to photoionize molecular oxygen upstream. The production of free electrons§ ahead of a plane shock wave in air by this "precursor radiation" is treated in some detail by Hammerling in Ref. 61. There is also evidence of such photoionization effect being observed during satellite re-entry.[65] In the present context, however, it is necessary only to estimate the cumulative effect of such upstream photoionization on the resultant ionization history behind the shock.

The electron density distribution behind the shock with and without consideration of upstream photoionization is shown in Fig. 17. The solid curve, with no upstream photoionization and identical to that shown in Fig. 13, was obtained by integrating the specific ionization rates behind the shock, with $[e] = 0$ at $x = 0$. The dotted curve, on the other hand, was obtained by integrating the same set of rate equations, but with $[e] = [O_2^+] = 10^{-4}$ at $x = 0$. This initial value of $[e]$ was obtained from a balance between the outgoing ionizing photon flux ($h\nu \geq 12.1$ ev) described by Eq. (21), and the flux of ion pairs convected back into the translational-rotational shock front per unit area due to the mass motion of the fluid. It represents an upper limit based on one-dimensional flow through a plane shock wave of infinite extent, so that the radiation overshoot region always subtends a solid angle of $2\pi$

§Precursor ionization ahead of shock waves in other gases has received extensive treatment in the literature, as for example in Ref. 64. Its possible importance in the early stages of ionization in shock-heated argon was discussed in Ref. 43.

steradians at any upstream point. In a shock tube experiment, of course, the upstream photoionization would be reduced since the photon flux originates behind a shock wave of finite area. Figure 17 shows that even though the neglect of upstream photoionization would lead to serious error in the very early part of the ionization history, the error becomes steadily smaller as the peak electron concentration is approached. Errors of this type would assume greater importance at higher shock velocity, where the electron impact ionization process would be expected to play a larger role.

The effects of electron diffusion on the theoretical ionization history must also be considered. However, even at $p_1 = 0.02$ mm Hg, the Debye shielding distance (see Eq. (8)) for the electron density profiles of Fig. 13 is always less than 0.05 cm. This is sufficient to ensure ambipolar diffusion[63, 66] and charge neutrality down to a distance scale smaller than $0.2\,\ell_I$. Since ambipolar diffusion is not much more rapid than momentum (or viscosity) diffusion, it should be negligible outside the translational-rotational relaxation zone. Gibson[67] has published calculations in which ambipolar diffusion is found to produce distortions of the ionization profile comparable with the upstream photoionization effect of Fig. 17.

Comparison with Experiment

Figures 18 through 21 show comparisons between calculations based on the above model and the experimental data of Ref. 51. Since no direct experimental determination of the positive ion distributions has yet been made, the comparison is restricted to features of electron density distribution.

The theoretical electron density profile in physical scales for the case $U_s = 6.9$ km/sec, $p_1 = 0.02$ mm Hg is plotted in Fig. 18 for comparison with the electron density profile obtained from four separate experiments taken at $U_s = 6.9 \pm 0.05$ km/sec, at the same initial air pressure. Curve (1) is identical to the normalized curve [e] shown in Fig. 13. Curves (2) and (3) are reference curves obtained by changing the absolute rate constant for the predominant ionizing reaction, (R1), by a factor of three in either direction (i.e., taking $\kappa_{R_1} = 9 \times 10^{-3}\,T^{-3/2}$ and $\kappa_{R_1} = 1 \times 10^{-3}\,T^{-3/2}$, respectively), while holding all other rate constants the same as those specified in the preceding section. The effect of upstream photoionization is hardly noticeable in this linear plot, since for $[e] = 10^{-4}$ the initial electron density amounts to only about $7 \times 10^{10}$ electrons/cm$^3$ just ahead of, and $4 \times 10^{11}$ electrons/cm$^3$ just behind the translational-rotational shock front.

It is seen that the front part of the experimentally-observed profile appears to fall between curve (1) and curve (2), but the rear part of the observed profile seems to fall systematically above the theoretical curves. Even though the apparent discrepancy of the order of 20 percent is still within the probable inaccuracy of the experiment in absolute electron density determination,[51] the distinct difference in slope between the theoretical and experimental profiles may be significant. However, in view of the fact that no reasonable adjustment of either the ionization or the chemical rate constants could reduce the theoretical rate of decrease of electron density behind the peak to the experimentally observed rate, it is believed that the slower observed rate of decrease of $n_e$ at large distances behind the shock front was due to the slight deceleration of the shock wave in the experimental situation.[68] In a decelerating shock wave, the fluid elements further away from the shock front are those that have been heated by a stronger shock wave, so that the resultant perturbation is in the direction of reducing the rate of decrease of translational temperature behind the shock.

The theoretical peak electron density and the maximum electron density gradient behind the shock at initial air pressure $p_1 = 0.02$ mm Hg are plotted in Figs. 19 and 20, as functions of the shock velocity (or the shock Mach number $M_s = U_s/a_1$, where $a_1$ is the speed of sound for air at room temperature). These may be compared with the corresponding measured quantities presented in Figs. 8 and 10 of Ref. 51. The theoretical curves (1), (2) and (3) are based on the electron density profiles calculated according to $\kappa_{R_1} = 3 \times 10^{-3}\ T^{-3/2}$; $9 \times 10^{-3}\ T^{-3/2}$; and $1 \times 10^{-3}\ T^{-3/2}$, respectively, as in Fig. 18. The electron density corresponding to the final equilibrium value far behind the shock is also plotted for comparison. It is seen that the amount of "ionization overshoot" observed in the experiment was perfectly reasonable according to the present theoretical model, even though the relative insensitivity of the peak electron density to the ionization rate makes this quantity a poor measure of the rate constant. The maximum electron density gradient shown in Fig. 20, on the other hand, is sensitive to the ionization rate and shows a clear preference for curves (1) and (2) over curve (3). It is also interesting to note that according to these calculations, the "ionization overshoot" phenomenon is expected to disappear at shock velocities below 4 km/sec and above 9 km/sec.

The theoretical "ionization incubation distance" is plotted in Fig. 21 for comparison with the corresponding experimental results presented in Fig. 12 of Ref. 51. This quantity is defined here as the distance for the local electron density to reach one-half the peak value behind the shock. The curves (1), (2) and (3) have the same meaning as before. It may be noted that the experimental points presented include measurements both from microwave reflection probes (hollow points) and from a magnetic induction probe (solid points). The microwave probes measured the electron density directly, but the magnetic probe measured only the dc electrical conductivity, which is related to the electron density by $\sigma_{dc} \approx n_e e^2/m_e \nu_e$. The electron collision frequency, $\nu_e$, may be calculated by using the theoretical temperature, density and chemical composition profiles, together with the electron momentum transfer cross sections of Fig. 4. Within the range of shock strength covered by the experiments, (i.e., $U_s \leq 7$ km/sec), it is found that $\nu_e$ does not vary by more than a factor of two with distance behind the shock front, while the electron density changes by orders of magnitude. Therefore, the measurements obtained by both methods may be compared directly with the theoretical curves. It is seen that the experimental points are in reasonably good agreement with the geometrical mean of curves (1) and (2).

The initial pressure dependence of the peak electron density and of the maximum electron density gradient behind the shock reported in Ref. 51 may be interpreted as a direct consequence of the predominance of binary collisions in the forward rate processes within the range of gas density under consideration. This is perfectly consistent with the theoretical results (see Figs. 11-14).

## Discussion

From the foregoing results, it is seen that all the experimental observations reported in Ref. 51 can be satisfactorily interpreted according to the present theoretical model. Even though many of the ionization rate constants are still uncertain, comparison of the specific rates based on tentative rate constants shows that only a few of them are pertinent. The apparent predominance of the atom-atom reactions throughout the range of shock velocity under consideration shows that only through an adjustment of the rate constant for reaction (R1), and to a lesser extent, the rate constants for reactions (R2) and (R3), can one effectively alter the theoretical ionization profile behind the shock to achieve agreement with experiment.

The reasonably close (i.e., within a factor of two) agreement between the experimental points and the theoretical curve (1) shown in Figs. 18 through 21 suggests that the "basic" values assigned to $\kappa_{R1}$, $\kappa_{R2}$ and $\kappa_{R3}$ are already quite satisfactory. Even though an upward revision of $\kappa_{R1}$ by a factor of two would bring the theoretical curve to near perfect agreement with the experimental points, such a fine adjustment is probably not justified in view of the existence of other uncertainties in the theoretical model which could be of comparable magnitude. These include, especially, uncertainties in the chemical composition profile caused by uncertainties in the high temperature dissociation rate constants, and in the model of coupling between vibrational relaxation and dissociation.

It has already been pointed out that the value

$$\kappa_{R1} = 3 \times 10^{-3} T^{-3/2} \quad cm^3/sec \tag{25}$$

is compatible with measurements made in flames and at room temperature. Several other relatively high temperature experiments provide additional information about the rate constants for reaction (R1). Eschenroeder, et al.[69] used microwave interferometric techniques to measure electron density at various stations in the nozzle of a shock tunnel. They were able to fit their data by using a rate constant for reaction (R1) in the form

$$\kappa_{R1} = (3.7 \pm 1.5) \times 10^{-3} T^{-3/2} \quad cm^3/sec$$

Eckerman and Stern[70] have measured electron recombination in the wakes of hypervelocity spheres, and are able to obtain reasonable theoretical agreement by using a similar value of $\kappa_{R1}$. Eckerman also quotes an unpublished recombination measurement of Laurmann, who obtained $\kappa_{R1} = 3 \times 10^{-8}$ cm$^3$/sec at 3000°K, in good agreement with Eq. (25). Other measurements obtained from studies of the forward process are discussed in Ref. 51, but none appear able to withstand critical evaluation.

The theoretical results presented in this section may be extended over a much wider range of initial air density without difficulty. The upper density limit is reached when the rates of dissociation and ionization due to three-body collisions (comparing only the forward rates) become comparable to those due to binary collisions. This is not expected to occur at densities much below liquid air density. For extension to lower densities, while there seems to be no theoretical limit, a practical limit should be set by the realizability of air samples without excessive contamination from the vapor pressure of the container, or from the environment, if the theoretical results were to be compared with experimental observations.

The theoretical approach can readily be extended to other initial $O_2/N_2$ mixture ratios, and also to predissociated oxygen-nitrogen mixtures for upper atmosphere application. With minor modification, the effects of various contaminants in the mixture can also be estimated. The model can also be used for lower shock velocities ($U_s < 4$ km/sec), provided attention is given to the increased role of impurities having low ionization potentials.

The present one-dimensional results can also be made applicable to three-dimensional problems. As is well known in fluid dynamics,[71] there is a one-to-one correspondence between the flow across an oblique shock wave and the flow across a normal shock wave. Therefore, the present results are directly applicable to plane oblique shock waves provided that only the normal component of the upstream flow velocity is used to determine the shock strength, and that the distance scale refers to the normal distance from the shock surface. With slight modifications, the model can also be used to follow the ionization history in the three-dimensional inviscid flow field of a hypersonic object[72] and in high-enthalpy nozzle flows.[69]

Extension of the present theoretical model to a wider range of shock strength, on the other hand, represents a more complicated problem. Apart from the uncertainties inherent in extrapolation of the chemical model to very high temperatures, several other effects must be considered. In the first place, when the degree of ionization approaches 0.01, the energy invested in ionization becomes an appreciable fraction of the flow energy, and the ionization processes become energetically coupled to the neutral histories. Furthermore, modifications to the chemical model are probably required, since appreciable dissociation is likely to occur from collisions between electrons and molecules. Most important, however, is the revision of the approximate scheme used above for handling ionization by electron impact, and this will require a formulation which uses separate temperatures for electrons and neutrals. Ionization will occur by way of excitation processes as in argon, through inelastic collisions with electrons, and the electronic state at which the effective "bottleneck" will be located can be calculated as in Refs. 46 or 49. As long as there are appreciable molecular concentrations, the energy lost by the electrons through collisions resulting in excitation will be replaced through collisions with molecules, which provide an efficient heating process even when the electron concentration is too low for coulomb interactions to be effective. However, it is possible that the predominance of electron-molecule collisions will effectively couple the electron temperature to the vibration temperature of the molecules, and from Fig. 11 it is clear that the latter could remain well below the translational temperature over a large portion of the relaxation zone. In conclusion, it appears that more experimental information will be required before an effective theoretical model can be formulated.

## Ionization at Higher Shock Velocities

The mechanisms which produce the ionization behind shock waves in noble gases and in air appear to be fairly well determined for shock velocities as high as 7 km/sec. However, several laboratory experiments with much faster shock waves have been reported, and their interpretation clearly depends strongly on understanding of the ionization phenomena involved. Thus, it is of interest to discuss briefly the extent of that understanding, and of the areas of ignorance exposed by these experiments.

Conventional shock tubes with combustion drivers perform satisfactorily at shock velocities up to ~ 7 km/sec. The shock speed attainable depends on both the pressure ratio across the diaphragm separating the high- and low-pressure gases in the shock tube, and also on the ratio of sound speeds in these two gases. For a given sound speed ratio, the shock speed approaches a limiting value as the pressure ratio is increased, so that improvements in shock tube performance are obtainable more readily by increase of sound speed in the driver. Significant increases in maximum shock speed have been obtained by means of arc-driven shock tubes, in which a capacitor bank is discharged into a driver section filled with helium, producing temperatures which are in the range 15,000 - 20,000 $^{\circ}$K prior to the bursting of the diaphragm. These temperatures contrast favorably with the 3,000 $^{\circ}$K range attainable in combustion drivers. Arc-driven shock tubes[73] have been operated successfully up to $U_s$ = 13 km/sec, and measurements have yielded adequate evidence for the existence of homogeneous gas samples, with apparent test times in general agreement with theoretical predictions. Convective heat transfer and radiation measurements obtained in such devices have been reported in the literature.[42, 74, 75] The equilibrium radiation phenomena for shock speeds in the range 9 - 12 km/sec are dominated by Kramers' radiation and Bremsstrahlung, and the intensities are compatible with equilibrium electron densities expected behind such strong shock waves. However, the details of the ionization production mechanisms have not yet been resolved.

At enthalpies somewhat higher than those attainable in an

arc-driven shock tube most gases become fully singly-ionized, and multiple ionization becomes significant. Electromagnetic means have been used to accelerate such plasmas to very high velocities, but the devices used do not necessarily produce a clean gasdynamic shock wave behind which the thermodynamic condition of the gas is well determined.[1, 76] Two key factors affect the production of a homogeneous gas sample behind an electromagnetically driven shock wave. First, it is essential that the shock be shielded from the driving magnetic field, so that interaction between the ionized gas sample and the field will not induce currents which could cause heating of the sample. Secondly, it is necessary to provide magnetic containment to inhibit diffusion of the plasma sample to the walls, and prevent drastic cooling of the gas or evaporation of a significant mass of wall material within the experimental test time. A fuller discussion of these factors is given in Ref. 1.

Experiments carried out in Magnetic Annular Shock Tubes have taken both these factors into consideration, and considerable progress in the production and diagnostics of plasma samples behind electromagnetically-driven shock waves in hydrogen has been achieved.[77-80] Some of the experimental conditions attained are shown on Fig. 22, together with theoretical conditions behind the shock calculated by means of the usual conservation relationships, with magnetic effects included. In general, the experiments were performed with an azimuthal magnetic field $B_1$ ahead of the shock wave. The important non-dimensional ratio is the Alfvén Mach number $M_A = U_s/b_1$, where $b_1 = (B_1^2/4\pi\rho_1)^{1/2}$ is the speed of an Alfvén wave in the undisturbed gas, in which the density is $\rho_1$. The data of Fig. 22 are obtained under conditions where $M_A = 2$. Over the higher shock speed range, the experimental measurements of magnetic field and emitted light intensity provide indication of the existence of homogeneous gas samples, with density and magnetic field consistent with the conservation relationships. Determinations of shock thickness have been made from measurements of the magnetic field compression across the shock wave, and of the rise in light intensity (due to Bremsstrahlung) associated with the ionization of the gas. These two events appear to occur almost simultaneously in the experiments.

For the argon and air shocks discussed in the main body of the paper, the initial jump in translational temperature and density at the shock front is produced by viscous dissipation, and occurs within a region which has thickness of only a few mean free paths ($\sim 3\,\ell_1$). The region in which chemical and ionization relaxation occurs is downstream of this narrow viscosity-controlled zone. For the MAST experiments in hydrogen, however, it is clear that the usual viscous dissipation plays a negligible role at such high velocity, since the momentum transfer cross sections for ion-ion or neutral-neutral collisions become too small at large relative energy. Cross sections which remain large at high energy are those for charge exchange between ions and neutrals, and for ionization by electron-neutral collisions. Charge exchange collisions can produce the effect of a frictional force between ions and neutrals, since the ions are essentially anchored to the field lines, whereas the neutrals can pass freely through the magnetic field. Ionizing electron-neutral collisions can also produce a dissipation effect, since the large energy of a neutral particle relative to the magnetic field is converted into thermal energy when the particle becomes ionized. The rate of ionization by this process, of course, would be controlled by the rate of transfer of energy to the electrons by elastic collisions. Gerry, et al.[80] discuss shock structure calculations based on a preliminary model which includes these dissipation mechanisms. They show that reasonable agreement with measured shock thickness can be obtained by using available cross section information, but they point out that they have neglected many effects which later investigation may show to be important. Their present calculations show the ionization process occurring after completion of the increase in density and magnetic field, which is analogous to the behavior at more moderate shock speeds with no magnetic field present, and in contrast to the measurements made in the MAST experiments.

The behavior of a MAST is clearly much more complex than that of the conventional shock tube, but improvements in diagnostics are leading to the unravelling of some of the complexities. The gas conditions produced behind the shock wave in this type of device are in the "collision-free" regime, which is not readily accessible to many laboratory experiments. The study of the ionization and dissipation phenomena within such strong shock waves raises many challenging problems, both theoretical and experimental, but it offers promise of significant contribution to basic physical understanding in this regime.

## References

[1] Kantrowitz, A. R., "Shock Tubes for High Temperature Gas Kinetics," Avco-Everett Research Laboratory Research Report 141 (October 1962); also Proceedings of the Fiftieth Anniversary Solvay Conference, Brussels, Belgium (November 1962) (to be published).

[2] Resler, E. L., Jr., Lin, S. C. and Kantrowitz, A. R., "The Production of High Temperature Gases in Shock Tubes," J. Appl. Phys. 23, 1390-1399 (1952).

[3] Hollyer, R. N., Jr., Hunting, A. C., Laporte, O. and Turner, E. B., "Luminosity Generated by Shock Waves," Nature 171, 395-396 (1953).

[4] Bond, J., "Structure of a Shock Front in Argon," Los Alamos Report LA-1693 (1954).

[5] Hilsenrath, J. and Beckett, C. W., "Tables of Thermodynamic Properties of Argon-Free Air to 15,000°K," AEDC-TN-56-12 (September 1956).

[6] Hilsenrath, J. and Klein, M., "Tables of Thermodynamic Properties of Air in Chemical Equilibrium including Second Virial Corrections from 1500°K to 15,000°K," AEDC-TDR-63-161 (August 1963).

[7] Lick, W. J. and Emmons, H. W., "Thermodynamic Properties of Helium to 50,000°K " (Harvard University Press, Cambridge, Mass., 1962).

[8] Drellishak, K. S., Knopp, C. F. and Cambel, A. B., "Partition Functions and Thermodynamic Properties of Argon Plasma," AEDC-TDR-63-146 (August 1963).

[9] Chapman, S. and Cowling, T. G., The Mathematical Theory of Non-Uniform Gases (Cambridge University Press, London and New York, 1952).

[10] Spitzer, L. and Härm, R., "Transport Phenomena in a Completely Ionized Gas," Phys. Rev. 89, 977-981 (1953).

[11] Lin, S. C., Resler, E. L., Jr. and Kantrowitz, A. R., "Electrical Conductivity of Highly Ionized Argon Produced by Shock Waves," J. Appl. Phys. 26, 95-109 (1955).

[12] Lamb, L. and Lin, S. C., "Electrical Conductivity of Thermally Ionized Air Produced in a Shock Tube," J. Appl. Phys. 28, 754-759 (1957).

[13] Shkarofsky, I. P., Bachynski, M. P. and Johnston, T. W., "Collision Frequency Associated with High Temperature Air and Scattering Cross Sections of the Constituents," Planetary and Space Sciences 6, 24-46 (1961).

[14] Neynaber, R. H., Marino, L. L., Rothe, E. W. and Trujillo, S. M., "Low-Energy Electron Scattering from Atomic Oxygen," Phys. Rev. 123, 148-152 (1961).

[15] Neynaber, R.H., Marino, L.L., Rothe, E.W. and Trujillo, S.M., "Low-Energy Electron Scattering from Atomic Nitrogen," Phys. Rev. 129, 2069-2071 (1963).

[16] Lin, S.C. and Kivel, B., "Slow Electron Scattering by Atomic Oxygen," Phys. Rev. 114, 1026-1027 (1959).

[17] Patrick, R.M. and Brogan, T.R., "One Dimensional Flow of an Ionized Gas through a Magnetic Field," J. Fluid Mech. 5, 289-309 (1959).

[18] Pain, H.J. and Smy, P.R., "Magnetic Field Interactions with Shock Ionized Argon," Proc. Phys. Soc. 76, 849-856 (1960).

[19] Dolder, K. and Hide, R., "An Experiment on the Interaction between a Plane Shock and a Magnetic Field," Nature 181, 1116-1118 (1958).

[20] Fishman, F., Lothrop, J., Patrick, R. and Petschek, H., "Supersonic Two Dimensional Magnetohydrodynamic Flow," The Magnetohydrodynamics of Conducting Fluids, edited by D. Bershader (Stanford University Press, 1959), pp. 90-119.

[21] Petschek, H.E., Locke, E.V. and Rose, P.H., "Experiments with Magnetohydrodynamically Supported Shock Layer," Bull. Am. Phys. Soc. II, 8, 436 (1963).

[22] Petschek, H.E., Rose, P.H., Glick, H.S., Kane, A. and Kantrowitz, A.R., "Spectroscopic Studies of Highly Ionized Argon produced by Shock Waves," J. Appl. Phys. 26, 83-95 (1955).

[23] Finkelnburg, W. and Peters, T., "Kontinuierliche Spektren," Handbuch der Physik, edited by S. Flügge (Springer-Verlag, Berlin, 1957), Vol. 28, pp. 79-204.

[24] Treanor, C.E., "Radiation at Hypersonic Speeds," ARS Series: Progress in Astronautics and Rocketry: Hypersonic Flow Research, edited by F.R. Riddell (Academic Press, New York, 1962), Vol. 7, pp. 255-280.

[25] Keck, J.C., Allen, R.A. and Taylor, R.L., "Electronic Transition Moments for Air Molecules," J. Quant. Spectr. Rad. Transf. 3, (to be published) (1963).

[26] Nardone, M.C., Breene, R.G., Zeldin, S.S. and Riethof, T.R., "Radiance of Species in High Temperature Air," General Electric TIS R63SD3 (June 1963).

[27] Meyerott, R. E., Sokoloff, J. and Nicholls, R. W., "Absorption Coefficients of Air," Geophys. Res. Papers 68, U.S.A.F. Res. Div. GRD-TR-60-277 (1960).

[28] Kivel, B. and Bailey, K., "Tables of Radiation from High Temperature Air," Avco-Everett Research Laboratory Research Report 21 (December 1957).

[29] Biberman, L. M. and Norman, G. E., "Calculations of Photoionization Absorption," Optika i Spektrosk. 8, 433-438 (1960); also Optics and Spectroscopy 8, 230-232 (1960).

[30] Taylor, R. L., "Continuum Infrared Radiation from High Temperature Air and Nitrogen," J. Chem. Phys. 39 (to be published) (1963).

[31] Boldt, G., "The Recombination and 'Minus' Continua of Nitrogen Atoms," Z. Physik 154, 330-338 (1959).

[32] Van der Noordaa, R. S. L., "Heat Transfer Measurements from Ionized Argon produced by Strong Shock Waves," Master of Aeronautical Engineering Thesis, Cornell University (1957).

[33] Bershader, D. and Rutowski, R. W., "Studies of an Argon Shock-Layer Plasma," ARS Paper No. 1998-61 (August 1961).

[34] Smiley, E. F., "The Measurement of the Thermal Conductivity of Gases at High Temperatures with a Shock Tube; Experimental Results in Argon at Temperatures between 1000°K and 3000°K," Ph. D. Thesis, The Catholic University of America (1957).

[35] Camac, M., Fay, J. A., Feinberg, R. M. and Kemp, N. H., "Heat Transfer from High Temperature Argon," Proceedings of the 1963 Heat Transfer and Fluid Mechanics Institute (June 1963), pp. 58-76.

[36] Kemp, N. H. and Camac, M., "A Multitemperature Boundary Layer," AIAA Preprint No. 63-460 (August 1963).

[37] Yos, J. M., "Transport Properties of Nitrogen, Hydrogen, Oxygen and Air to 30,000°K," Avco Corporation Research and Advanced Development Division Technical Memorandum RAD-TM-63-7 (March 1963).

[38] Cohen, N., "Boundary Layer Similar Solutions and Correlation Equations for Laminar Heat-Transfer Distribution in Equilibrium Air at Velocities up to 41,000 Feet Per Second," NASA Technical Report R-118 (1961).

39 Hoshizaki, H., "Heat Transfer in Planetary Atmospheres at Super-Satellite Speeds," ARS J. 32, 1544-1552 (1962).

40 Pallone, A. and Van Tassell, W., "The Effects of Ionization on Stagnation-Point Heat Transfer in Air and Nitrogen," Phys. of Fluids 6, 983-986 (1963).

41 Fay, J.A. and Kemp, N.H., "Theory of Stagnation-Point Heat Transfer in a Partially Ionized Diatomic Gas," AIAA J. (to be published).

42 Rose, P.H. and Stankevics, J.O., "Stagnation-Point Heat Transfer Measurements in Partially Ionized Air," AIAA J. (to be published).

43 Petschek, H. and Byron, S., "Approach to Equilibrium Ionization behind Strong Shock Waves in Argon," Annals of Physics 1, 270-315 (1957).

44 Maier-Leibnitz, H., "Excitation Efficiency of Slow Electron Collisions with Inert Gas Atoms," Z. Physik 95, 499-523 (1935).

45 Bray, K.N.C., "Electron-Ion Recombination in Argon Flowing through a Supersonic Nozzle," High Temperature Aspects of Hypersonic Flow, edited by W.C. Nelson (Pergamon Press, New York, 1963, to be published).

46 Byron, S., Stabler, R.C. and Bortz, P.I., "Electron-Ion Recombination by Collisional and Radiative Processes," Phys. Rev. Letters 8, 376-379 (1962).

47 Bates, D.R., Kingston, A.E. and McWhirter, R.W.P., "Recombination between Electrons and Atomic Ions. I. Optically Thin Plasmas," Proc. Roy. Soc. A 267, 297-312 (1962).

48 Gryzinski, M., "Classical Theory of Electronic and Ionic Inelastic Collisions," Phys. Rev. 115, 374-383 (1959).

49 Makin, B. and Keck, J.C., "Variational Theory of Three-Body Electron-Ion Recombination Rates," Phys. Rev. Letters 11, 281-282 (1963); also "Phase Space Calculation of Three-Body Electron-Ion Recombination Rates," Proceedings of the Third International Conference on the Physics of Electronic and Atomic Collisions, London (July 1963) (to be published).

50 Camac, M. and Feinberg, R. M., "Thermal Conductivity of Argon at High Temperatures," Avco-Everett Research Laboratory Research Report (to be published).

51 Lin, S. C., Neal, R. A. and Fyfe, W. I., "Rate of Ionization behind Shock Waves in Air. I. Experimental Results," Phys. of Fluids 5, 1633-1648 (1962).

52 Wray, K. L. and Teare, J. D., "A Shock Tube Study of the Kinetics of Nitric Oxide at High Temperatures," J. Chem. Phys. 36, 2582-2596 (1962).

53 Wray, K. L., "Chemical Kinetics of High Temperature Air," ARS Series: Progress in Astronautics and Rocketry: Hypersonic Flow Research, edited by F. R. Riddell (Academic Press, New York, 1962), Vol. 7, pp. 181-204.

54 Lin, S. C. and Teare, J. D., "Rate of Ionization behind Shock Waves in Air. II. Theoretical Interpretation," Phys. of Fluids 6, 355-374 (1963).

55 Bates, D. R. and Massey, H. S. W., "Slow Inelastic Collisions between Atomic Systems," Phil. Mag. 45, 111-122 (1954).

56 Kasner, W. H., Rogers, W. A. and Biondi, M. A., "Electron-Ion Recombination Coefficients in Nitrogen and in Oxygen," Bull. Am. Phys. Soc. II, 7, 131 (1962).

57 Utterback, N. G. and Miller, G. H., "Ionization of Nitrogen Molecules by Nitrogen Molecules," Phys. Rev. 124, 1477-1481 (1961).

58 Utterback, N. G., "Ionization of Nitrogen and Oxygen Molecules by Nitrogen and Oxygen Molecules," Phys. Rev. 129, 219-224 (1963).

59 Massey, H. S. W. and Burhop, E. H. S., Electronic and Ionic Impact Phenomena (Oxford University Press, 1952).

60 Camm, J., Kivel, B., Taylor, R. and Teare, J. D., "Absolute Intensity of Nonequilibrium Radiation in Air and Stagnation Heating at High Altitudes," J. Quant. Spectr. Rad. Transf. 1, 53-75 (1961).

61 Hammerling, P., "Ionization Effects of Precursor Radiation from Shocks in Air," Avco-Everett Research Laboratory Research Report 98 (June 1960).

[62] Weissler, G. L., "Photoionization in Gases and Photoelectric Emission from Solids," Handbuch der Physik, edited by S. Flügge, (Springer-Verlag, Berlin, 1956), Vol. 21, pp. 304-382.

[63] Brown, S. C., Basic Data of Plasma Physics (M. I. T. and John Wiley and Sons, New York, 1959).

[64] Ferrari, C. and Clarke, J. H., "Photoionization Upstream of a Strong Shock Wave," Brown University Report CM-1020 (January 1963).

[65] Lin, S. C., "Radio Echoes from a Manned Satellite during Re-entry," J. Geophys. Res. 67, 3851-3870 (1962).

[66] Loeb, L. B., Basic Processes of Gaseous Electronics (University of California Press, 1955).

[67] Gibson, W., "The Effect of Ambient Dissociation and Species Diffusion on Nonequilibrium Shock-Layers," IAS Paper No. 63-70 (January 1963).

[68] Lin, S. C. and Fyfe, W. I., "Low-Density Shock Tube for Chemical Kinetic Studies," Phys. of Fluids 4, 238-249 (1961).

[69] Eschenroeder, A. Q., Daiber, J. W., Golian, T. C. and Hertzberg, A., "Shock Tunnel Studies of High-Enthalpy Ionized Airflows," High Temperature Aspects of Hypersonic Flow Dynamics, edited by W. C. Nelson (Pergamon Press, New York, 1963, to be published).

[70] Eckerman, J. and Stern, R. G., "Electron-Ion Recombination in High Temperature Air," AIAA Preprint No. 63-445 (August 1963).

[71] Liepmann, H. W. and Roshko, A., Elements of Gasdynamics (John Wiley and Sons, New York, 1957) p. 85.

[72] Gibson, W. E. and Marrone, P. V., "Correspondence between Normal-Shock and Blunt-Body Flows," Phys. of Fluids 5, 1649-1656 (1962).

[73] Camm, J. C. and Rose, P. H., "Electric Arc-Driven Shock Tubes," Phys. of Fluids 6, 663-678 (1963).

[74] Allen, R. A., Rose, P. H. and Camm, J. C., "Non-equilibrium and Equilibrium Radiation at Super-Satellite Re-entry Velocities," IAS Preprint 63-77 (January 1963).

75 Gruszczynski, J.S. and Warren, W.R., Jr., "Experimental Heat Transfer Studies of Hypervelocity Flight in Planetary Atmospheres," AIAA Preprint 63-450 (August 1963).

76 Cloupeau, M., "Interpretation of Luminous Phenomena Observed in Electromagnetic Shock Tubes," Phys. of Fluids 6, 679-688 (1963).

77 Patrick, R.M., "The Production and Study of High Speed Shock Waves in a Magnetic Annular Shock Tube," Phys. of Fluids 2, 589-598 (1959).

78 Kemp, N.H. and Petschek, H.E., "Theory of the Flow in the Magnetic Annular Shock Tube," Phys. of Fluids 2, 599-608 (1959).

79 Patrick, R.M. and Camac, M., "Experimental Investigation of Collision-Free Shocks and Plasmas," Plasma Hydromagnetics, edited by D. Bershader (Stanford University Press, 1962), pp. 97-123.

80 Gerry, E.T., Patrick, R.M. and Petschek, H.E., "Magnetohydrodynamic Shock Structure in a Partially Ionized Gas," Proceedings of 6th International Conference on Ionization Phenomena in Gases, Paris (July 1963) (to be published).

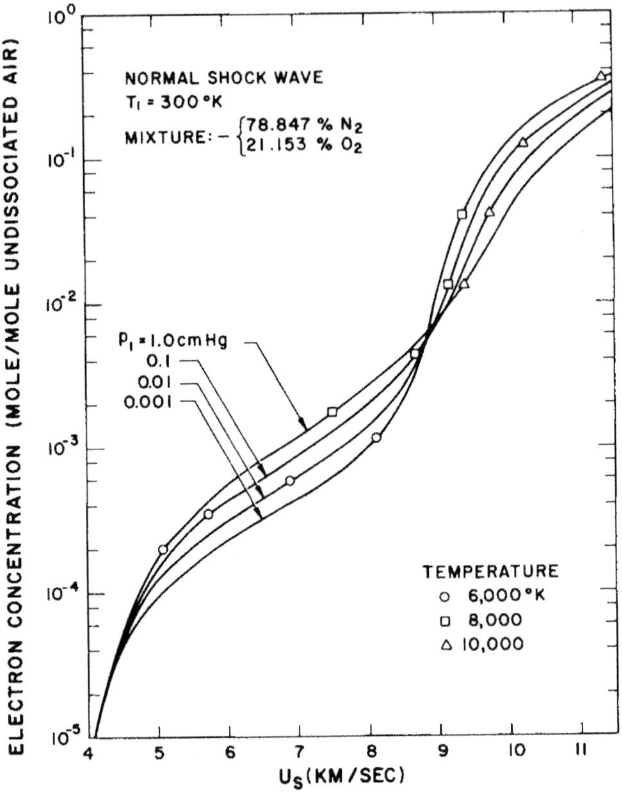


Fig. 1 Equilibrium degree of ionization behind a normal shock wave in argon-free air as a function of shock velocity for a range of values of initial pressure, $p_1$, in the undisturbed gas. The variation of equilibrium temperature with shock speed is also shown.

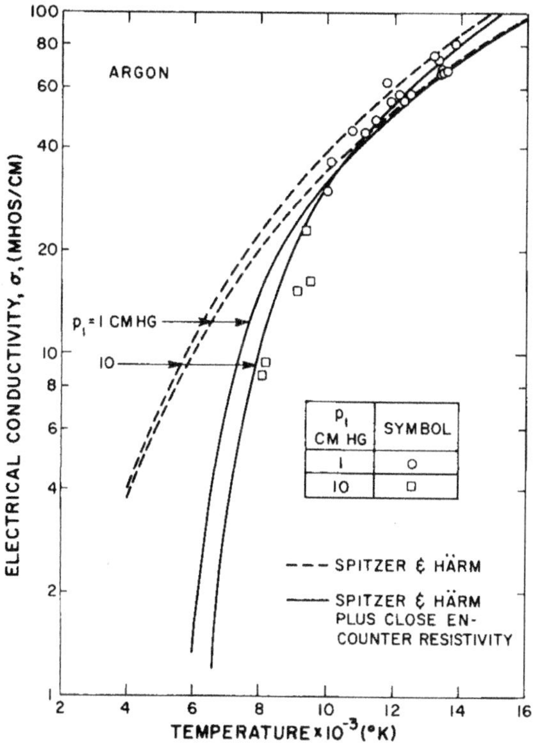


Fig. 2 Equilibrium electrical conductivity in shock-heated argon as a function of equilibrium temperature behind the shock wave, for initial pressures of 1 cm and 10 cm Hg. The dashed curves include only the effects of distant encounters between electrons and ions, according to the theory of Ref. 10. The solid curves include the contribution from electron-atom collisions, using the argon cross section of Fig. 4. The data points are taken from Ref. 11.

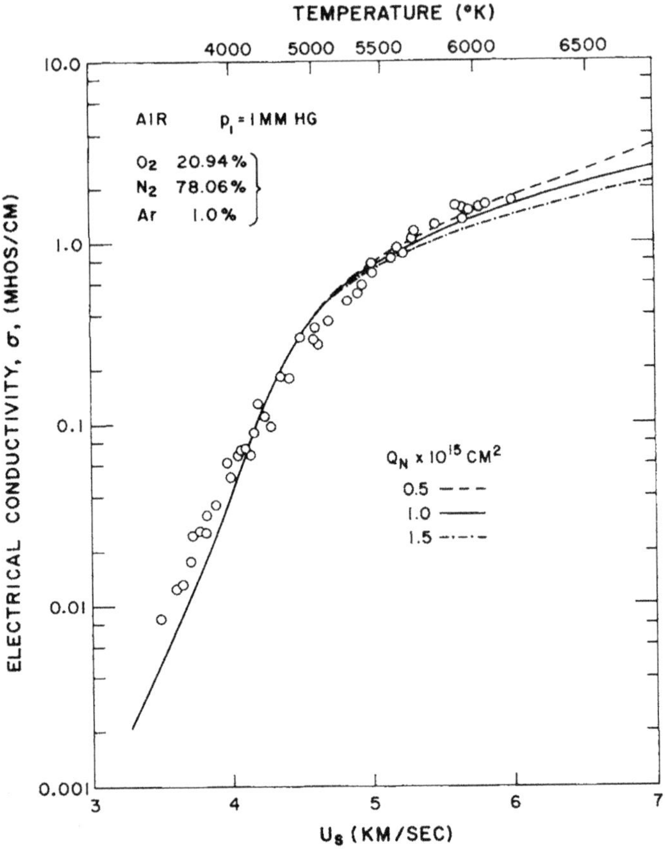


Fig. 3 Equilibrium electrical conductivity behind a normal shock wave in air as a function of shock velocity, for initial pressure of 1 mm Hg. The solid line is calculated using the cross sections of Fig. 4, and the dashed curves show the effect of ± 50% variation in the assumed cross section, $Q_N$, for collisions between electrons and nitrogen atoms. The data points are taken from Ref. 12.

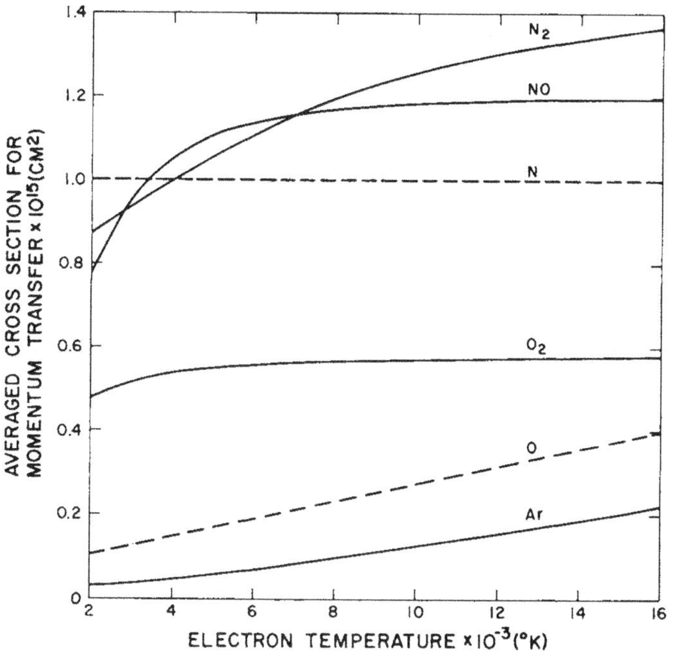


Fig. 4 Velocity-averaged cross sections for momentum transfer between electrons and the various neutral components of high temperature air, as functions of electron temperature. The solid curves are obtained from Fig. 7 of Ref. 13, after integration over a Maxwellian distribution. The atomic oxygen line is arbitrarily drawn through a value of $1.5 \times 10^{-16}$ cm$^2$ at 4,000°K (from Ref. 16), and is shown as increasing with electron temperature, for compatibility with higher energy measurements of Ref. 14. The atomic nitrogen line is arbitrarily located at a level intermediate between the values quoted in Refs. 13 and 15.

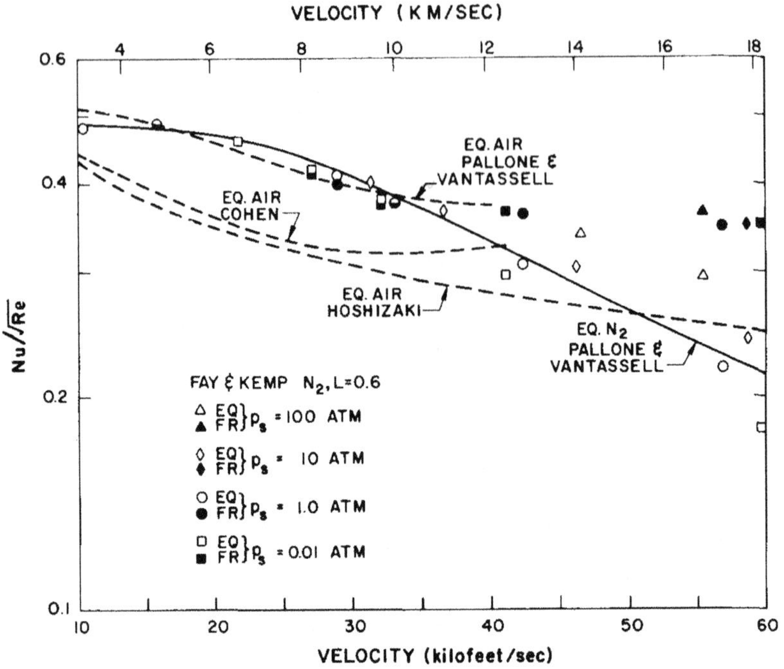


Fig. 5 Convective heat transfer at the stagnation point as a function of flight velocity. The figure shows calculations of Cohen[38] and Hoshizaki,[39] (both for equilibrium air), of Pallone and Van Tassell[40] (equilibrium air and equilibrium $N_2$) and of Fay and Kemp[41] (equilibrium $N_2$ and frozen $N_2$). The ordinate is the ratio of Nusselt number to (Reynolds number)$^{1/2}$.

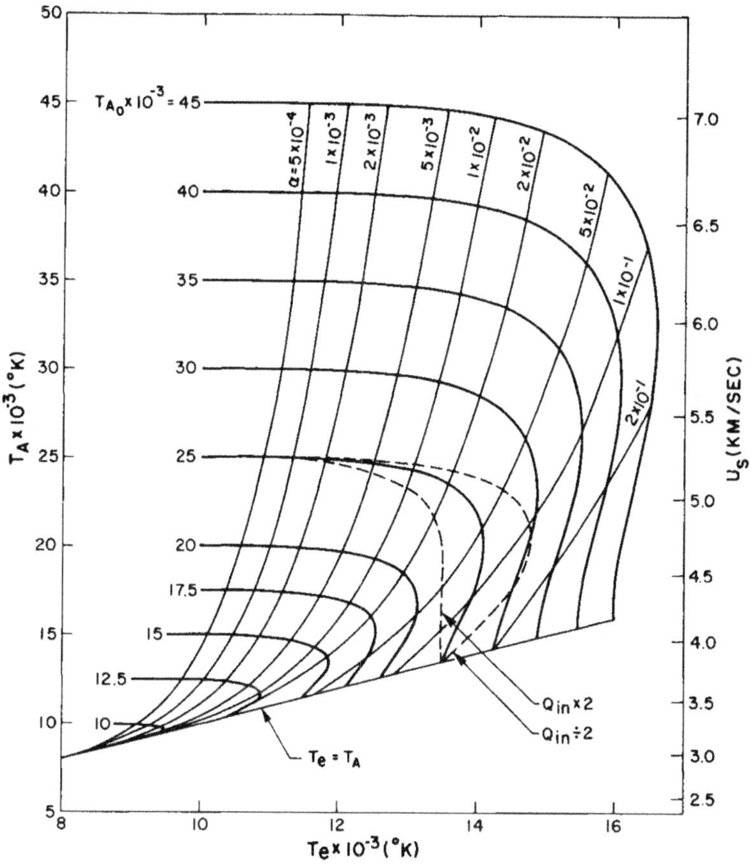


Fig. 6 Variation of electron temperature, $T_e$, with atom temperature, $T_A$, throughout the ionization process behind shock waves in argon. (Figure is similar to Fig. 2 of Ref. 43.) Each heavy solid line describes the interrelation between $T_A$, $T_e$ and degree of ionization, $\alpha$, for a given shock strength characterized by the initial atom temperature, $T_{A_0}$, with corresponding shock speed given on the right hand scale. The calculations assume a balance between the rates at which electrons lose energy through inelastic collisions, Eq. (12), and gain energy due to elastic collisions with ions, Eq. (14). The solid curves are computed with an inelastic cross section, $Q_{in}$, given by Eq. (10), and the effect on $T_e$ of factor-of-two changes in $Q_{in}$ is shown by the dashed curves.

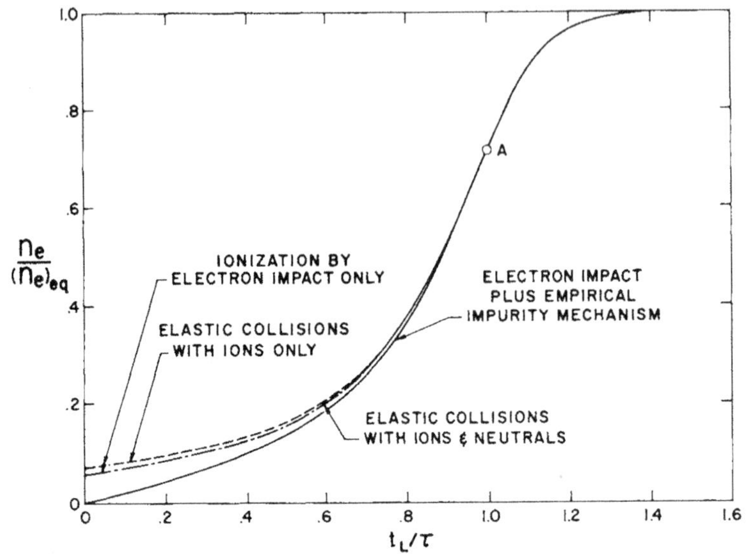


Fig. 7 Variation of electron concentration during the approach to equilibrium behind a strong shock wave in argon. The calculations are for $U_S = 5.25$ km/sec ($T_{A_0} = 25,000^\circ K$) and $p_1 = 1$ cm Hg, but the shape of the solid curve in this nondimensional plot is insensitive to the shock wave parameters. Time, $t_L$, is measured in the laboratory coordinate system (distance behind shock front divided by $U_S$), and $\tau$ is defined as the time at which $n_e^2 = 0.5\ (n_e)^2_{eq}$, corresponding to the point A. The solid line uses the empirical rate equation of Eq. (16) to permit integration from zero electron concentration at the shock front, while the broken curves are obtained by integrating backwards from A with the electron impact mechanism only.

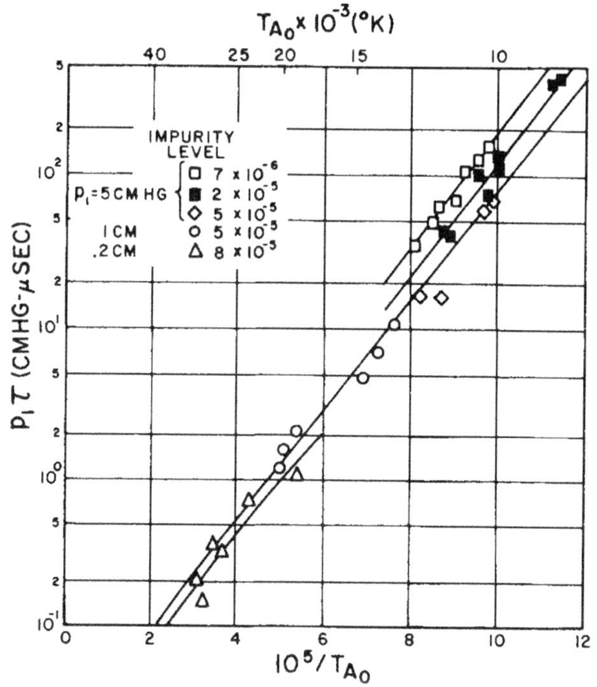


Fig. 8 Ionization time, $\tau$ , (defined as in Fig. 7) as a function of reciprocal initial temperature, $T_{A_0}$, for various values of the impurity level, $X_{imp}$. Initial pressure, $p_1$, is used as a scaling factor for the ordinate, in order to correlate data obtained with different values of $p_1$. The theoretical lines are obtained by integrating the sum of Eqs. (11) and (16). (Reproduced from Ref. 43.)

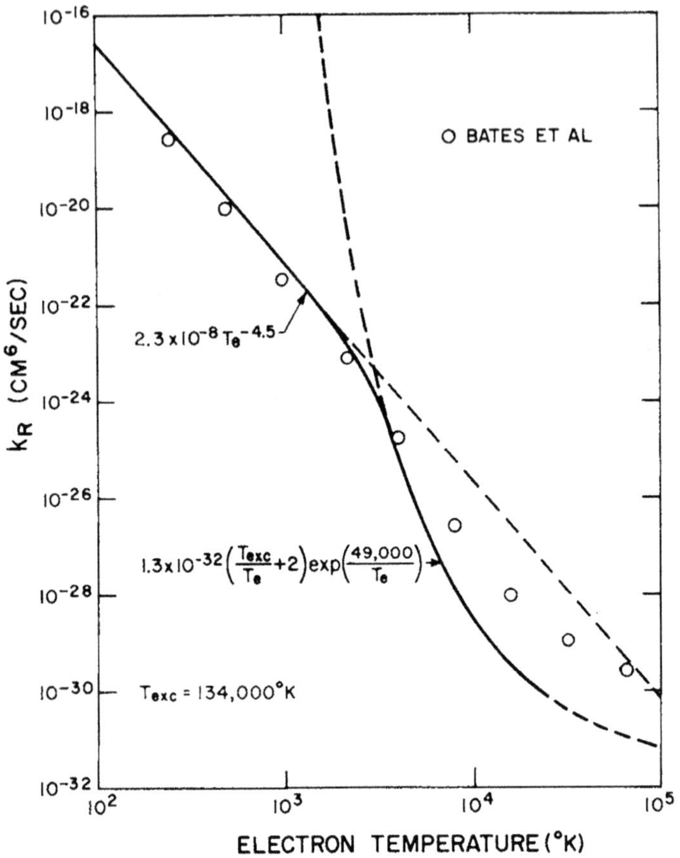


Fig. 9 Three-body recombination rate, $k_R$, for ionized argon as a function of electron temperature, $T_e$. (Reproduced from Ref. 50.) The open circles are taken from Ref. 47.

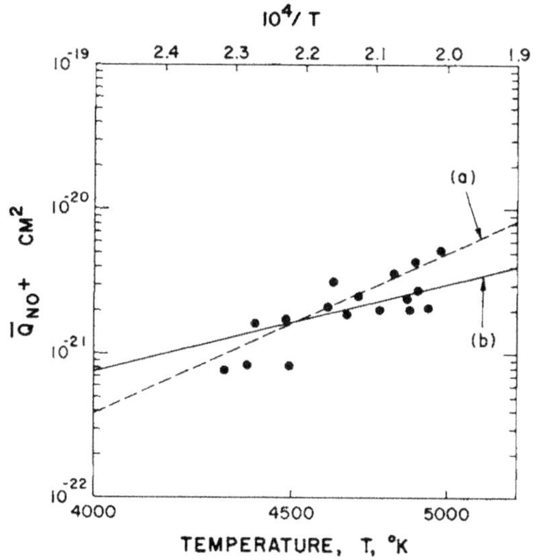


Fig. 10 Velocity-averaged ionization cross section for N-O collisions deduced from observation of the ionization rate behind shock waves in gas mixtures containing a small percentage of $O_2$ in $N_2$ (see Ref. 54). The solid line uses the minimum activation energy, 2.8 ev, for the reaction $N + O \rightarrow NO^+ + e$. The dashed line uses an activation energy of 4.8 ev.

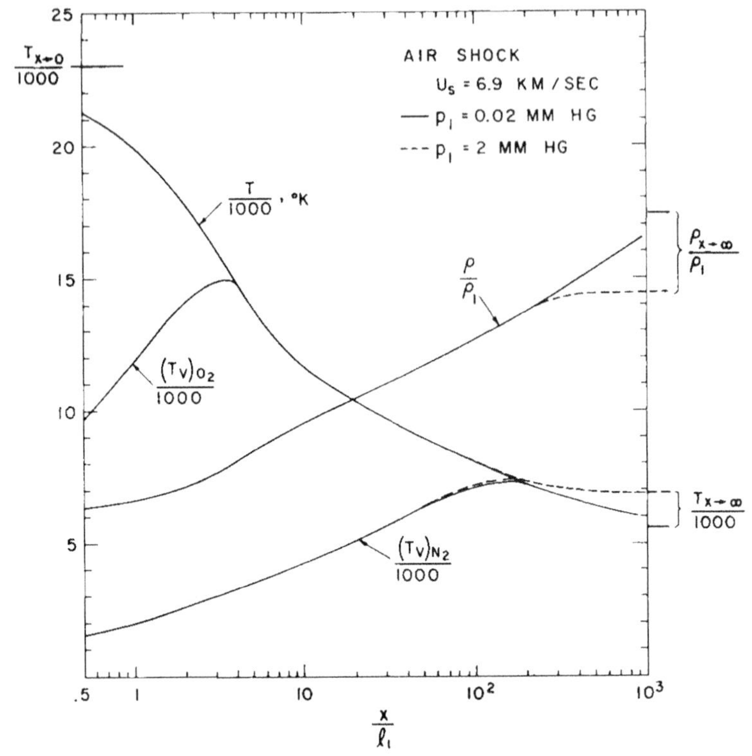


Fig. 11 Typical temperature and density distribution behind a normal shock wave in air calculated from coupled vibrational excitation and chemical reactions. $x/\ell_1$ is the distance behind the translational-rotational shock front divided by the upstream mean free path.

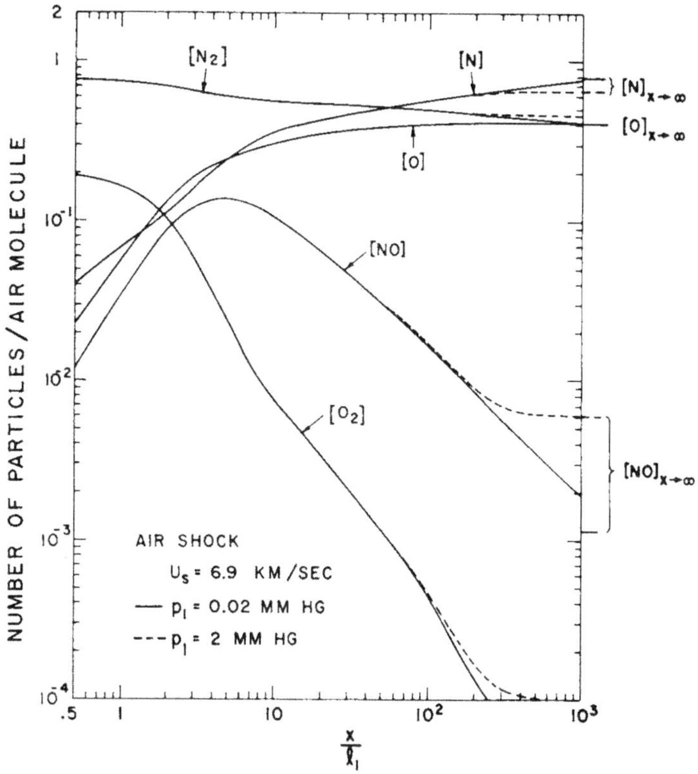


Fig. 12 Typical neutral particle distributions behind a normal shock wave in air calculated from coupled vibrational excitation and chemical reactions. [X] is the total number of type X particles at all internal energy states per unit volume divided by the equivalent number of air molecules within the same volume.

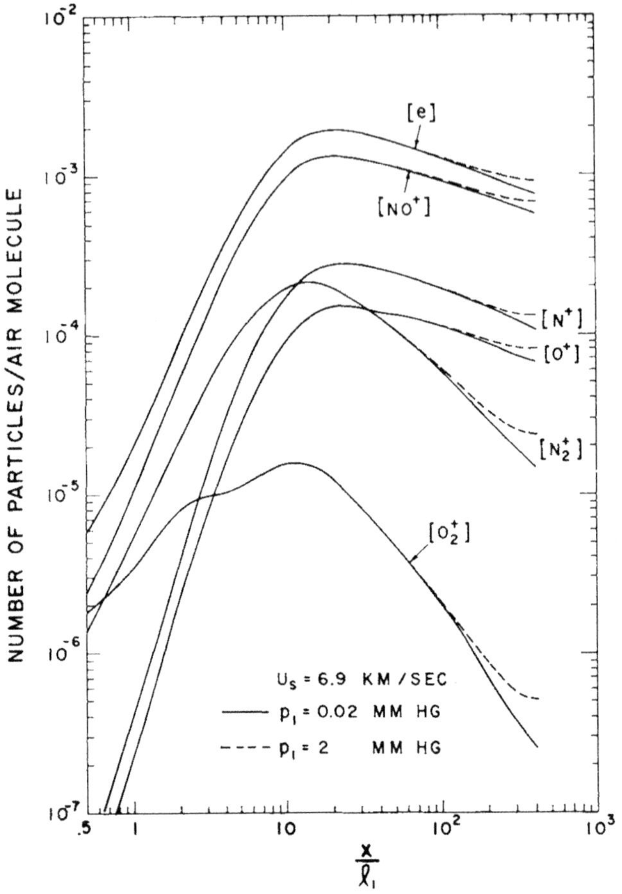


Fig. 13 Distribution of free electrons and various positive ions behind a normal shock wave in air, as calculated with the "basic" rate constants for the ionization processes considered. The effects of diffusion and upstream photoionization have been neglected.

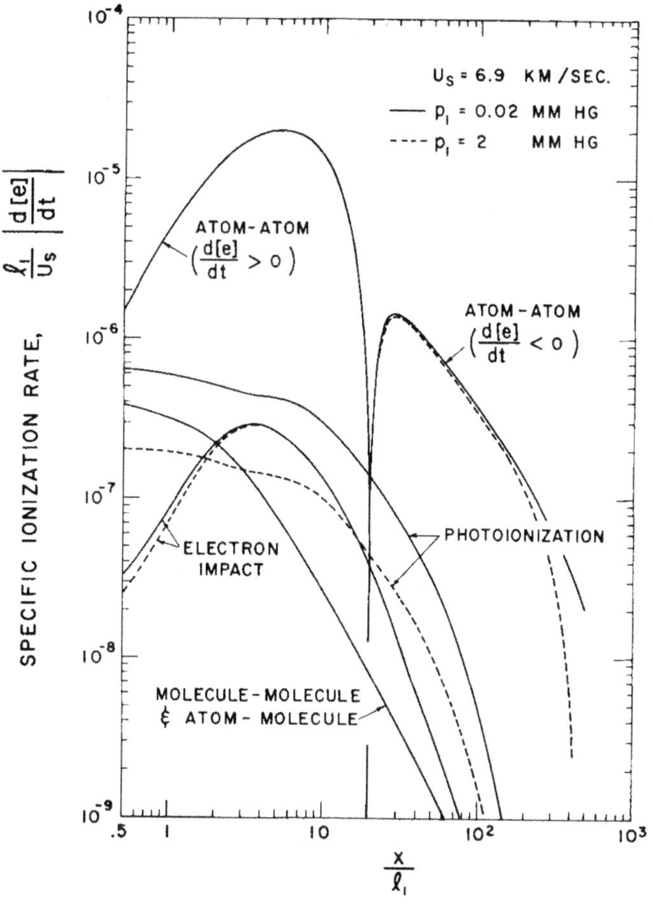


Fig. 14 Specific ionization rates for the various groups of ionization processes as functions of distance behind the shock. The "basic" rate constants are used.

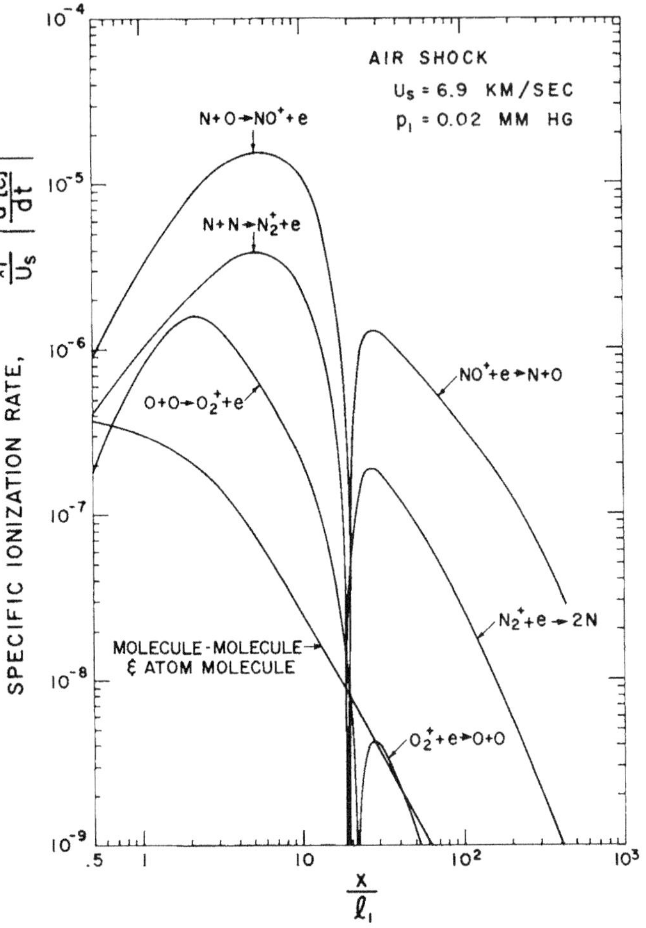


Fig. 15 Detail breakdown of the specific ionization rates for the reactions contributing to the atom-atom process of Fig. 14.

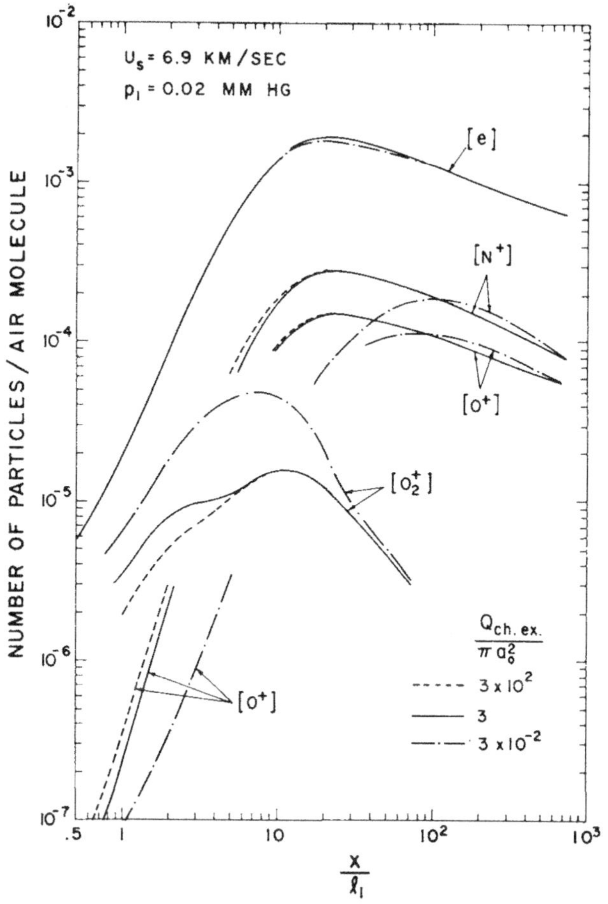


Fig. 16 Effects of varying the "basic" exothermic charge exchange cross section by a factor of 100 in either direction on the distributions of free electrons and $N^+$, $O^+$ and $O_2{}^+$ ions behind the shock for one of the cases shown in Fig. 13. The corresponding effects on the distribution of $NO^+$ and $N_2{}^+$ ions not shown are between those of the electrons and of the other three positive ions.

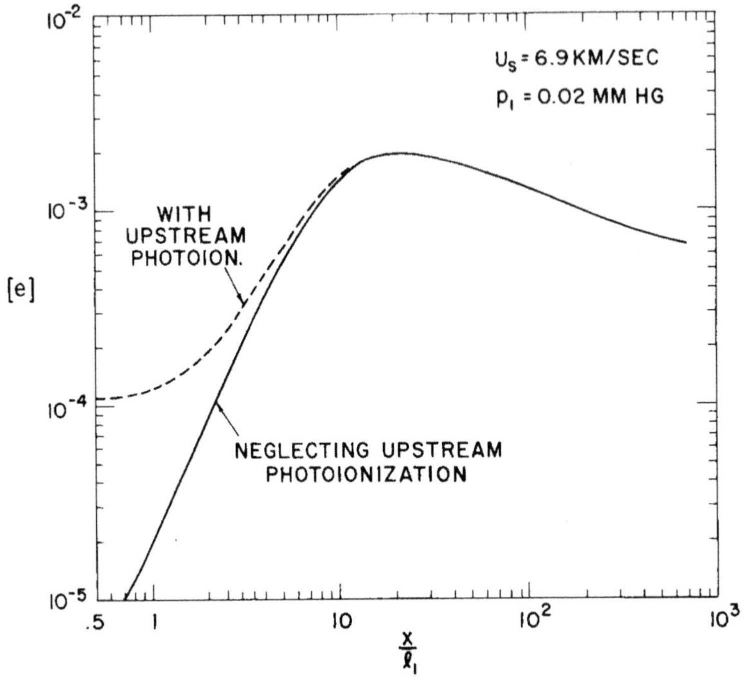


Fig. 17 Effect of inclusion of upstream photoionization on the distribution of free electrons behind the shock. The solid curve is taken directly from Fig. 13.

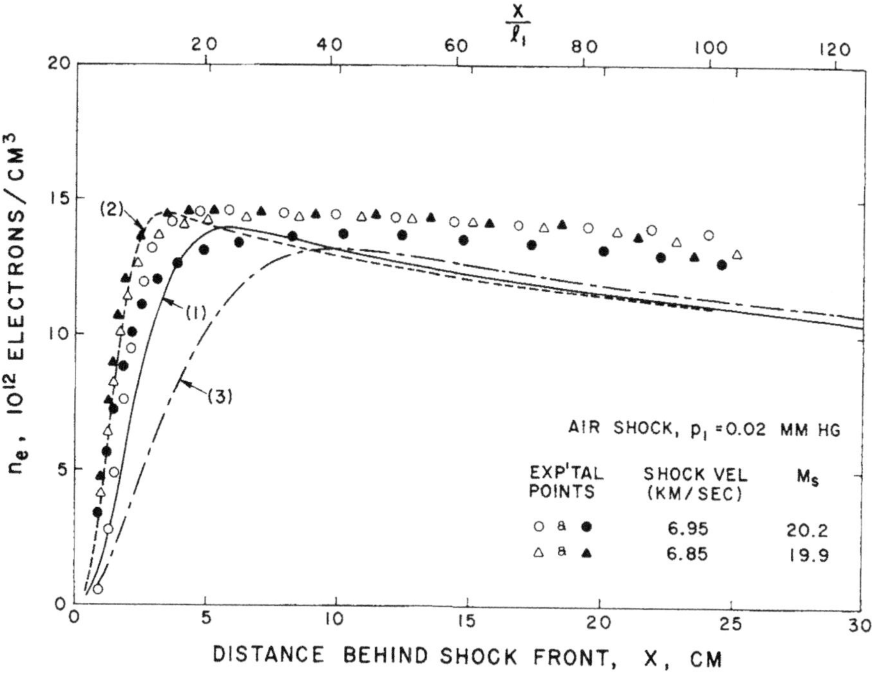


Fig. 18 Comparison between calculated and observed electron density distributions behind the shock for the case $U_s = 6.9 \pm 0.05$ km/sec and $p_1 = 0.02$ mm Hg. The theoretical curve (1) is based on the "basic" set of rate constants described in the text, and the curves (2) and (3) are obtained by varying the predominant ionization rate constant $\kappa_{R1}$ by a factor of 3 in either direction while keeping all other rate constants unchanged. Upstream photoionization has not been included in the theoretical curves since its effect is smaller than the experimental scatter.

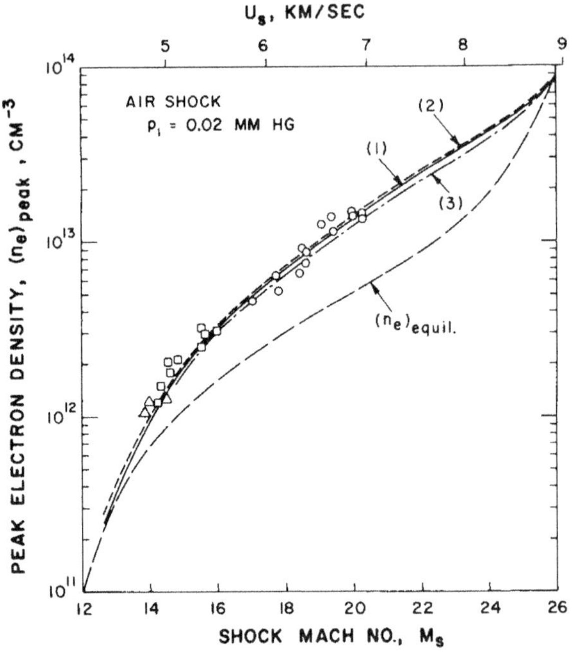


Fig. 19 Comparison between calculated and observed peak electron density behind the shock as a function of shock velocity at $p_1$ = 0.02 mm Hg. The theoretical curves (1), (2) and (3) have the same meaning as in Fig. 18. The curve $(n_e)$ equil. gives the theoretical electron density corresponding to the final adiabatic equilibrium state far behind the shock. The experimental points are taken directly from Fig. 8 of Ref. 51.

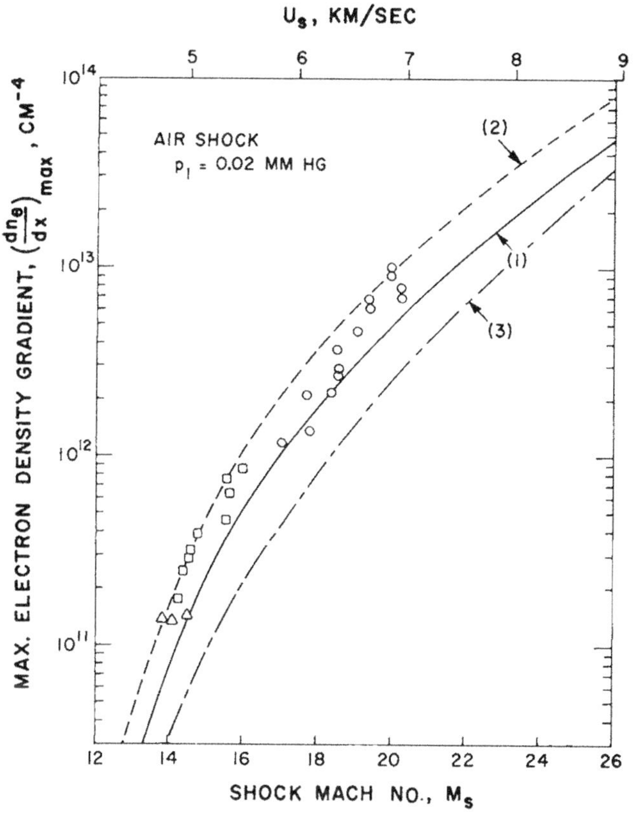


Fig. 20 Comparison between calculated and observed maximum electron density gradient behind the shock as a function of shock velocity at $p_1$ = 0.02 mm Hg. The theoretical curves (1), (2) and (3) have the same meaning as in Fig. 18. The experimental points are taken directly from Fig. 10 of Ref. 51.

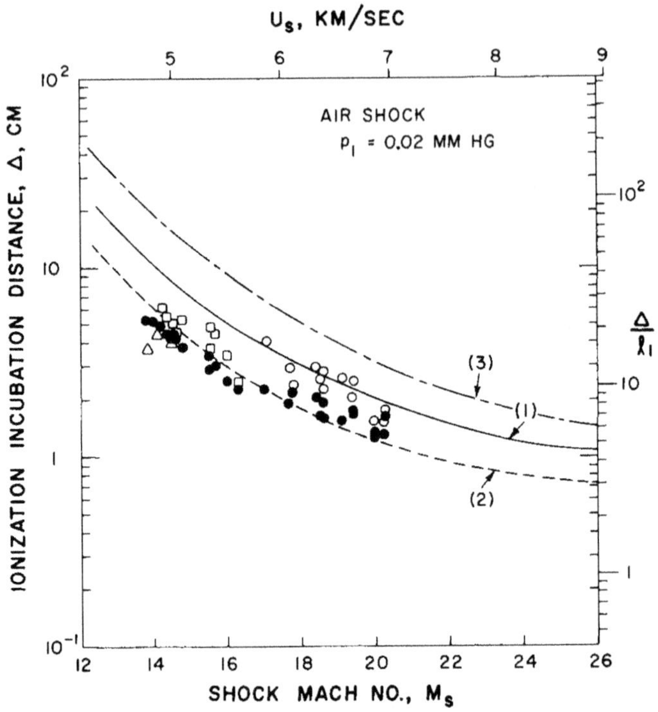


Fig. 21 Comparison between calculated and observed ionization incubation distance behind the shock as a function of shock velocity at $p_1$ = 0.02 mm Hg. The theoretical curves (1), (2) and (3) have the same meaning as in Fig. 18. The experimental points are taken directly from Fig. 12 of Ref. 51.

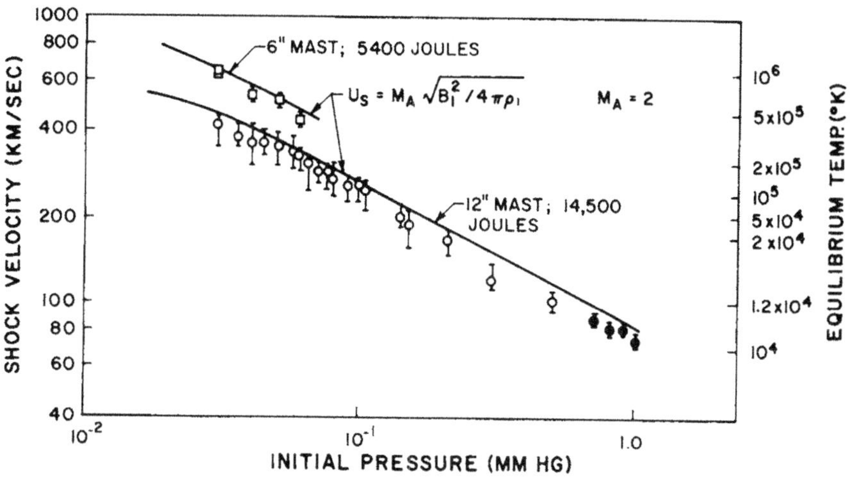


Fig. 22 Measurements of shock velocities in hydrogen obtained in Magnetic Annular Shock Tubes as a function of initial pressure. Data are taken with constant driving energy and constant Alfvén Mach number $M_A = 2$. The solid points indicate conditions for which no evidence of a homogeneous gas sample was obtained. The theoretical shock velocities are obtained from Ref. 78, and the corresponding temperatures indicated would be applicable if the gas behind the shock were in thermal equilibrium. Agreement between calculated and experimental velocities is of the same order as that obtained in conventional shock tubes.

# IONIZATION IN GASEOUS DETONATION WAVES

Tau-Yi Toong*

Massachusetts Institute of Technology, Cambridge, Mass.

## Abstract

The literature on ionization in gaseous detonations is reviewed. Possible ionization mechanisms within detonation waves are discussed in the light of findings obtained in the corresponding studies of flames and shocks. Also included is a brief account of work on electromagnetic effects on exothermal and endothermal waves.

## Introduction

A combustible gas is known to support, in general, two steady exothermal waves, each propagating at its characteristic speed.[1-3] A flame propagates at a subsonic speed relative to the unreacted gas upstream, whereas a detonation wave propagates at a supersonic speed. Early theoretical work on gaseous detonation was mainly restricted to the use of gasdynamical theory (the classical Chapman-Jouguet theory) to predict the steady propagation speed without consideration of the detailed structure. A detonation wave was simply assumed to consist of a shock wave followed closely by a zone of chemical reaction, the combination being treated as a one-dimensional discontinuity. However, such a model is not adequate for the study of ionization within the wave, which includes problems such as the determination of the electrical-conductivity profile, the mechanism of ionization, the rates of ion formation and recombination, and others.

Submitted for publication to AIAA August 1, 1963. The assistance of E. Plett and R. L. Alpert in the preparation of this review is gratefully acknowledged. This work was supported by the U. S. Army Research Office (DURHAM) under Grant DA-ARO(D)-31-124-G190 and the National Science Foundation under Grant G 20880.

*Professor of Mechanical Engineering.

Recent research on the structure and stability of gaseous detonations indicates that the waves are definitely not one-dimensional and are possibly inherently unstable. Although the mechanism of coupling between shock and reaction zone remains unresolved, it is important to note that a strong coupling exists in a self-sustained detonation wave. As ionization in detonations is intimately associated with their structure, interested readers should be familiar with the pertinent literature dealing with the latter problem, which has been reviewed recently by Gross and Oppenheim,[4] Oppenheim,[5] Fay,[6] and Oppenheim, Manson, and Wagner.[7]

This paper will review the literature on ionization in gaseous detonations. Also included is a brief account of the work on electromagnetic effects on exothermal and endothermal waves, which is currently receiving much attention.

## Flames, Shocks, and Detonations

During past years, ionization in flames and shock waves has been the subject of many intensive investigations, both theoretical and experimental.† Ionization in detonation waves, however, has not received nearly as much attention. As a self-sustained detonation wave may be considered to consist of a shock wave and a flame, which are coupled to each other in varying degrees, the study of ionization in detonation waves may be aided by the results obtained in the corresponding studies of flames and shocks. One should note, however, in this connection, that the predominant mechanisms of ionization in a particular case may differ, depending on such factors as the composition, the presence of seeding material of low ionization potential, and the range of pressure and temperature involved.

The two main mechanisms of ionization known to be important in flames and shocks are thermal ionization and chemi-ionization. In general, the former is in thermodynamic equilibrium‡ and becomes more important at higher temperatures, whereas the latter is the result of chemical reaction and is thus essentially a nonequilibrium process.

---

†See, for example, the pertinent articles in this book.

‡The degree of ionization can be computed in this case by the Saha equation.[8]

It is now well established that chemi-ionization is the predominant mechanism in many flames, thus accounting for the fact that their electrical conductivities are much higher than those predicted from the equilibrium theory. Until recently, ionization in shock waves was believed to be due to purely thermal effects. For example, Lin, Resler, and Kantrowitz[9] found that the maximum conductivity of highly ionized argon behind incident shocks at temperatures higher than 8000° to 10,000°K agreed within 10% with the value predicted on the basis of the equilibrium theory, whereas at lower temperatures, the measured value was lower than the computed equilibrium value, presumably because the gas did not have time to reach thermal equilibrium in the region bounded by the shock front and the contact discontinuity. However, recent measurements of Lin, Neal, and Fyfe[10] of ionization rates behind strong normal shock waves in air (at low initial pressures, 0.02 to 0.2 mm Hg) showed that the electron density rose rapidly behind the shock front to a transient peak value, two or three times higher than that corresponding to the final equilibrium value, thus indicating the possible presence of nonequilibrium effects. Although such a peak is to be expected to exist behind strong shocks in diatomic and polyatomic gases whenever the ionization process is sufficiently fast compared with the dissociation process,[11] the nonequilibrium effect has been detected here due to the high spatial resolution made possible by the use of a low-density shock tube.

It would be interesting to point out in this connection the importance of spatial resolution demonstrated in the study of ionization in flames. Shuler and Weber[12] indicated that their measurements in premixed hydrogen-oxygen and acetylene-oxygen flames at atmospheric pressure (by means of a microwave attenuation technique) pertained to an ionization density averaged over a rather extensive region of the flame, which included both the reaction zone and parts of the hot combustion gases. It is thus possible that their measured values agreed well with the thermal-equilibrium values due to contribution from the hot gases, which are essentially in thermal equilibrium. On the other hand (as it was pointed out earlier), results showing local ion concentrations throughout the reaction zone indicate that chemi-ionization is the predominant mechanism in flames.

## Ionization in Detonation Waves

### 1. Experimental

Using a d. c. probe, Basu and Fay[13] measured the electrical conductance behind stoichiometric oxy-hydrogen (with 2/3 mole %

nitrogen) detonation waves at initial pressures of 0.1 to 1 atm and oxy-acetylene detonation waves at 1 atm initial pressure with mixture composition varying from 30 to 80 % acetylene. By comparing their measured values with theoretical results computed on the basis of thermodynamic equilibrium considerations, they concluded that equilibrium ionization was achieved in detonation waves in the mixtures tested. It is important to note, however, that their probe signal built up from zero to a steady value in about 10 μsec or less, corresponding to a distance behind the detonation wave front of about 3 cm. The probe did not seem to have enough spatial resolution to detect any nonequilibrium effects, if present, within the wave. Their conclusion regarding equilibrium ionization is thus valid only with respect to the detonation products.

Electrical conductivities in <u>seeded</u> equimolal oxy-acetylene detonations at 0.1-atm initial pressure were determined by Basu[14] by a magnetic induction method as used by Lin et al.[9] The measured maximum values were compared with the results of thermodynamic equilibrium calculations, as shown in Fig. 1, which is taken from Fig. 6 of Ref. 14. They were found to agree with curve A within a factor of two and to have the same dependence on the mole fraction of potassium (from the seed material, potassium acetylide), which was varied from 0.1 to 10 %. The conductivity was a maximum (2.7 mho/cm) at about 1 mole % potassium.§ Ionization was also found to be essentially complete within about 40 μsec behind the wave front.

In view of the general agreement between theory and experiment, Basu concluded (as in his previous work) that equilibrium ionization// was achieved in the product gases (at 3500° to 4000°K) behind seeded detonation waves in the mixtures tested, within about 40 μsec after the passage of the wave front. It is interesting to note, however, that his search-coil responses showed two peaks in most experiments, the second peak being on the average about 25% of the first peak and trailing it by 7 to 25 cm of wave travel. He indicated

§The conductivity behind unseeded detonation waves in the same mixture at 1-atm initial pressure was measured by Basu and Fay[13] to be about $10^{-3}$ mho/cm.

//In this conjunction, one should note the corresponding work in flames by Belcher and Sugden[15] which showed that, in coal-gas/air flames, sodium and potassium gave the expected relative ionization from the Saha equation.

that the first peak corresponded to the passage of the detonation front and the second peak was probably due to waves similar to those observed by Cher and Kistiakowsky.[16] Thus, the final conductivity of the product gases seemed to have been achieved in two steps. The reason for the existence of these double waves is not clear. However, they seem to imply the possible presence of nonequilibrium effects within the detonation wave.

Nonequilibrium effects have indeed been observed by Hand and Kistiakowsky[17] in their study of ionization in shock waves in acetylene-oxygen-argon mixtures (97% A, 1.2% $C_2H_2$, and 1.8% $O_2$ at an initial pressure of 1 Torr and a final temperature of 1500° to 2800°K). It should be noted that these waves were not (self-sustained) detonation waves but shock waves accompanied by exothermic chemical reactions. After an induction period, the ionization (measured by means of a Langmuir probe) rose exponentially and then decayed[#], as shown in Fig. 2, taken from Fig. 2 of Ref. 17. The maximum free-electron concentrations were found to be several orders of magnitude higher than those at thermodynamic equilibrium, thus suggesting again the possible presence of chemi-ionization. This behavior of ionization is similar to that of the emission of short ultraviolet radiation (due to chemi-luminescence) from shock waves in acetylene-oxygen-argon mixtures described by Kistiakowsky and Richards.[19]

In comparing the results of Basu and Fay,[13] Basu,[14] and Hand and Kistiakowsky,[17] one should note the following main differences: 1) in the former two cases, the waves were self-sustained detonations, whereas in the last case, the waves were shocks accompanied with exothermic reaction; 2) the temperature was higher in the former two cases (3500° to 4200°K) than in the last case (1500° to 2800°K); and 3) seeding material of low ionization potential was used in Basu's experiments.[14] These differences might account for the varying degrees of importance of different ionization mechanisms in these cases. However, it is interesting to point out that the nonequilibrium effects were detected by Hand and Kistiakowsky mainly because of the adequacy of spatial resolution of their probe as the result of rather long reaction time. The question remains whether chemi-ionization or other nonequilibrium mech-

[#]The ionization was found to decay according to a second-order rate equation over a considerable range of currents, just like that behind the downstream edge of the luminous zone in a flame.[18]

anism plays a predominant role within detonation waves just as in flames or in shocks. This problem is now being studied by the author's research group at the Massachusetts Institute of Technology.

Nonequilibrium effects seemed to have been observed also by Cook, Bracke, and Keyes[20] in their study of the effect of pressure on ionization in gaseous detonations in $2H_2 + O_2 + 0.002N_2$, $2H_2 + O_2 + N_2$, and $2CO + O_2$ (up to 350 atm in detonation pressures). The electron densities were determined from d. c. electrical-conduction measurements between probes (of 1.6-mm-diam copper wire) mounted diametrically opposite, 1.5 mm apart. The conductance rose to a maximum in 5 $\mu$sec or less, dropped sharply, and then rose again to a smaller peak, resembling somewhat the response observed by Basu.[14] The probe calibration factor was chosen to bring to agreement the measured maximum conductivity at low detonation pressures and the theoretical value based on thermodynamic equilibrium considerations, as justified by the findings of Basu and Fay[13] (although the justification seems to be rather questionable, especially because the probe signals were quite different in shape in the two cases). With the low-pressure values thus brought into agreement, they noted a relatively abrupt rise with pressure of the experimental (maximum) values above the theoretical ones (calculated on the basis of $H_2O^+$, $O_2^+$, and $NO^+$ formation) at about 20 and 50 atm for the $2H_2 + O_2 + 0.002\ N_2$ and the $2H_2 + O_2 + N_2$ systems, respectively. The difference between observed and computed values increased until at 250 atm (detonation pressure) the measured maximum values were about 10 times greater than the computed ones.

Two explanations for the excess ionization at high pressures described in the foregoing were attempted by Cook, et al.[20] The first was based on the formation of the hydrated ion, $H_3O^+$, which has been found from mass spectrometric data of Knewstubb and Sugden[21] and Deckers and van Tiggelen[22] to be a dominant ion in hydrocarbon-oxygen-nitrogen flames. Although the contribution to the total ionization by the formation of $H_3O^+$ increased with pressure, the increase was negligibly small and could not account for the observed pressure dependence of the electrical conductivity. The second explanation was based on the quasi-lattice model of Cook and McEwan.[23] According to this model, provided that the ion density is sufficiently large, the ions and electrons assume a quasi-metallic-like lattice structure, in which the electrons move through the positive-ion lattice in correlated patterns. Such a situation leads to a cohesive energy, which would effectively result in a lower ionization potential in the Saha equation for the computation of the fraction ionized. This decrease in the effective ionization potential as the result of increase in cohesive energy with rising pressure seemed to account for the observed pressure dependence of the electrical conductivity.

Despite good agreement between experiment and theory (on the basis of thermodynamic equilibrium and quasi-lattice structure), the possible presence of nonequilibrium effects cannot be ruled out in view of the many assumptions used in bringing about this agreement and the shape of the probe signal obtained in this work.

## 2. Theoretical

Theoretical work on ionization in gaseous detonations published in the literature is restricted to the computation of the degree of ionization and of the electrical conductivity of the product gases in thermodynamic equilibrium. In addition to the theoretical calculations cited in part 1 of this section for comparison with the corresponding experimental measurements, extensive work has been published by Gross and his associates.[24,25] Chinitz, Eisen, and Gross[24] presented pressure, temperature, chemical composition, and electrical conductivity of gas mixtures behind strong air shocks and detonations of hydrogen-air and acetylene-air mixtures up to Mach 20. Their results also included the initial gas composition at which the electrical conductivity is a maximum for a given shock or detonation strength. In another publication, Eisen, Gross, and Rivlin[25] presented Chapman-Jouguet, strong and weak detonation properties for a five-atom (C, H, O, N, and A) system up to Mach 10. Equilibrium and frozen sound speeds were also calculated for the hot products. It is to be noted that, in the temperature range considered in both of these references, NO is the only ionized component that appears in appreciable quantities because of its low ionization potential. Any impurities with low ionization potentials (such as the alkali metals), if present, would significantly change the reported values of electrical conductivity.

Figure 3, taken from Fig. 11 of Ref. 24, shows the electrical conductivity of the gases behind air shocks and stoichiometric hydrogen-air and acetylene-air detonations (at an initial pressure and temperature of $10^{-3}$ atm and 298°K, respectively) at different Mach numbers. In the same figure is shown also the effect of the presence of a "seed" of low ionization potential. The very small amount of cesium increases appreciably the electrical conductivity of gases behind air shocks. In this conjunction, one should note the corresponding conductivity increase (by three orders of magnitude) behind seeded detonation waves as measured by Basu.[14]

In the discussion of the work on ionization in flames, shocks, and gaseous detonations, it has been suggested that nonequilibrium chemi-ionization effect may be important within a detonation wave. However, depending on the pressure, temperature, composition, and presence of seed material or impurities of low ionization potential, thermal ionization may assume a role just as important as chemi-ionization. In the case of seeded detonation waves, one would expect that the contribution to electrical conductivity from thermal ionization of the seed material (in some part of the wave) may be several orders of magnitude higher than that from chemi-ionization (as estimated from results in flame studies). On the other hand, for unseeded detonation waves, one would expect that the contributions from the two mechanisms may be of equal importance in the major part of the wave. Preliminary calculations of Plett[26] based on a structure of a detonation wave obtained by Hirschfelder and Curtiss[27] seemed to support these expectations. Experimental work is now in progress in the author's group at Massachusetts Institute of Technology to confirm these findings.

## Electromagnetic Effects

A new class of problems in which ionization plays an important role — the study of electromagnetic effects on exothermal and endothermal waves — has become of great interest in recent years. Much of this study is theoretical in nature and predicts many interesting interactions. However, definitive experimental work needed to further understanding of these interactions is still lacking.**

As a critical survey of the literature on electromagnetic effects is beyond the scope of this review paper, only a brief account of some of these studies will be presented below.

In an early note, Gross[28] made a study of the one-dimensional motion of a plasma of infinite electrical conductivity in a transverse magnetic field, involving exchange of thermal energy with its surroundings. In a later paper, Gross, Chinitz, and Rivlin[29] examined some magnetogasdynamic effects on exothermal, plane, supersonic waves. In the mean time, Larish and Shekhtman[30] studied the effects of a transverse magnetic field on

**Such experimental program is now in progress in the author's group at Massachusetts Institute of Technology.

the propagation of detonation waves for the relativistic and nonrelativistic cases. Lyubimov[31] and Helliwell[32] considered the case in which a detonation wave was propagating into a nonconducting gas in a transverse electromagnetic field, with the conductivity of the gas behind the wave assumed to be infinite. Barmin[33] presented a discontinuity analysis for an infinitely conducting medium and for the general case of an oblique field. More recently, Kunkel and Gross[34] have examined the characteristics of hydromagnetic ionizing waves and discussed their similarities to detonation waves.

Nicholls, et al.[35] studied the magnetic interaction zone behind a detonation wave (treated as a discontinuity) propagating into a nonconducting gas. Fong, Bollinger, and Edse[36] also presented results for the magnetic interaction zone and for an ionization-inducing exothermal discontinuity.

Demutskii and Polovin[37, 38] investigated possible shock-ionization, detonation, and combustion regimes in the presence of a perfectly conducting, moving piston. In these two papers, the piston velocity, the Alfvén velocity, and the reaction energy were assumed to be small. In a later paper, Demutskii[39] examined the propagation of detonation and combustion waves in a magnetohydrodynamic medium in a tube from its closed end (i.e., a motionless piston). The Alfven velocity was still assumed to be small (compared with sound velocity) and the energy of reaction large (compared with the square of sound velocity).

It would be of interest to study the interactions between electromagnetic fields and exothermal or endothermal waves, with the wave structure taken into consideration. Toong[40] has conducted a theoretical study of a steady, one-dimensional flow of a chemically reactive gaseous mixture with variable electrical conductivity in a transverse magnetic field. This study further indicates the need of definitive experimental work to aid in the interpretation of the theoretical results, which have been obtained for widely different models.

## References


1 Courant, R. and Friedrichs, K. O., Supersonic Flow and Shock Waves (Interscience Publishers, Inc., New York, 1948), pp. 204-234.

2 Emmons, H. W., "Fundamentals of gas dynamics", High Speed Aerodynamics and Jet Propulsion (Princeton University Press, Princeton, N. J., 1958), Vol. III, pp. 574-686.

[3] Lewis, B. and von Elbe, G., Combustion, Flames and Explosions of Gases (Academic Press, Inc., New York, 1961), 2nd. ed., pp. 511-554.

[4] Gross, R. A. and Oppenheim, A. K., "Recent advances in gaseous detonations," ARS J. 29, 173-179 (1959).

[5] Oppenheim, A. K., "Development and structure of plane detonation waves," Fourth AGARD Combustion and Propulsion Colloquium (Pergamon Press, Oxford, 1961), pp. 186-258.

[6] Fay, J. A., "The structure of gaseous detonation waves," Eighth Symposium (International) on Combustion (Williams and Wilkins Co., Baltimore, Md., 1962), pp. 30-40.

[7] Oppenheim, A. K., Manson, N., and Wagner, H. G., "Recent progress in detonation research and its relation to rocket propulsion," AIAA Preprint 63-227 (1963).

[8] Saha, M. N. and Saha, N. K., A Treatise on Modern Physics (Indian Press, Allahabad and Calcutta, 1934), p. 630.

[9] Lin, S. C., Resler, E. L., and Kantrowitz, A., "Electrical conductivity of highly ionized argon produced by shock waves," J. Appl. Phys. 26, 95-109 (1955).

[10] Lin, S. C., Neal, R. S., and Fyfe, W. I., "Rate of ionization behind shock waves in air. I. Experimental results," Phys. Fluids 5, 1633-1648 (1962).

[11] Lin, S. C. and Teare, J. D., "Rate of ionization behind shock waves in air. II. Theoretical interpretation," Phys. Fluids 6, 355-375 (1963).

[12] Shuler, K. E. and Weber, J., "A microwave investigation of the ionization of hydrogen-oxygen and acetylene-oxygen flames," J. Chem. Phys. 22, 491-502 (1954).

[13] Basu, S. and Fay, J. A., "Ionization in detonation waves," Seventh Symposium (International) on Combustion (Butterworth Scientific Publications, London, 1959), pp. 277-282.

[14] Basu, S., "Ionization in seeded detonation waves," Phys. Fluids 3, 456-463 (1960).

[15] Belcher, H. E. and Sugden, T. M., "Studies on the ionization produced by metallic salts in flames. II. Reactions governed by ionic equilibrium in coal-gas/air flames containing alkali metal salts," Proc. Roy. Soc. (London) A202, 17-39 (1950).

[16] Cher, M. and Kistiakowsky, G. B., "Gaseous detonations. XI. Double waves," J. Chem. Phys. 29, 506-511 (1958).

[17] Hand, C. W. and Kistiakowsky, G. B., "Ionization accompanying the acetylene-oxygen reaction in shock waves," J. Chem. Phys. 37, 1239-1245 (1962).

[18] Calcote, H. F., "Mechanisms for the formation of ions in flames," Combust. Flame 1, 385-402 (1957).

[19] Kistiakowsky, G. B. and Richards, L. W., "Emission of vacuum ultraviolet radiation from the acetylene-oxygen and the methane-oxygen reactions in shock waves," J. Chem. Phys. 36, 1707-1714 (1962).

[20] Cook, M. A., Bracke, P. P., and Keyes, R. T., "The effect of pressure on ionization in gaseous detonation," Inst. Metals and Explosives Res. Univ. Utah, ASTIA AD 258 247 (1961).

[21] Knewstubb, P. F. and Sugden, T. M., "Mass spectrometry of the ions present in hydrocarbon flames," Seventh Symposium (International) on Combustion (Butterworths Scientific Publications, London, 1959), pp. 247-253.

[22] Deckers, J. and van Tiggelen, A., "Ion identification in flames," Seventh Symposium (International) on Combustion (Butterworths Scientific Publications, London, 1959), pp. 254-255.

[23] Cook, M. A. and McEwan, W. S., "Cohesion in plasma," J. Appl. Phys. 29, 1612-1613 (1958).

[24] Chinitz, W., Eisen, C. L., and Gross, R. A., "Aerothermodynamic and electrical properties of some gas mixtures to Mach 20," ARS J. 29, 573-579 (1959).

[25] Eisen, C. L., Gross, R. A., and Rivlin, R. J., "Theoretical calculations in gaseous detonation," Combust. Flame 4, 137-147 (1960).

[26] Plett, E., "Structure of detonation waves," unpublished report, Mech. Eng. Dept., Mass. Inst. Tech. (1963).

[27] Hirschfelder, J. O. and Curtiss, C. F., "Theory of detonations. I. Irreversible unimolecular reaction," J. Chem. Phys. 28, 1130-1147 (1958).

[28] Gross, R. A., "A note on one-dimensional plasma motion," J. Aerospace Sci. 25, 788-789 (1958).

[29] Gross, R. A., Chinitz, W., and Rivlin, T. J., "Magnetohydrodynamic effects on exothermal waves," J. aerospace Sci. 27, 283-290 (1960).

[30] Larish, E. and Shekhtman, I., "Propagation of detonation waves in the presence of a magnetic field," J. Exptl. Theoret. Phys. (USSR) 35, 203-207 (1958); transl. in Soviet Phys. — JETP 35(8), 139-143 (1959).

[31] Lyubimov, G. A., "The effect of an electromagnetic field on the detonation regime," Dokl. Akad. Nauk SSSR 126, 532-533 (1959); transl. in Soviet Phys. —Doklady 4, 526-528 (1960).

[32] Helliwell, J. B., "Gas-ionizing shock and combustion waves in magnetogasdynamics," J. Fluid Mech. 14, 405-419 (1962).

[33] Barmin, A. A., "Surfaces of discontinuity with release or absorption of energy in magnetic hydrodynamics," Dokl. Akad. Nauk SSSR 138, 77-80; transl. in Soviet Phys. — Doklady 6, 374-376 (1961).

[34] Kunkel, W. B. and Gross, R. A., "Hydromagnetic ionizing waves," Plasma Hydromagnetics, Sixth Lockheed Symposium on MHD, edited by D. Bershader (Stanford University Press, Stanford, Calif., 1962), pp. 58-82.

[35] Nicholls, J. A., Dabora, E. K., Morrison, R. B., Adamson, T. C.,Jr., Phillips, R. L., Ranger, A. A., Lelli, G. J., and Cosens, G. L., "An experimental and theoretical study of stationary gaseous detonation waves," Dept. Aeronaut. Astronaut. Eng., Univ. Mich., U. S. Air Force Office Sci. Res. TN 1764 (1961).

[36] Fong, M. C., Bollinger, L. E., and Edse, R., "Magnetohydrodynamic effects on exothermal waves," Ohio State Univ., ASTIA AD 269 280 (1961).

[37] Demutskii, V. P. and Polovin, R. V., "Shock ionization and detonation in magnetohydrodynamics," Zh. Tekhn. Fiz. 31, 419-427 (1961); transl. in Soviet Phys. — Tech. Phys. 6, 302-307 (1961).

[38] Polovin , R. V. and Demutskii, V. P., "Magnetohydrodynamic combustion," J. Exptl. Theoret. Phys. (USSR) 40, 1746-1754 (1961); transl. in Soviet Phys. — JETP 13, 1229-1234 (1961).

[39] Demutskii, V. P., "Detonation and combustion in magnetohydrodynamics," Sh. Tekhn. Fiz. 32, 445-449 (1962); transl. in Soviet Phys. — Tech. Phys. 7, 324-327 (1962).

[40] Toong, T. Y., "Steady, one-dimensional, magnetogasdynamics of chemically reacting systems with variable electrical conductivity," unpublished report, Mech. Eng. Dept., Mass. Inst. Tech. (1961).

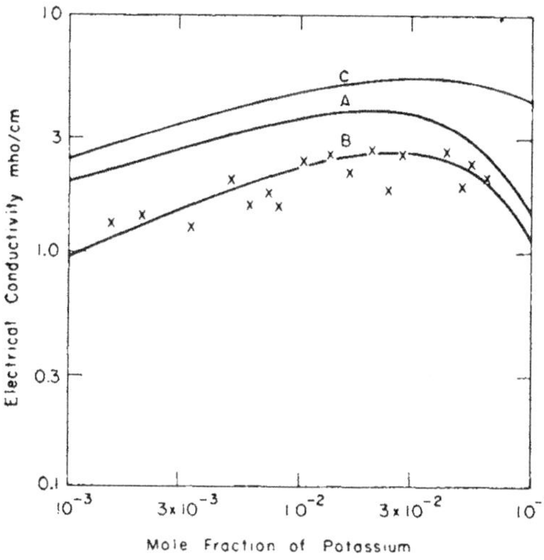


Fig. 1 Comparison between experimental and theoretical electrical conductivity as a function of mole fraction of potassium in the product gases. Curve A corresponds to experimental conditions with $Q_g = 10^{-15}$ cm$^2$ and $Q_K = 4 \times 10^{-14}$ cm$^2$. Curve B corresponds to experimental conditions with $Q_g = 2.5 \times 10^{-15}$ cm$^2$ and $Q_K = 4 \times 10^{-14}$ cm$^2$. Curve C is a constant temperature line corresponding to 4000°K. x denotes an experimental point.

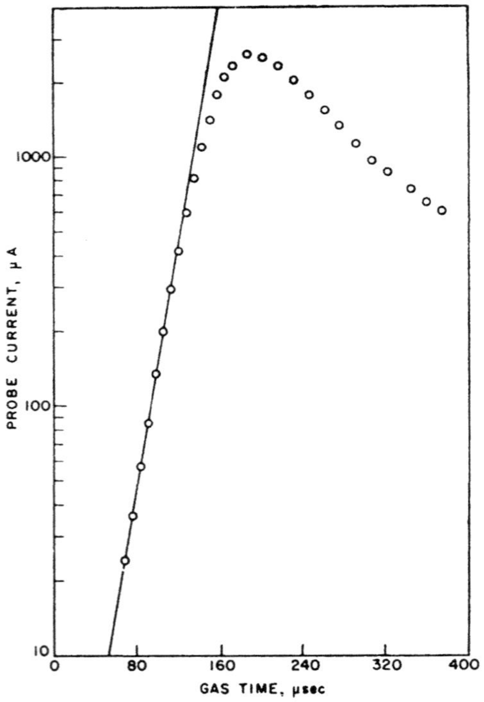


Fig. 2 Log *current* vs *time for* +5.00 v probe. Shock temperature 2575°K.

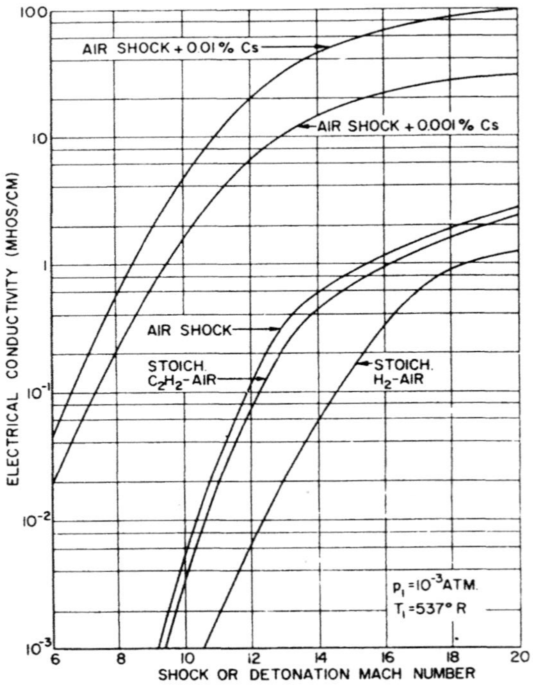


Fig. 3 Electrical conductivity vs shock or detonation Mach number.

# IV
# IONIZATION IN ROCKET EXHAUSTS

# CHEMISTRY OF IONIZATION IN ROCKET EXHAUSTS

Felix T. Smith* and Carole R. Gatz†

Stanford Research Institute, Menlo Park, Calif.

## Abstract

Theoretical and experimental studies contributing to the understanding of ionization in rocket exhausts are reviewed. Equilibrium thermal ionization of atoms and solid particles is readily calculable, at least in principle, and nonequilibrium effects in the rapidly expanding exhaust gases can be taken into account to some extent. Interactions between the exhaust gases and the ambient atmosphere which can produce electrons include chemionization, particularly in association with afterburning, ionization in collision by transfer of translational or electronic energy, and photoionization and photodetachment. Electrons are removed by recombination and attachment processes.

## Introduction

It has long been known that the flight of some rockets is accompanied by electromagnetic effects which interfere with communications and guidance by radio and radar. One of the commonest types of interference is due to free electrons in the neighborhood of the rocket which absorb, reflect, diffract, and scatter electromagnetic radiation. These effects may attenuate or distort communication and guidance signals. They may also influence detection and tracking of a missile by increasing its apparent cross section. The electrons causing these effects may originate and disappear in various ways. They may emerge as part of the hot gases of the rocket exhaust,[1] or they may be created by interactions of the rocket exhaust with the external environment.[2] They may be lost by attachment or recombination or may diffuse until their concentration is insignificant.

---

Presented at the ARS Ions in Flames and Rocket Exhausts Conference, Palm Springs, Calif., October 10-12, 1962. This work was sponsored by the Ballistic Systems Division, U.S. Air Force, under Contracts AF 04(647)-221, AF 04(647)-751, and AF 04(694)-128.

*Senior Physical Chemist, Chemical Physics Division.

†Physical Chemist, Chemical Physics Division.

Unfortunately we must admit at the beginning that the variety of possible mechanisms is so great and the chemical rate constants are so uncertain that it is impossible to make detailed and reliable calculations in most cases. However, it is often possible to make rough order-of magnitude estimates. These at least illustrate some of the possible effects that could be encountered in practice. Even these order-of-magnitude estimates should be of some value in pointing out situations in which communication and guidance problems may be expected. If these situations are known early enough, it is often possible to design equipment to overcome them. In this article we shall review the most important mechanisms of electron production and loss that seem to be well established at the present time. It must be kept in mind that the introduction of novel propellant materials, as well as changes in temperature and other parameters of the exhaust, may make still other mechanisms important in the future.

The ionization in the exhaust may be divided into two classes: the primary ionization, which is that of the exhaust gases themselves as they emerge, and the secondary ionization due to interaction between the rocket exhaust and the external environment. The principal factors in the primary ionization which we shall consider are the following: thermal ionization of atoms and molecules, thermal ionization of solid particles, and nonequilibrium effects due to rapid expansion of the exhaust in the nozzle. The principal sources of secondary ionization are chemical interactions between components of the exhaust and the atmosphere, principally afterburning; interactions in high-speed collisions and shock waves, which may be briefly referred to as impact ionization; and interaction with sunlight, which may produce photoionization and photodetachment of the electrons from negative ions.

## Primary Ionization

The thermal ionization of atoms and molecules at the temperature of the rocket exhaust is necessarily dominated by species such as sodium and potassium which have a low ionization potential. In most solid propellant materials these alkali metals are almost inevitably encountered in amounts of the order of a few parts per million, and these amounts are sufficient to cause a very important ionization problem. In the simplest case, this ionization can be calculated as a function of temperature and pressure by the use of the Saha equation. In practice, of course, the chemistry is more complex, and a number of competing equilibria may have to be considered at the same time. In particular, substances with a high electron affinity, such as chlorine, may reduce the over-all electron density by attaching electrons. Also, some substances may be important as ions, although they are insignificant as neutrals—for instance, clustered ions such as $Na^+ \cdot H_2O$. Provided that all the important species are taken into account, the equilibrium level of ionization can be computed by the usual methods of thermodynamics, and the calculations, although complicated, are of a fairly routine sort.

When solid particles are present, an additional source of ionization becomes possible, since at high temperatures various solids become thermionic emitters. This problem was first considered with regard to carbon particles in sooty flames, [3,4] but it is also important in rocket exhausts containing large amounts of alumina resulting from the combustion of aluminized propellants. Here again, the level of ionization is strongly temperature-dependent and depends also on the work function of the solid material. When this work function is low, solids may become an important source of electrons. A high particle concentration may reduce the total electron density in cases where electrons are also produced by alkali metals. If the work function of a solid is known, as well as the size and concentration of particles, it is possible to make a thermodynamic calculation in which the competing effects of the solids and of the ionizing atoms and molecules can all be taken into account. In practice, the state of subdivision of the solid is not well known, and, even worse, the work function of the solid may not be well known. This is unfortunately the case with alumina, for which the work function is, in fact, very inadequately known. There is an old value in the literature of 3.7 ev for the work function of alumina on platinum[5] ; this is extremely low and would make alumina a dominating source of electrons in some rocket exhausts. It is possible that this value is not reliable for the alumina to be found in rocket exhausts, which may be full of impurities--indeed, its effective work function cannot be considered to be known at present, and it may not be reproducible from one situation to another. It can at least be shown from the work of Pearson et al.[6] on ionization of alumina in acetylene-oxygen flames that the work function of alumina is much less than 6 ev.

In the actual rocket exhaust, further complications arise because the assumption of thermal equilibrium is a very poor one for the rapidly expanding gases in and beyond the nozzle. In the vicinity of the throat, where temperature and pressure are high, a shifting equilibrium is more or less maintained. As the expansion continues, the chemical reactions can no longer keep up with it, and ultimately a frozen composition is to be expected. As a result, the ionization at the exit plane tends to be much higher than equilibrium calculations would predict. There is no simple way of making a reliable prediction of the final level of ionization, because the calculation would require detailed knowledge of the mechanisms and of the rate constants for electron production and loss as functions of temperature, density, and composition. For the complicated mixtures that occur in real rocket exhausts, this information is simply unavailable, and we must content ourselves with the results of artificially simplified calculations which illustrate the type of thing which is to be expected.

The transition from shifting equilibrium to frozen flow for a single recombination reaction in a conical nozzle has been solved by numerical integration.[7] When only gaseous ionization is involved, the composition is generally frozen by the time the exit plane is reached, and this composition will remain unchanged during further expansion into the plume. If expansion beyond the throat is rapid, it is often assumed that equilibrium under throat conditions is suddenly frozen and that this composition is maintained unchanged. This is a useful first approximation even when it is not reliable in detail. In addition to conditions where equilibrium is attained or approached, a knowledge of equilibrium properties is often useful as a measure of the direction and ultimate goal of kinetic processes.

When solid particles are present, they may serve as very effective energy dissipators and would then provide an especially effective mechanism for the recombinations of electrons and ions. Insofar as these particles govern the level of ionization, there are also kinetic effects connected with the cooling of the particles by radiation and the rate of their emission and reabsorption of electrons. Even during the expansion inside the rocket nozzle, the temperature and velocity of solid particles will lag considerably behind that of the gas. As the gas continues to expand, the electron density decreases and the particles emit electrons more rapidly than they recombine with them. On the other hand, the particles also undergo radiative cooling, as a result of which the rate of electron emission falls, and the ultimate process will be recombination of particles and electrons.

An added complication is that the motion of the particles will not, in general, follow that of the gas. Emerging from the nozzle, they may form a core much smaller than the total exhaust diameter. Ultimately, they will follow diverging paths carrying them out much further than the gas plume boundary. When the particles are hot, the core is probably a region of high electron density, from which electrons may diffuse into the plume. Both the rate of radiative cooling and the electron emission of the particles will depend on the distribution of particles in the core. Much work needs to be done before we shall be able to predict the resulting electron density distribution. It is clear, however, that the electron density will be expected to vary strongly as a function of radial and longitudinal position in the plume.

The equations for calculating equilibrium ionization in the cases previously discussed are given below. Table 1 gives some sample calculations of electron densities in equilibrium with sodium and potassium atoms for conditions at the throat and at the exit plane of a rocket motor. Table 2 gives some samples involving the presence and absence of sodium atoms and of a solid with a work function of 3.7 or 4.5 ev. It will be seen that when temperatures are high and the work function is low the solid material dominates. In other conditions, when the particles are plentiful, there is a competition between them and the alkali metal which leads to an intermediate value for the resulting electron density.

## Thermal Ionization of Atoms and Molecules

Thermal ionization of atoms or molecules in a gas can be represented by the equilibrium:

$$A \rightleftarrows A^{+} + e \qquad (1)$$

The equilibrium constant $K_A$ is given by Saha's equation:

$$K_A = \frac{[A^{+}]n_e}{[A]} = \left(\frac{2\pi m_e}{h^2}\right)^{3/2} (kT)^{3/2} e^{-w/kT} \frac{g_{A^+}g_e}{g_A} \qquad (2)$$

where $n_e$ = electron density ($cm^{-3}$), $m_e$ = mass of the electron, $h$ = Planck's constant, $k$ = Boltzmann's constant, $T$ = absolute temperature, $w$ = ionization potential of $A$, and the $g$'s are ground state electron multiplicities.

In a mixture of several ionizing species, the equilibrium electron density is given by

$$n_e = (\sum_i K_i [A_i])^{1/2} \qquad (3)$$

Frequently the ionization is dominated by one or two species with low ionization potential, and other terms in Eq. (3) need not be included. A species may control the ionization even though it is present in amounts too small to be chemically important. A few parts per million of alkali metal, for example, are usually sufficient to swamp other atomic ionizers.

Other equilibria might also be important in determining $n_e$. Association reactions, such as

$$K + OH \rightleftharpoons KOH \tag{4}$$

can remove ionizing species from the gas. Electron attachment

$$Cl + e \rightleftharpoons Cl^- \tag{5}$$

can provide an additional sink for electrons. Equilibrium constants for such processes can be calculated by the methods of statistical mechanics. With the conditions of mass balance and charge balance, a complete set of equations is obtained which is in principle solvable for $n_e$.

Effect of Solid Particles

It is a well-known result of the theory of thermionic emission that the saturation electron density $n_s$ in equilibrium with a pure conducting solid at the temperature $T$ is

$$n_s = 2(2\pi m_e kT/h^2)^{3/2}\, e^{-v/kT} = 4.83 \times 10^{15} T^{3/2}\, e^{-v/kT} \tag{6}$$

where $v$ is the work function of the solid. This value is attained when enough particles are present compared with the number of electrons they are to supply. If the particles are too few, they must be multicharged, and the increased Coulomb force will displace the equilibrium in the direction of a lower concentration of free electrons.[3,4] Under these circumstances, the electron density becomes very sensitive to the state of dispersion of the particles. The dependence of $n_e$ on the average particle radius $r_p$ and on the concentration of particles $N_p$ enters through the quantity

$$\phi = 4\pi\epsilon_0 r_p e^{-2} N_p = 3\epsilon_0 e^{-2} \rho^{-1} r_p^{-2} M_p \tag{7}$$

where $\epsilon_0 = 8.854 \times 10^{-12}$ farad/m is the dielectric constant of space, $e$ is the electronic charge, $\rho$ is the density of the solid material composing the particles, and $M_p$ is the mass of solid material per unit volume in the gas. The term $n_e/\phi$ is a correction to the work function representing the additional work to remove an electron from the field due to the positive charge already on the particle. The electron concentration is then given by

$$n_e = 4.83 \times 10^{15} T^{3/2} \exp\left(-\frac{v + (n_e/\phi)}{kT}\right)$$

$$= n_s \exp\left(\frac{-n_e}{\phi kT}\right) \qquad (8)$$

When the solid is not a conductor but a semiconductor or an insulator, Eqs. (6) and (8) must be modified by inserting the appropriate effective electron mass $m^*$ in place of $m_e$.[8] $m^*$ is generally smaller than $m_e$, often by a factor of 10 or 100. For alumina, there is some evidence that has been interpreted to indicate that $m^*/m_e = 1/9$.[9]

If the particle size is unknown, Eq. (6) will give an upper bound for the electron concentration provided that the temperature and work function are known.

When electrons are contributed by both gaseous species and particles, in many cases one source of electrons will dominate and buffer out the other. There are transition cases which can be handled also. A general expression for such ionizing mixtures is

$$U\left(\frac{n_s}{n_e}\right) + W \ln\left(\frac{n_s}{n_e}\right) - V\left(\frac{n_e}{n_s}\right) = 0 \qquad (9)$$

In this equation, the effect of small positive ions enters through

$$U = (1/n_s^2)\Sigma k_i [A_i]$$

the effect of small negative ions through

$$V = 1 + \sum_j [(B_j]/L_j)$$

where

$$L_j = [B_j] n_e / [B_j^-]$$

and the effect of large particles through

$$W = \sum_k (a_k/2f_k n_s)$$

Here $a_k$ is the concentration of particles of the $k$th species, independent of charge, and

$$f_k = e^2/2c_k kT$$

where $e$ is the electronic charge and $c_k$ is the electrostatic capacitance of the isolated particle; for spherical particles at 2500°K,

$$f_k = 33.4/r_k(\mathring{A})$$

Various cases of Eq. (9) are shown in Fig. 1. At low particle concentrations (small $W$), the electron density is determined by atomic ion formation. As $W$ increases, $n_e$ passes through a transition region and approaches $n_s$.

Numerical Examples

We have made sample calculations of equilibrium atom ionization for a hypothetical rocket motor with assumed values of temperature, pressure, and average molecular weight, at the exit plane and at the throat, as given in Table I. A sodium contamination was assumed of 40 ppm by weight and a potassium contamination of 300 ppm—the average values found by photometric analyses of several batches of a solid propellant. Table 1 lists the calculated values of the electron density at exit plane and throat conditions for three cases, assuming that the equilibria controlling the electron density are

Case 1: $Na \rightleftharpoons Na^+ + e$

Case 2: $Na \rightleftharpoons Na^+ + e$

$K \rightleftharpoons K^+ + e$

Case 3: $Na \rightleftharpoons Na^+ + e$

$K \rightleftharpoons K^+ + e$

$KOH \rightleftharpoons K + OH$

Table 1 Equilibrium electron densities for gaseous ionization

| A) Experimental parameters | Exit plane | Throat |
|---|---|---|
| Temp., °K | 1300 | 3000 |
| Pressure, erg/cm$^2$ | $4 \times 10^5$ | $2 \times 10^7$ |
| Avg. mol. wt | 27 | 27 |
| *Na* content, ppm | 40 | 40 |
| *K* content, ppm | 300 | 300 |
| **B) Calculated electron density (cm$^{-3}$) at the exit plane** | **Equil. at exit conditions** | **Equil. frozen at throat** |
| Case 1 | $10^7$ | $2 \times 10^{12}$ |
| Case 2 | $10^9$ | $2 \times 10^{13}$ |
| Case 3 | $10^7$ | $1 \times 10^{13}$ |

For throat conditions, potassium rather than sodium is the dominant source of ionization, as would be expected from its lower ionization potential. At the lower temperature of the exit plane, however, formation of potassium hydroxide lowers the equilibrium concentration of potassium atoms to such an extent that sodium becomes the dominant ionizing species.

The actual electron density at the exit plane, which has been measured by microwave attenuation during test firings of real motors, would probably be about $10^{10}$ to $10^{11}$ electrons/cm$^3$ for this hypothetical motor. This concentration is bracketed by the equilibrium concentrations at the throat and exit plane and can be described, at least empirically, as an equilibrium density frozen at some intermediate stage of nozzle expansion.

In the above calculation the effect of solid particles on equilibrium ionization was not taken into account. As shown by Eq. (9), this effect depends in a complicated way on temperature; particle size, concentration, and work function; concentration and ionization potential of atomic ionizing species; and concentration of electron-attaching species.

Instead of trying to illustrate the role of each of these parameters by means of a large number of sample calculations, we limit our examples to a fixed concentration of sodium atoms, $10^{14}$ cm$^{-3}$, and a fixed concentration of particles, $10^7$ cm$^{-3}$, with the particles assumed to be spherical and of uniform size with radius = $3\mu$. Results at 1300° and 1800°K with a particle work function of 3.7 or 4.5 ev are given in Table 2.

Table 2 Calculated electron densities for ionizing particles and atoms

| Temperature, °K | $n_{Na}$, cm$^{-3}$ | $v$, ev | $n_s$, cm$^{-3}$ | $n_e$, cm$^{-3}$ $V = 1$ |
|---|---|---|---|---|
| 1300 | $2 \times 10^7$ | 3.7 | $1.6 \times 10^6$ | $2 \times 10^6$ |
| | | 4.5 | $7.5 \times 10^2$ | $7.5 \times 10^4$ |
| 1800 | $9.5 \times 10^9$ | 3.7 | $1.5 \times 10^{10}$ | $1.2 \times 10^{10}$ |
| | | 4.5 | $9 \times 10^7$ | $2 \times 10^9$ |

The equilibrium electron density at each temperature due to sodium ionization in the absence of particles, calculated from Eq. (3), is given in Table 2 as $n_{Na}$. The saturation electron density for particles in the absence of sodium, $n_s$, calculated from Eq. (6), is given for each value of work function and temperature. Finally, electron density, $n_e$, in equilibrium with sodium atoms and particles, was obtained from Fig. 1 for each set of conditions (with electron attachment assumed to be negligible).

Comparing the values of $n_{Na}$, $n_s$, and $n_e$, we concluded, at least for the specific conditions of this calculation, that the electron density is controlled by particle ionization if the particle work function is 3.7 ev.

If the work function of the particle is as high as 4.5 ev, the particles tend to buffer the electron density, reducing it slightly below that due to sodium ionization alone.

## Secondary Ionization

We shall first consider afterburning which involves combustion in the atmosphere of the incompletely burned materials in the rocket exhaust. These reactions usually involve oxygen atoms or molecules and are similar to ordinary combustion reactions.

It is well known that combustion of hydrocarbons with oxygen leads to high nonequilibrium electron concentrations in the combustion zone of laboratory flames. Similar electron production in the mixing zone between the rocket exhaust and the atmosphere may also lead to high electron densities in this region. It is probable that a number of chemical reactions may be involved in this sort of electron production. Calcote[10] has suggested a number of chemionization processes which may occur in hydrocarbon flames. Sugden and co-workers[11] have shown that the primary ionization processes in hydrocarbon flames is

$$CH + O \rightarrow CHO^+ + e \tag{10}$$

This reaction, first suggested by Calcote, is approximately thermoneutral and has a rate constant $\sim 3 \times 10^{-12} cm^3\text{-}sec^{-1}$. It is probably the most important chemionization reaction in rocket exhaust afterburning.

This result can be applied to an afterburning rocket exhaust by assuming a fractional concentration of $10^{-6}$ for reactive species such as CH, as may be the case for laboratory flames. At an altitude of 100 km, where the populations of $O_2$ and O are both $\sim 10^{12}$ $cm^{-3}$, reaction (10) can lead to electron production rates on the order of $10^6$ or $10^7/cm^3/sec$. The accuracy of such an estimate is at present limited by our knowledge of the populations of minor exhaust constituents such as CH and of the details of turbulent mixing of the exhaust with the atmosphere.

With respect to these afterburning reactions, there are two regions of flight which may have to be considered. At low altitudes, where the only oxygen species is molecular, it is possible that the rocket exhaust may not ignite when it encounters the air. The ionization and the luminosity are expected to be correlated, and it is sometimes possible to eliminate this form of ionization by controlling the ignition. On the other hand, at altitudes where atomic oxygen is abundant (its maximum concentration occurs at about 100 km), afterburning reactions that produce ionization may occur without any specific ignition event.

Electronically excited species produced in highly exothermic afterburning reactions may have enough energy to produce ionization when they collide with species of low ionization potential. The few parts per million of alkali metal commonly present in the exhaust gases are sufficient to account for electron densities significantly higher than background. Excited atoms and molecules present in the atmosphere may participate in such energy-transfer collisions and may also be involved in chemionization processes which are not energetically feasible for ground state species. Excited states of O, $O_2$, and $N_2$ are of possible importance, but neither their populations nor their reactions are well known.

Another important interaction which creates secondary ionization is the impact ionization that may occur at high elevations when even the rocket exhaust has a very large forward velocity with respect to the ambient atmosphere. Basically this interaction is the same as the collisional type of ionization which can be observed in shock waves. Comparatively little is yet known quantitatively about this form of ionization, but some information is now being obtained from shock wave studies[12] and from studies with fast molecular beams.[13]

A couple of generalizations can be made which seem to be supported by such information as is available. First, the rare gas atoms have much smaller cross sections for collisional ionization than substances whose electrons are fairly loosely bound and have large orbits. Johnston[12] has, in fact, found that the ionization of the rare gases in shock waves is predominantly a two-step or many-step process; he finds the first step to be the excitation of one atom to the first excited state (a process with a very low cross section), followed by a much more rapid collisional ionization of the excited species whose electron is in a much larger orbit. Experiments by Dukelsky and others[13] on the ionization of alkali metals have only been carried on at considerably higher energies, but they show that the cross section depends very strongly on the other partner in the collision. Again the rare gases are particularly inefficient, while substances which can capture the electrons (such as $Cl_2$ and $O_2$) are particularly effective (it may be noted that even $N_2$ forms a short-lived metastable $N_2^-$ which may participate in the collisional ionization). Since free radicals and excited species may similarly be expected to have comparatively high cross sections for ionization, this process may be expected to be fairly important in the interaction between the rocket exhaust and the atmosphere.

Some consideration has been given at various times to photoionization or photodetachment as a source of electrons in rocket exhaust in the daytime. In most cases, the cross sections appear to be too small for this to be a very important source of electrons. Nevertheless, exceptional cases in which this effect may occur should be kept in mind.

## Electron Loss Processes

Although diffusion and bulk flow affect electron distribution, the only processes that can reduce the total number of free electrons in the exhaust are recombination and attachment. These processes are important in the case of primary ionization—the rate of electron loss determines the extent to which shifting equilibrium can be maintained during nozzle expansion. In the exhaust plume, the electron density is determined by the rates of the completing electron production and loss processes.

For example, let us consider those conditions for which we previously estimated that reaction (10) might produce electrons at a rate $\sim 10^7/cm^3/sec$. In this case, the predominant loss process for electrons is dissociative recombination

$$CHO^+ + e \rightarrow CO + H \tag{11}$$

which occurs at a rate = $\alpha[CHO^+]n_e$. The recombination coefficient, $\alpha$ can be estimated as $10^{-7}$ $cm^3$-$sec^{-1}$, by analogy with other reactions of this type. If $[CHO^+] = n_e$, then in the steady state, when production rate equals loss rate, the electron density will be about $10^7$ $cm^{-3}$. The time scale for attaining this steady state is on the order of tenths of a second, corresponding to distances in the exhaust plume on the order of a kilometer.

If the predominant positive ions in the exhaust plume are monatomic ions, such as $Na^+$, dissociative recombination cannot occur, and electrons and ions probably recombine by radiative recombination (a typical rate constant is $10^{-12}$ $cm^3$-$sec^{-1}$):

$$M^+ + e \rightarrow M + h\nu \tag{12}$$

In this case, a production rate of $10^7$ $cm^{-3}$-$sec^{-1}$would lead to a steady-state electron density of $\sim 10^9$ $cm^{-3}$, but with a time scale on the order of several hundred seconds.

We can also estimate the effect of attachment on steady-state electron densities. Oxygen molecules can attach electrons by a three-body process:

$$2O_2 + e \rightarrow O_2^- + O_2 \tag{13}$$

with rate constant $10^{-30}$ $cm^3$-$sec^{-1}$. The maximum daytime rate constant for photodetachment is 0.4 $sec^{-1}$. The $O_2^-$ ions may undergo mutual neutralization with positive ions

$$O_2^- + M^+ \rightarrow O_2 + M \tag{14}$$

with a rate constant perhaps $10^{-8}$ $cm^3$-$sec^{-1}$. On the basis of these three processes, the steady-state concentration of $O_2^-$ is given by

$$[O_2^-] = \frac{10^{-30}[O_2]\, n_e}{0.4 + 10^{-8}[M^+]} \qquad (15)$$

Under most conditions of interest, the concentration ratio $[O_2^-]/n_e$ is less than unity, which means that attachment by $O_2$ is not directly important in reducing the electron density. Attachment may increase the effective rate of recombination of monatomic ions and electrons, since the rate of charge loss by reaction (14) may be one or two orders of magnitude larger than the rate predicted for reaction (12).

However, this effect is probably less important than these calculations predict, because several seconds are required to reach the steady-state concentration of $O_2^-$.

A similar calculation indicates that electron attachment by atomic oxygen is not important, except perhaps near 100 km, where the atomic oxygen concentration has its maximum value of about $10^{12}$ $cm^{-3}$.

One may also conclude that exhaust constituents, such as Cl, which can attach electrons, are probably present in amounts too small to affect the electron loss rate to any significant extent, unless they are able to attach electrons much more rapidly than can oxygen.

The rate constants used in the above calculations were based on reviews of experimental and theoretical work on recombination and attachment processes of importance in the ionosphere.[14]

## Conclusion

We have surveyed existing knowledge of electron production and loss processes and ways in which this information can be used to synthesize simple models for rocket ionization. The goal of these crude models is primarily a general description of the types of processes which dominate the ionization, secondarily a rough estimation of the ionization levels that can occur.

More information is needed, particularly with regard to mechanisms and rate constants for chemionization processes and the role of particles in charge production and recombination. In addition, further applications of existing knowledge to the practical case of the rocket exhaust is necessary. Improved descriptions of the mixing and flow of exhaust gases in the ambient atmosphere would be particularly valuable.

As more of this type of information becomes available, models can become more sophisticated, leading to more accurate and detailed predictions of rocket exhaust ionization and increased ability to control it.

References

[1]Gatz, C. R., Rosser, W. A., and Smith, F. T., "Study of radar beam attenuation in rocket exhaust gases. Part 2: The chemistry of ionization in rocket exhausts," Air Force Ballistic Systems Div. TR AFBMD-TR-61-39, Part 2 (February 1961), pp. 9-30.

[2]Smith, F. T. and Gatz, C. R., "Ionization in afterburning rocket exhausts," Air Force Ballistic Systems Div. TR BSD-TR-61-76 (November 1961), pp. 33-41, 65-73.

[3]Einbinder, H., "Generalized equations for the ionization of solid particles," J. Chem. Phys. **26**, 948-953 (1957).

[4]Smith, F. T., "On the ionization of solid particles," J. Chem. Phys. **28**, 746-747 (1958); also "Ionization and the structure of carbon particles at high temperatures," *Proceedings of 1957 Conference on Carbon* (Pergamon Press, London, 1957), pp. 419-424.

[5]*International Critical Tables* (McGraw-Hill Book Co. Inc., New York, 1929), Vol. VI, p. 54.

[6]Pearson, M. D., Larsen, O. W., Linder, W. J., and Bowlby, C. W.,"Electron density in aluminium oxide contaminated flames" Bull. Am. Phys. Soc. **4**, 362 (1959).

[7]Smith, F. T., "On the analysis of recombination reactions in an expanding gas stream," *Seventh Symposium on Combustion* (Butterworths Scientific Publications, London, 1958), pp. 93-97; also Rosner, D. E., "Estimation of electrical conductivity at rocket nozzle exit sections," ARS J. **32**, 1602-1605 (1962).

[8]Kittel, C., *Introduction to Solid State Physics* (John Wiley and Sons Inc., New York, 1956 ) 2nd ed., Chap. 13.

[9]Fisher, J. C., and Giaever, I., "Tunnelling through thin insulating layers," J. Appl. Phys. **32**, 172 (1961).

[10]Calcote, H. F., "Ions in flames," Air Force Ballistic Systems Div. TR AFBMD-TR-61-54 (December 1960), pp. 89.

[11]Bascombe, K. N., Green, J. A., and Sugden, T. M., *Joint Symposium on Mass Spectrometry, ASTM and Institute of Petroleum* (Pergamon Press, Oxford, 1962); also Green, J. A. and Sugden, T. M., "Some observations on the mechanism of ionization in flames containing hydrocarbons," *Ninth Symposium on Combustion* (Academic Press Inc., New York, 1963), pp. 607-621.

[12]Johnston, H. S. and Kornegay, W., "Rate and mechanism of the thermal ionization of xenon," Trans. Faraday Soc. **57**, 1563-1577 (1961).

[13]Utterback, N. G. and Miller, G. H., "Ionization of nitrogen molecules by nitrogen molecules," Phys. Rev. **124**, 1477-1481 (1961); also Utterback, N. G., "Ionization of nitrogen and oxygen molecules by nitrogen and oxygen molecules," Bull. Am. Phys. Soc. **7**,487 (1962); also Sluyters, T. J. M., de Haas, E., and Kistemaker, J., "Interactions between 5-24 kev ions or neutrals and gas molecules," *Proceedings of the Fourth International Conference on Ionization Phenomena in Gases* (North-Holland Publishing Co., Amsterdam, 1960), Vol. I, pp. 60-64; also Bydin, Y. F. and Bukhteev, A. M., "Ionization of fast Na, K, Rb, and Cs atoms in collisions with $H_2$, $D_2$, $N_2$ and $O_2$ molecules," Soviet Phys.-Tech. Phys. **5**, 512-519 (1960); also Bukhteev, A. M., Bydin, Y. F., and Dukelsky, V. M., "Electron capture by $O_2$ and $Cl_2$ molecules bombarded with fast alkali-metal atoms," Soviet Phys.-Tech. Phys. **6**, 496-499 (1961); also Dukelsky, V. M., Bydin, Y. F., and Bukhteev, A. M., "Ionization of high velocity alkali atoms by collisions with atoms of the inert gases," *Proceedings of the Fourth International Conference on Ionization Phenomena in Gases* (North-Holland Publishing Co., Amsterdam, 1960), Vol. I, pp. 65-68.

[14]Massey, H. S. W., "Recombination of gaseous ions," Advan. Phys. **1**, 395-426 (1952); also Bortner, M. H. and Baulknight, C. W., "Deionization kinetics," DASA 12-44, Contract No. AF 19 (604)-8820 (July 1961); also Nawrocki, P. J. and Papa, R., *Atmospheric Processes* (Prentice-Hall, Inc., Englewood Cliffs, N. J. 1963), Chap. 3.

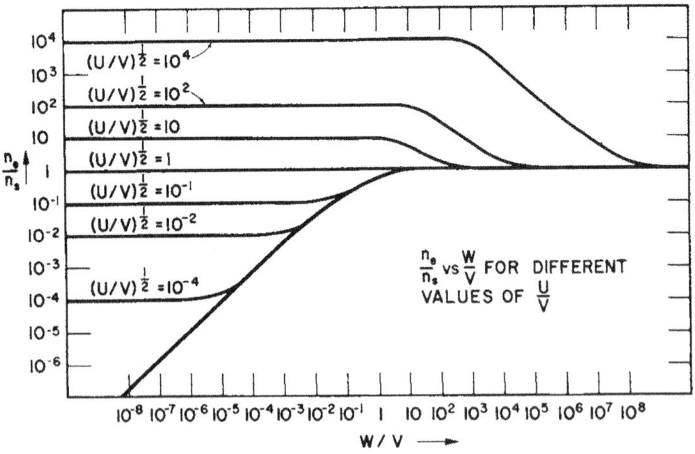


Fig. 1 Electron density in a mixture.

# V

# ELECTRON GENERATION BY SEEDING

# ELECTRICAL PROPERTIES OF SEEDED COMBUSTION GASES

Thomas R. Brogan*

Avco-Everett Research Laboratory,
Everett, Mass.

## Abstract

There are important cases where it is desirable and necessary that combustion gases be good conductors of electricity, and where steps are taken to enhance the conductivity. The MHD generator is an important example of a case where enhancement ot conductivity is required. The process of enhancing the conductivity is known as seeding, and consists of adding to the gas a relatively small amount of easily ionizable impurity called seed, usually an alkali salt. At typical flame temperatures the conductivity can be increased by roughly two orders of magnitude by seeding. The paper considers the various factors which influence the conductivity of seeded gases and describes methods for predicting and measuring it. It is shown that measurement and prediction are largely in agreement. The formation of heavy ions and reduction in conductivity due to the presence of electronegative species is discussed. Finally, because seeding is of importance for magnetohydrodynamic applications, Hall effects in seeded gases are discussed briefly.

## Introduction

In contrast with some of the problems where the fact that combustion gases conduct electricity is often bothersome to various degrees, there are important applications where conductivity is desirable and necessary and where steps are taken to enhance it. The MHD generator is an important example of a case where enhancement of conductivity is required in order to achieve important and desirable effects. Power outputs as high as 1.5 MW have been achieved in MHD generators driven by seeded combustion gases.[13] Another application requiring conductivity enhancement is the use of MHD to effect deflections for the purpose of attitude control of rockets.

---

§Submitted for publication to AIAA June 14, 1963.
*Principal Research Scientist.

The process of enhancing the conductivity of gases is called seeding[1] and consists of adding to the gas a relatively small amount of easily ionizable impurity called seed. In practice, the seed is usually one of the alkali metals that have low ionization potentials; these, when added to the gas, can produce sufficient conductivity to cause appreciable MHD effects to occur at temperatures between 2000° and 3000°K, depending on the size and geometry of the device, the conditions under which it is operated, and the magnetic field applied. If the gas is combustion products, the seed usually is added in the form of a potassium or cesium salt that is vaporized, decomposed, and ionized in the flame. For reasons to be discussed, the halides cannot be used. At typical flame temperatures, the electrical conductivity of flame gases can be increased by roughly two orders of magnitude using the seeding technique.

Explanation of the conductivity of unseeded gases is complicated by the fact that the ion often cannot be identified. This difficulty is not present in the seeded case where very nearly all the electrons are due to ionization of alkali, whose ionization potential is well known. The electron density can be predicted using the Saha equation, and various secondary effects such as electron attachment to heavy particles, i. e., OH radicals, and partial dissociation of the seed can be taken into account. Using the computed electron densities together with values for cross sections for the various flame species, various investigators have succeeded in explaining the observed conductivity. These studies are reviewed briefly in this paper.

The presence of a halogen is a serious impediment to the success of the seeding process because of the high electron affinity of these elements in their atomic state. Here again, experimental work and computations are largely in agreement. The effect of attachment to chlorine on a typical MHD device is discussed.

Gases, like solids, exhibit Hall effects but generally to a much greater degree. The importance of the Hall effect is determined by the geometry, magnetic field, gas density, and the details of the electron-heavy particle collision processes of the gas. These Hall effects are particularly important for the case of seeded gases that are used in MHD devices. The Hall effect is discussed briefly.

It will be shown that the range of temperatures required to produce significant values of electrical conductivity in seeded combustion gases is above 2000°K and is thus most typical of temperatures common to rocket exhaust products.

## Simplified Theory of Electrical Conductivity of Ionized Gases

Ionized gases at the temperatures of interest here have been dealt with ever since the advent of the use of electricity. The common carbon arc lamp, gaseous rectifier tubes, fluorescent lights, and so on all employ conducting gases. The theory of electrical conductivity of seeded gases will be considered here briefly, but only to the extent necessary to illustrate the parameters that are important. Ample detailed treatments of the theory of electrical conductivity in gases are contained in the literature; a good review of these is given by Lin.[2,3]

It first should be noted that in all the calculations it is assumed that the gas is thermally ionized, that is to say that the degree of ionization is in thermodynamic equilibrium with the temperature of the molecules and atoms. The normal source of heat, such as the combustion, produces the ionization. This is to be contrasted with the situation in, say, a fluorescent lamp, where the electrical discharge itself is responsible for the ionization. Such methods of nonequilibrium ionization may have some MHD applications. It does not seem likely that these applications will include combustion gases. It is important to note that in the case of thermal ionization the gas is a conductor independent of the electrical discharge being passed through it, as contrasted with the nonequilibrium case of, for instance, the fluorescent light, where the presence of the discharge is required if the gas is to conduct.

The conductivity of an ionized gas is caused by a drift of electrons under the influence of an applied electric field. If an electron of charge e is subjected to an electric field E, it is given an acceleration, a, of

$$a = \frac{e\,E}{m_e}$$

where $m_e$ is the mass of an electron. Now this acceleration persists until the electron undergoes a collision with a heavy particle. Assuming no persistence of velocity, the mean collision time $\tau$ is equal to the mean free path for electrons in the gas divided by the electron thermal velocity. So, in a collision time and under the acceleration described in the foregoing, the electron gains an average drift velocity $v_e$ in the direction of the electric field which is given by

$$v_e = \frac{e\,E\,\tau}{m_e\,2}$$

Now, if the electron density is $n_e$, the current density j is given by

$$j = n_e \, e \, v_e$$

The coefficient of E is defined as the electrical conductivity $\sigma$, so that

$$\sigma = \frac{n_e \, e^2 \, \tau}{2 \, m_e}$$

The mean free time $\tau$ may be expressed in terms of an average over the number density of each species times the appropriate electron collision cross section $Q_{ie}$ which an atom or molecule of the $i^{th}$ species presents to the electron of velocity c:

$$\tau = \frac{1}{n_o} \int_o^x \frac{f(c)\, dc}{\Sigma \frac{n_e}{n_o} \, C \, Q_{ie}}$$

where f (c) dc is the electron thermal velocity distribution function, $n_o$ is the total particle density of atoms, molecules, and ions, and $n_e/n_o$ the mole fraction of the $i^{th}$ species.

In practice, the coefficient in the foregoing equation differs somewhat from one half when the electron-heavy particle collisions are considered in more exact detail. However, the foregoing expression serves to illustrate the important parameters. Aside from physical constants, the conductivity may be expressed as

$$\sigma \sim \frac{n_e}{\overline{Q_e}}$$

where $\overline{Q_e}$ is some appropriate average cross section.

It first is noted that the conductivity is directly proportional to electron density (although this may not be true if the gas is at a slightly higher temperature than is being considered), and second that the conductivity is inversely proportional to an appropriately averaged electron-heavy particle collision cross section.

Let the latter parameter be considered first. The collision cross section is a function both of temperature and of the particular gas being used. Variations of a factor of some 10,000 in its value can be expected between a gas like argon and an alkali metal vapor. In addition, there are large variations with temperature for the individual species.

Combustion products of primary concern here have rather large cross sections, as gases go. This is caused primarily by the very strong effect of water vapor. Although these large cross sections, of course, tend to reduce the conductivity of combustion products, it may be expected that impurities in combustion products will have little effect on the conductivity.

Now consider the electron density $n_e$. If a number of seed atoms $n_o$ per unit volume are added to a gas, then the number of electrons and ions produced at a given temperature may be expressed as

$$\frac{n_e\, n_i}{(n_o - n_e)} = K\,(T)$$

where $n_i$ is the ion density, $n_e$ the electron density, and K (T) the equilibrium constant. Now the gas as a whole is electrically neutral, so that $n_e = n_i$. The value of K (T) is found from statistical mechanics, and finally the electron density (for $n_e/n_o$ small) may be written approximately as

$$n_e^2 = A\, n_o\, T^{3/2}\, e^{-\frac{T_i}{T}}$$

for small degrees of ionization. A is a constant, and $T_i$ is an ionization temperature proportional to the ionization potential of the seed. For potassium $T_i$ has the value of about 50,000°K, whereas T, the gas temperature, is about 3000°K. The exponential term is extremely sensitive to the temperature, so that the conductivity is a very strong function of gas temperature in the range being considered. The conductivity is proportional only to the square root $(n_o)^{1/2}$ of the seed density.

At this laboratory, conductivity experiments have been carried out using the combustion products of JP4 (a mixture of kerosene and gasoline) and oxygen seeded with potassium.[4,5] This mixture gives rise to combustion products not unlike those of a medium fuel oil and was used because it is simple to handle, and its combustion properties are well known. To illustrate how the theoretical methods can be applied, the expected conductivity will be calculated for this mixture.

Figure 1 is a plot of the equilibrium constant for the ionization of potassium vs temperature. Potassium has an ionization potential of 4.32 ev, that is to say, the energy to strip an electron from a potassium atom is equivalent to the energy the atom itself would have if heated to about 35,000°K.

Fortunately, however, it is not necessary to heat the potassium to this temperature in order to get sufficient ionization so that the combustion products will be good conductors at the flame temperature. As a general rule, a gas must be heated to about one-tenth the ionization temperature before appreciable ionization occurs, so that, in the case of potassium, temperatures above 2000°K will produce ionization in sufficient degree to induce appreciable conductivity.

Since the conductivity is proportional to the electron density, the importance of a high combustion temperature is evident from the extremely rapid variation of the equilibrium constant with temperature.

There are some possible sources of error in the calculation of the equilibrium constant. First, very slight variations of the ionization potential can make substantial changes in the theoretical electron density. The calculation of the equilibrium constant involves a specification of the state of the potassium ion after ionization, and here, too, there is some uncertainty. Also, the calculation assumed that the electron is either in the ground state in the potassium atom or free, and no account is taken of the fact that the various other states of the atom may be populated. These considerations add up to something like a possible error of a factor of 2 in the electron density.

The composition of the combustion products vs temperature and pressure also can be calculated using the methods of statistical mechanics. Figure 2 shows the combustion of the JP4-oxygen flame at stoichiometric proportions at 1 atm pressure. At low temperatures the reaction has gone nearly to completion, and the products are merely water vapor and carbon dioxide. At higher temperatures, dissociation sets in, and these final products are broken down into molecules and atoms. For the calculation of the theoretical conductivity, the cross sections of all of these species must be taken into consideration in the correct ratio. Figure 3 is a plot of the cross sections for these various species vs temperature. Note that the cross section of the water vapor is large and moreover increases rapidly as the temperature is decreased. In the case of $CO_2$, the increase in $Q_{ei}$ as T decreases is because the electron, as it nears the molecule, induces a dipole moment that exerts long-range forces on the electron. For the water vapor, the effect is even more pronounced. Water vapor is an asymmetrical molecule and possesses a permanent dipole moment.

Figure 4 shows the contribution of each of the species to the total cross section along with the total cross section. The water vapor cross section is dominant for all of the temperatures considered. Also included here is the cross section due to the potassium atoms and ion. Although the mole percent of these species is small, their cross sections are very large. In the case of the atom, this is because the outer shell electron is relatively far from the nucleus. The ion has an electrical charge and appears extremely large to the electron and is coupled to the electron by long-range Coulumb forces. The total cross section also is shown in the figure. At temperatures higher than have been considered, where the potassium is ionized further, the conductivity will not rise so rapidly, since for each new electron produced an ion of extremely large cross section is created and the ratio of electron density to cross section increases very slowly. This effect is only beginning to appear at the highest temperature in Fig. 4. The same is true of increased seed concentration. Because the alkali metal seed atoms have a very large cross section compared even with $H_2O$ molecules, the total cross section $Q_e$ increases as seed concentration is increased so that the conductivity actually rises somewhat slower than the square root of the seed concentration and, in fact, exhibits a maximum when the mole fraction of seeding is equal to $Q_{ea}/Q_{es}$, where $Q_{ea}$ and $Q_{es}$ are the averaged cross sections for gas and seed atoms. For combustion gases, typical values of optimum seed concentration are between 2 and 10%. It should be noted also that increased seed addition depresses combustion temperature and thus further reduces $\sigma$. Figure 5 is a plot of the calculated electrical conductivity of the products vs temperature and includes the combined effect of electron density and cross section. Shown on the figure are the experimental points from measurements to be described shortly. It is believed that the close agreement between experiment and theory is partially fortuitous.

Using the theoretical methods outlined in the foregoing, the conductivity of simple gases can be predicted accurately. For combustion products, however, the complex nature of the gas and the inherent errors in the calculation of electron density and cross section make it difficult to predict the conductivity to better than a factor of 2. For accurate results, experimental work is needed.

A further effect has not been discussed as yet. Some of the products of combustion, such as oxygen atoms and hydroxyl radicals, have an affinity for electrons, that is, they tend to attach the electrons produced by ionization to form a negative ion, thus reducing the density of electrons. The chlorine atom is a particular offender in this regard, and this is important because potassium chloride is the cheapest potassium salt. Preliminary calculations and experiments have been made on the effect of electron attachment, and it has been found that the presence of the chlorine atom is damaging, whereas the other electronegative species present in the gas have little effect on the conductivity. These effects are discussed in more detail later.

## Experimental Determination of the Conductivity of Seeded Combustion Gases

Several techniques are available for determining the electrical conductivity of gases in the range of interest for MHD work; these are described briefly.

a. Distortion of Magnetic Field by Conducting Gas in Shock or Detonation Tube

This technique is due originally to Lin,[2] who used it to study the electrical properties of argon produced in a shock tube.[9] The experiment is shown schematically in Fig. 6. A shock wave is driven through a round glass tube, and the coil around the tube produces a magnetic field with a radial component. As the shock wave and the hot conducting gas behind the wave enter the magnetic field, currents are induced in the gas as shown. A small pickup coil is used to detect the change in magnetic field due to currents in the gas. The apparatus can be calibrated by shooting a metal slug of known electrical conductivity through the field at known velocity.[2] The technique is extremely sensitive, and the experiment has been carried out over a conductivity range of $10 \times 10^6$.[2,10]

There are some difficulties in using this technique for the study of seeded conducting gases, however. These difficulties arise from the problems inherent in suspending the seed uniformly in the shock or detonation tube prior to firing. Basu[11] has suspended potassium acetylide in a vertical $H_2$-$O_2$ detonation tube and measured the electrical conductivity behind the detonation wave. Some of his results are shown in Fig. 7. The results are, on the whole, in good agreement with predictions.

b. The Q Coil[12]

This is a steady-state technique and is more adaptable for

use with seeded flame gases. It involves a geometry in which the conducting gas is closely coupled to a resonant RF coil. The enclosed fields in the gas give rise to dissipation that may be detected either as a change in the Q of the resonant circuit or as a shift in resonant frequency. A typical geometry is shown in Fig. 8.[12] Here the seeded gases are passed down a duct that has a coil embedded in one wall. Success and accuracy with the technique usually is dependent to a large degree on careful attention to limitation of extraneous effects such as those caused by heating of the coil by the hot gas. The experiment can be calibrated by coupling the coil to a water salt solution of known conductivity.

c. Voltage-Current Characteristics [12]

In this most straightforward method, the conductivity is determined from the voltage-current characteristics of a discharge (either a.c. or d.c.) through the conducting gas, together with the geometry of the particular experiment. Although the experiment is simple in principle, good accuracy demands a thorough understanding of the geometry used, in particular, knowledge of the mode of operation of the electrodes. The determined conductivities can vary by a factor of up to 3, depending on whether electrode current density is uniform as with thermionic emission or concentrated as in an arc with cold metal electrodes.

The measurement of conductivity from the V-I characteristic has developed generally along two more or less distinct typical experiments that may be described as the high-power or channel flow experiment, and the low-power or cell experiment. Both have received the attention of several investigators.

High-power or channel flow experiment[5, 12, 13]

Here, a high-speed flow of the seeded combustion gases is passed down a channel to which potentials are applied; apparatus typical of such experiments[4, 5, 17] is shown schematically in Fig. 9 and a photograph in Fig. 10. A typical V-I characteristic is shown in Fig. 11. The voltage intercept at zero current is an indication of the electrode drop; in this case, cold graphite electrodes were in use, and a high drop was experienced. The use of cold electrodes, of course, makes it necessary to apply a correction in order to infer the conductivity from the apparent resistance between electrodes as determined from the voltage-current characteristics. These corrections take into account the effect of current concentration on spots in the cold electrodes and may amount to as much as a factor of 3. Such effects can be eliminated largely by the use of probes in the gas.

The high-power experiment has the advantage that the high energy in the flow precludes the possibility that the discharge can change the electrical properties of the gas. Moreover, the geometry resembles that of an MHD generator in which such gases have a prime application, and, in fact, such measurements commonly can be carried out in an experimental generator by applying potentials with the field off.[13] The experiment generally is not sensitive to variation in gas flow or other parameters.

The experiment has the disadvantage of requiring a sizable piece of apparatus which quite often is rather noisy. Typical results from this experiment are shown in Fig. 5.

Low-power or cell experiment[6,7]

In this experiment (Fig. 12) a seeded combustion gas at low-speed flow (nearly quiescent by comparison with the high-power experiment) is passed through a cell containing concentric electrodes that are well insulated thermally and so, after a time, become heated nearly to the temperature of the gas being studied. Ideally, the electrodes are at a temperature equal to that of the gas so that thermoelectric effects are eliminated. Also, at such temperatures, the current density can be supplied at the cathode by thermionic emission so that the electrode drop is zero and a linear V-I characteristic is obtained. Typical results from such an experiment are shown in Fig. 13 for propane flows seeded with cesium and potassium salts.[7]

The experiments fit a theoretical prediction based on an average electron collision cross section of $1 \times 10^{-15}$ cm$^2$. The data of Fig. 3 and 4 for similar gases would indicate that cross sections of several times this amount would be more appropriate, and the good agreement here between these measurements and the predictions using the low cross section is not well understood by the author. It well may be that the extraneous effects to which this experiment is quite sensitive (thermoelectric, sheath drop, etc.) may contribute to the discrepancy. More recently, similar V-I experiments have been carried out using a Meker-type burner as the gas source.[14]

In summary, several techniques are available for measuring the conductivity of seeded combustion gases. The experimental results are in reasonably close agreement with predicted values.

## Electronegatives

The cases described in the foregoing have considered the use of potassium carbonate (or sometimes potassium hydroxide) as the seed material. With such materials, the experimentally determined conductivity has been found to agree quite well with the expected value.

In computing the expected conductivity, attachment of electrons to electronegative species in the gas has been ignored. Some of these species are present in combustion products. When electrons become attached to electronegatives to form a heavy negative ion, the gas conductivity is reduced since the ion mobility is very low compared to that of the electron. The combustion products of JP4-$O_2$ contain electronegative substances, the strongest of which are the oxygen atom and hydroxyl (OH) radical, both of which have an electron affinity of about 2.2 ev. Calculations show that attachment to these species introduce errors of the order of 10% in the computed conductivity, an error within the experimental and/or theoretical accuracy.

Potassium chloride (KCl) has been considered as a seed material because of its low cost. But chlorine atoms are strongly electronegative with an electron affinity of 3.8 ev. At the temperatures of interest, most of the chlorine liberated in the dissociation of KCl will be in atomic form, so that it is important in this case to determine the effect of the chlorine on conductivity.

Figure 14 shows the result of electron attachment calculations for a typical fuel oil-oxygen-nitrogen mixture seeded with 0.4% mole fraction K.[4] The ordinate $\alpha'$ is the ratio of the electron density to the value that would be obtained if there were no attachment, and the abscissa is the gas temperature. The calculations are carried out for several pressures as indicated and for two cases of attachment. In the first (upper curves), attachment to the oxygen atoms and OH present in the combustion is considered, whereas for the second set of calculations 0.4% mole fraction of Cl atoms is added to the gas. (It is assumed that KCl is completely dissociated, and this is not entirely correct; partial dissociation of KCl would make the effect more striking.)

The dashed lines indicate the path of the gas state (pressure vs temperature) in a typical MHD generator. With no chlorine present, $\sigma$ is reduced to some 75% of the ideal value at the generator inlet, whereas at the exit the figure is 90%. With chlorine present, however, the electron attachment is much more pronounced, $\sigma$ (proportional to $\alpha'$) being reduced to 25% of the ideal value at the inlet and 15% of the ideal value at the exit.

Using the equipment and technique described previously, the conductivity of the JP-4-$O_2$ flame at 1 atm pressure seeded with 1% mole fraction KCl has been measured, and the results are shown in Fig. 15 compared with the normal $K_2CO_3$ data.[4] As before, $\sigma$ is proportional to the inverse of the slope of the curves. From these data, $\sigma$ using KCl is a factor of 6 lower than $\sigma$ using $K_2CO_3$. Although the gas used in the experiment is somewhat different from the gas used for the calculations displayed in Fig. 14, the experimental results are in essential agreement with the theory. The differences are probably due to partial dissociation of KCl and a different partial pressure of KCl in the calculations and the experiment.

The high electron affinity of chlorine may mean that a combination of another potassium seed combined with an effective seed-recovery system will be necessary for an economically practical commercial MHD generator. Alternatively, other impurities present in fuel may combine with Cl atoms and remove them from the flame. For such a reaction to occur, in preference to the formation of negative ions, the heat of reaction per chlorine atom would need to exceed 3.8 ev.

Similar experiments to those described above have been carried out by other workers.[6,7] Other experiments have been carried out at lower electron densities.[8] These experiments confirm the deleterious effect of halogen on conductivity of seeded gases.

## Hall Effect and Ion Slip

In the foregoing, the electrical properties of seeded combustion gases in an applied electric field have been of concern, and for this purpose, the conductivity $\sigma$ is the only property of interest. However, a prime use of seeding is to enhance the conductivity of working fluids for MHD devices, particularly power generators. The presence of the magnetic field drastically alters the electrical characteristics of the gas due to Hall effect and ion slip. Although a thorough discussion of these effects is outside the scope of this paper, it is necessary to describe these phenomena briefly, since they are of profound importance for the MHD application of the seeded gases.

These effects are important, first because of the desire to minimize the length (and length/diameter ratio) of an MHD device in order that the influence of wall effects (friction and heat transfer) may be kept within bound, and second because of the desire to exert significant MHD forces to as low a temperature and conductivity as is possible. The relative magnitude of the MHD forces to inertial forces may be inferred from the value of the so-called interaction parameter $s = \sigma B^2 \ell / \rho U$, where B is the field strength, $\ell$ the length, $\rho$ the gas density, and U the typical gas velocity. For significant MHD effect, s should be of order unity or greater. Thus, the magnitude of the MHD effect is proportional to the square of the field strength divided by the gas density.

Since the ionizing species, its concentration, and its properties are known quantities, it is possible to predict the conductivity with good accuracy, and experiment and theory are in substantial agreement. Several techniques are available for measuring the electrical properties of these gases.

The importance of seeded combustion gases in the operation of MHD devices makes it very important to understand how the electrical properties are affected by a magnetic field. It is found that significant Hall effects and ion slip appear in practical situations and must be dealt with effectively if MHD devices are to operate efficiently. The presence of nonuniformities in the gas is particularly damaging when operating MHD devices at significant values of Hall coefficient.

Consider the interaction shown in Fig. 16. The gas velocity U is directed along the x axis, whereas the magnetic field B is directed along the y axis; an induced EMF of magnitude UB thus is produced along the positive z axis in which there is measured an electric field E in the laboratory system. In the absence of Hall effect, Ohm's law would read

$$j = \sigma (E + UB) = \sigma UB (1 - \alpha) \tag{1}$$

where $\alpha = - E/UB$ is a "loading factor" such that $0 \leq \alpha \leq 1$ if power is being extracted from the gas (generator) and $\alpha > 1$ if power is being added to the gas. As has been shown,

$j = n_e ev_e$, where $n_e$ is the electron (and ion density), e the electron charge, and $v_e$ the electron drift velocity. It has been assumed that the ion drift velocity is negligible; it also will be assumed that there is no current in the x direction. Then, because of the drift of the electrons across field lines, there will be an electric field $E_{Ho}$ (the Hall field) induced in the x direction given by

$$E_{Ho} = v_e B = \frac{j}{n_e e} B = \frac{\sigma UB^2 (1 - \alpha)}{n_e e}$$

$$= \left[\left(\frac{e B}{m_e}\right) T_e\right] (UB)(1 - \alpha) = (\omega_e \tau_e)(UB)(1 - \alpha)$$

The quantity $eB/m_e$ is the cyclotron frequency $\omega_e$ of the electron in the applied field B. The product of $\omega_e \tau_e$ is called the Hall coefficient and is a measure of the importance of the Hall effect. It measures the number of times an electron orbits in the field between collisions with heavy gas particles. When $\omega_e \tau_e$ becomes of order unity, it may be assumed that appreciable Hall effects are present.

It will be noted that $\omega_e \tau_e \sim B/\rho$. Thus, the desirable situation for producing an MHD device of minimum length leads directly to a high value of the Hall coefficient.

At high values of field strength, the heavy ions will begin to orbit between collision with other heavy particles. This corresponds to a drift of the heavy ions relative to the gas and represents a fundamental limitation on the maximum magnitude of the MHD effect locally.

The case has been examined where no current flows due to the Hall field. The completely general case has been studied previously,[15] with the result that the vector Ohm's law should be modified to read

$$\vec{j} = \sigma \vec{E} - \frac{\omega_e \tau_e}{B} \vec{j} \times \vec{B} + \frac{\omega_e \tau_e \, \omega_i \tau_i}{B^2} (\vec{j} \times \vec{B}) \times \vec{B} \qquad (2)$$

where $\vec{E}$ is the sum of all electric fields and motional emf's measured in the coordinate system of the observer (or, alternatively, the electric field measured in the coordinate system of the gas), $\sigma$ is the scalar conductivity of the gas, B is the magnetic field strength, $\omega_e$ and $\omega_i$ are the electron and ion cyclotron frequencies, and $\tau_e$ and $\tau_i$ are the electron and ion mean free times. The second term on the right gives rise to the Hall effect, and the third term introduces ion slip.

Assume the coordinate system of Fig. 13 with B along the y axis and consider currents and electric fields in the x, z plane. The components of the vector Ohm's law perpendicular and parallel to the planes of nonuniformity are found to be[15]

$$j_x = \Sigma E_x + \Omega j_z \tag{3a}$$

$$j_z = \Sigma E_z - \Omega j_x \tag{3b}$$

where two new symbols have been introduced:

$$\Sigma = \frac{\sigma}{1 + \omega_i \tau_i \omega_e \tau_e} \tag{4a}$$

$$\Omega = \frac{\omega_e \tau_e}{1 + \omega_i \tau_i \omega_e \tau_e} \tag{4b}$$

It can be seen that $\Sigma \approx \sigma$ and $\Omega \approx \omega_e \tau_e$ until ion mobility effects appear, making the third term in Eq. (2) significant. This does not in general happen until $\omega_e \tau_e$ has exceeded a value of 10 or more, since

$$\frac{\omega_e \tau_e}{\omega_i \tau_i} \approx \sqrt{\frac{m_i}{m_e}} \approx 100 \text{ to } 1000$$

where $m_e$ and $m_i$ are the electron and ion masses.

For efficient operation of the MHD devices, it is most desirable that no current be allowed to flow in the x (flow direction) due to the Hall field. The effect of Hall current ($j_x$) flow may be seen by putting $Ex = 0$, in which case

$$j_z = \frac{\sigma UB (1 - a)}{1 + (\omega_e \tau_e)^2}$$

where ion slip has been neglected. Thus, the result is effectively a reduction of the conductivity by the factor $1 + (\omega_e \tau_e)^2$. To prevent a short circuit of the Hall potential, it is necessary to segment the device in the axial direction; this has been accomplished successfully in seeded combustion gases for values of $\omega_e \tau_e$ up to 3.[13]

Further complications and deterioration of performance[16] of MHD devices occur if the gas is nonuniform for any reason. These effects are illustrated in Fig. 17, where both Hall effect and ion slip have been taken into account for various values of $\omega_e \tau_e / \omega_i \tau_i$. The case computed is for layers of conductivity $\sigma_1$ and $\sigma_2$ in the x-y plane of Fig. 16, and the serious reduction in performance when nonuniformities are present or when ion slip becomes appreciable is evident.

## References


[1] Rosa, R. J., "Shock wave spectroscopy (Part I), Engineering magnetohydrodynamics (Part II)," PhD Thesis, Cornell Univ.

[2] Lin, S. C., Resler, E. L., and Kantrowitz, A. R., "Electrical conductivity of highly ionized argon produced by shock waves," J. Appl. Phys. 26, 96-109 (1955).

[3] Lin, S. C., "Electrical conductivity of an ionized gas produced by strong shock waves," PhD Thesis, Cornell Univ. (1952).

[4] Brogan, T. R., Kantrowitz, A. R., Rosa, R. J., and Stekly, Z. J. J., "Progress in MHD power generation," Avco-Everett Research Lab. AMP 55 (March 1961); also Manuel, C. and Mather, N., Engineering Aspects of Magnetohydrodynamics (Columbia University Press, New York, 1962).

[5] Brogan, T. R., Rosa, R. J., Louis, J. F., and Stekly, Z. J. J., "A review of recent MHD generator work at the Avco-Everett Research Laboratory," Avco-Everett Research Lab. AMP 74 (March 1962); also presented at the 3rd Symposium on Engineering Aspects of Magnetohydrodynamics, Rochester, N. Y. (March 1962).

[6] Mullaney, G. J., Kydd, P. H., and Dibelius, N. R., J. Appl. Phys. 32, 668 (1961).

[7] Dibelius, N. R., Luebke, E. A., and Mullaney, G. J., Proceedings of the 2nd Symposium on the Engineering Aspects of Magnetohydrodynamics, March 1961 (Columbia University Press, New York, 1962), pp. 307-326.

[8] Sugden, T. M., "Microwave studies in the ionization of alkali metals in flame gases and related phenomena," Fifth Symposium (International) on Combustion (Reinhold Publishing Corp., New York, 1955), p. 406.

[9] Resler, E. L., Lin, S. C., and Kantrowitz, A. R., "The production of high temperature gases in shock tubes," J. Appl. Phys. 23 (1952).

[10] Lamb, L. and Lin, S. C., "Electrical conductivity of thermally ionized air produced in a shock tube," Avco-Everett Research Lab. Research Rept. 5 (February 1957); also J. Appl. Phys. 28 (1957).

[11] Basu, S., Phys. Fluids 3, 456-463 (1960).

[12] Rosa, R. J., "Physical principles of magnetohydrodynamic power generation," Avco-Everett Research Lab. Research Rept. 69 (January 1960); also Phys. Fluids 4, (1961).

[13] Louis, J. F., Lothrop, J., and Brogan, T. R., "Studies of fluid mechanics using a large, combustion driven MHD generator," Avco-Everett Research Lab. Research Rept. 145 (March 1963).

[14] Lapp, M. and Rich, J. A., "Electrical conductivities of seeded flame plasmas in strong electric fields," General Electric Research Lab. Rept. 62-RL-3178G (December 1962).

[15] Schluter, v., "Dynamik des Plasmas I: Grundgleichungen des Plasma in Gekreuzten Feldern," Z. Naturforch. 5a, 72-78 (1950); "Dynamik des Plasmas II: Plasma mit Neutralgas," Z. Naturforch. 6a, 73-78 (1951).

[16] Rosa, R. J., "Hall and ion slip effects in nonuniform gases," Avco-Everett Research Lab. Research Rept. 121 (December 1961); also Phys. Fluids 5, 1081-1090 (1962).

[17] Jones, M. S., Jr. and McKinnon, C. N., Symposium on Magnetoplasmadynamic Electrical Power Generation, Newcastle upon Tyne (September 6-8, 1962).

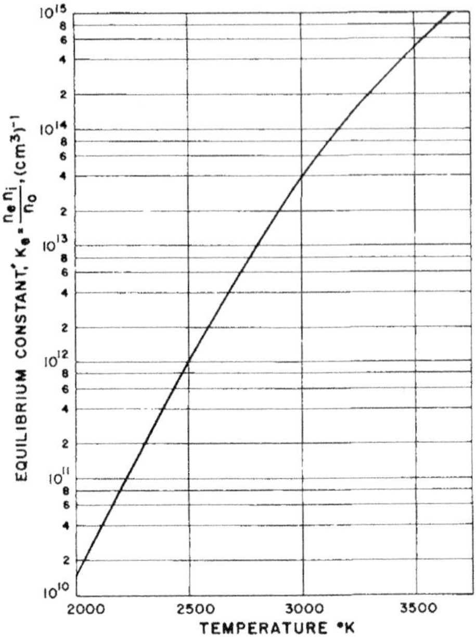


Fig. 1 Equilibrium constant for the ionization of potassium

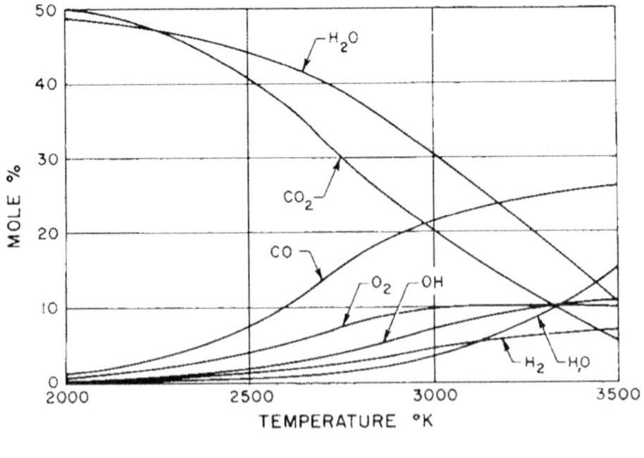


Fig. 2 Composition of stoichiometric JP-4-$O_2$ combustion gases vs temperature at atmospheric pressure

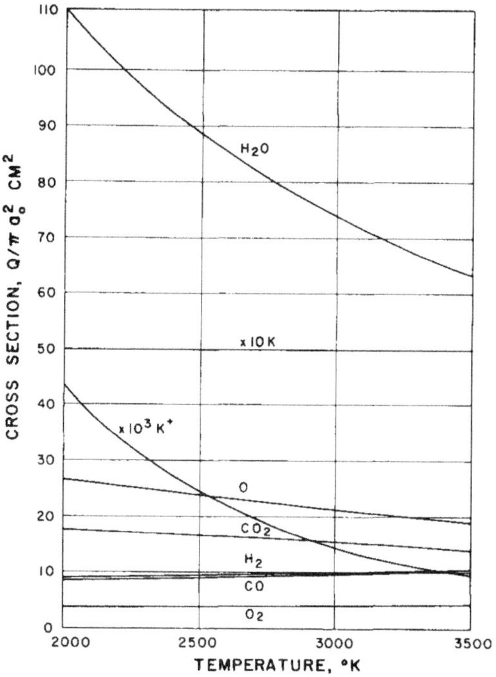


Fig. 3 Electron collision cross section for the various species in JP-4-$O_2$ combustion products as a function of temperature

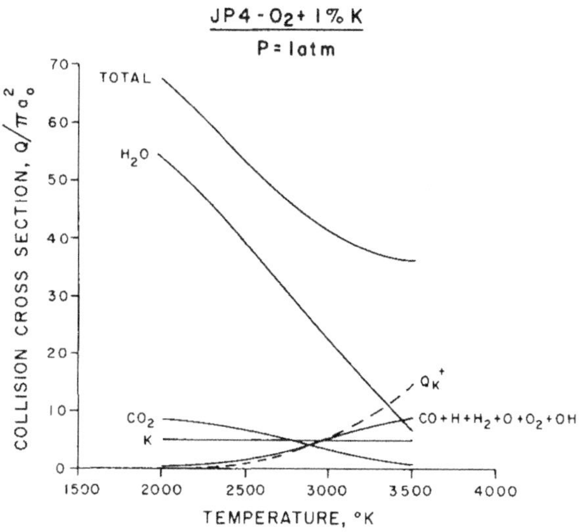


Fig. 4 Total electron collision cross section and the contribution to the total cross section by the various species as a function of temperature

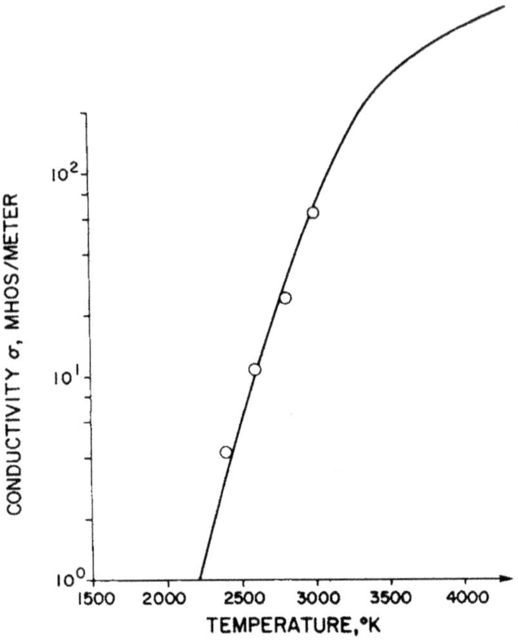


Fig. 5 Measured and predicted conductivity of the JP-4-$O_2$ products as a function of temperature

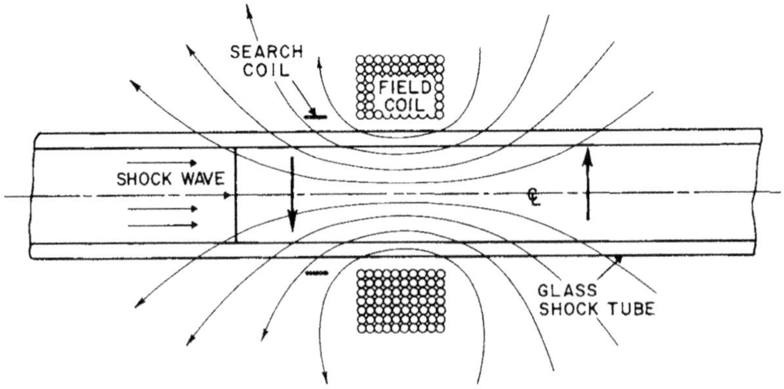


Fig. 6 Sketch of the shock tube conductivity experiment

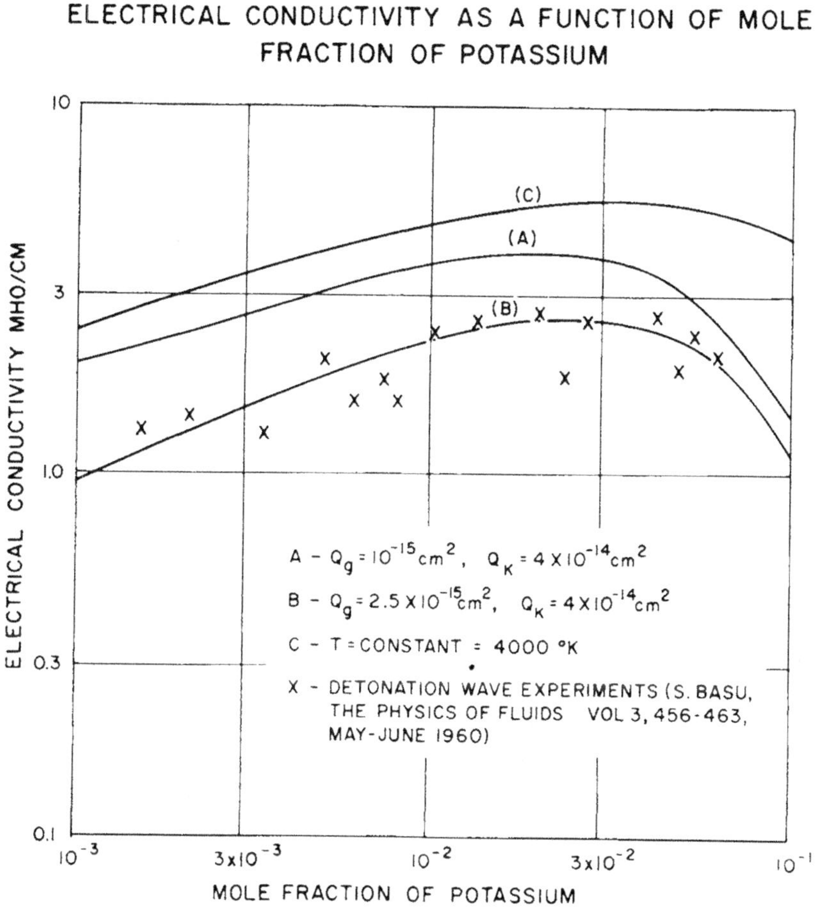


Fig. 7 Comparison between experimental and theoretical electrical conductivity as a function of mole fraction of potassium in the product gases. Curve A corresponds to experimental conditions with $Q_g = 10^{-15}$ cm$^2$ and $Q_K = 4 \times 10^{-14}$ cm$^2$. Curve B corresponds to experimental conditions with $Q_g = 2.5 \times 10^{-15}$ cm$^2$ and $Q_K = 4 \times 10^{-14}$ cm$^2$. Curve C is a constant temperature line corresponding to $4000^\circ$ K. x denotes an experimental point

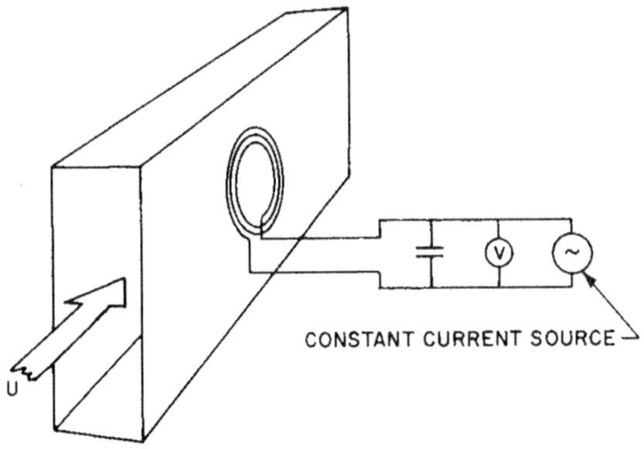


Fig. 8 Sketch of the "Q coil" conductivity measurement

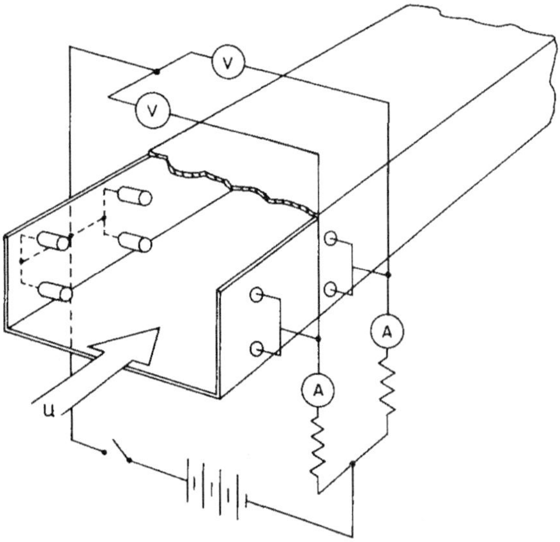


Fig. 9 Sketch of the V-I characteristic conductivity measurement

Fig. 10 Apparatus for the V-I measurement in a channel flow

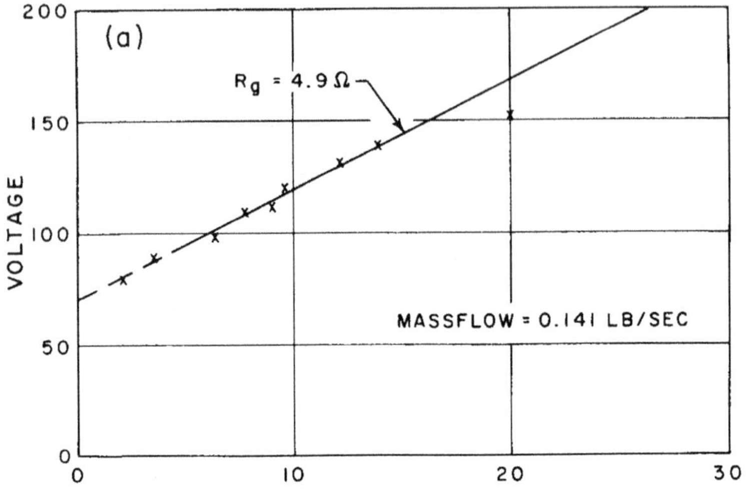


Fig. 11 Typical V-I characteristic; cold electrodes

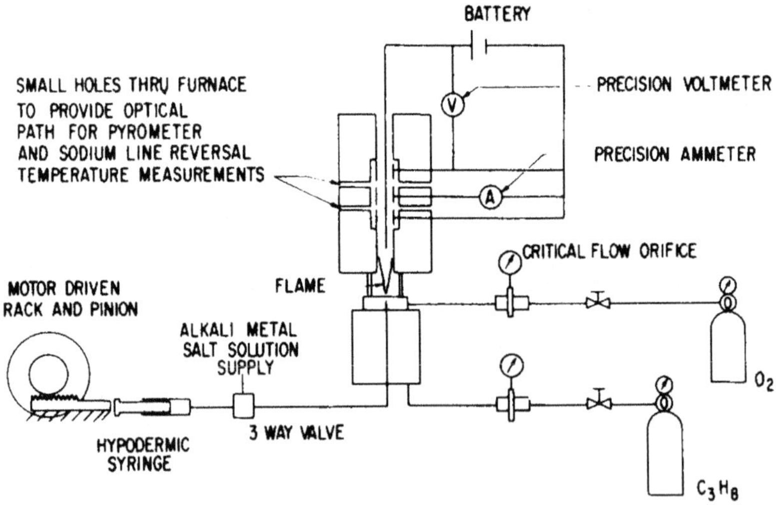


Fig. 12 Schematic of the hot cell V-I measurement apparatus

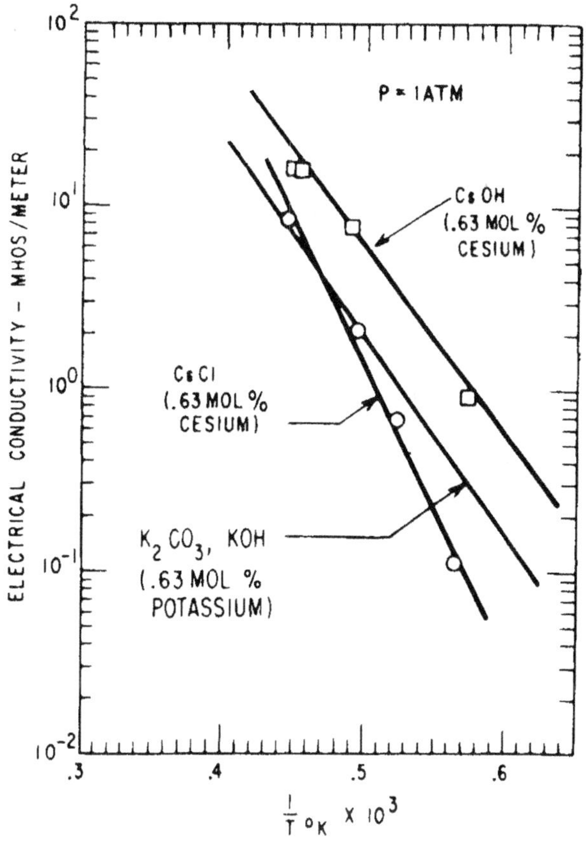


Fig. 13 Some results of the hot cell measurement

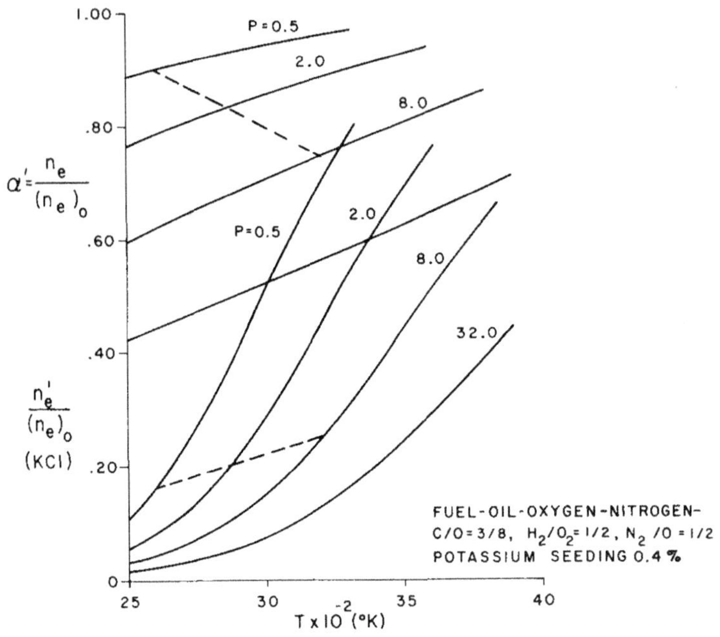


Fig. 14 The effect of electron attachment to combustion products (upper curves) and to chlorine atoms

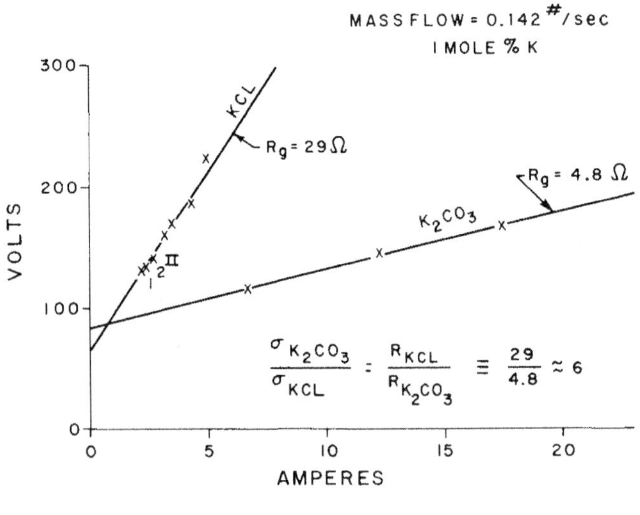


Fig. 15 V-I characteristic with equal amounts of $K_2CO_3$ and KCl seed

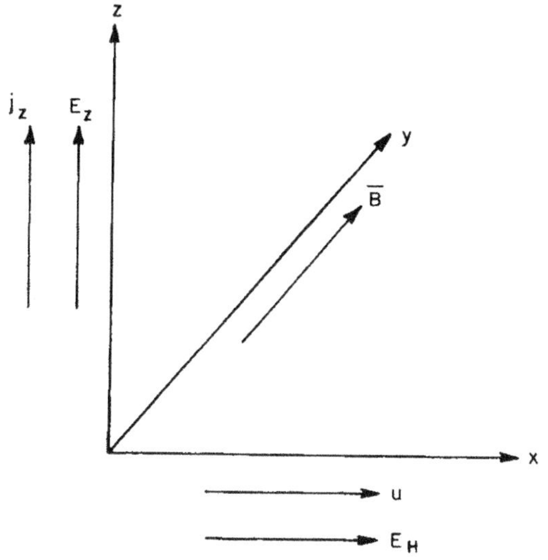


Fig. 16 Coordinate system for description of the Hall effect

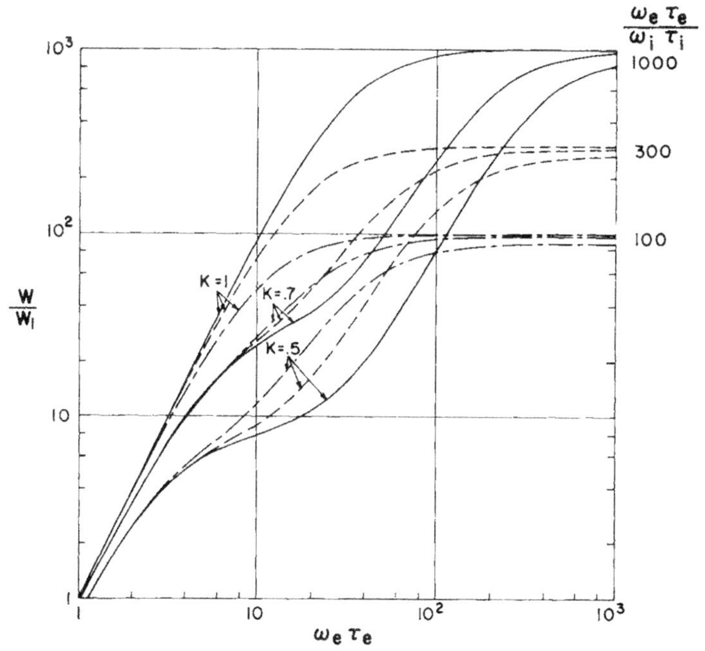


Fig. 17 Effect of nonuniformity on the performance of a cross-field MHD channel flow device

# EXPERIMENTAL STUDIES OF SOME ELECTRICAL PROPERTIES OF SEEDED FLAME GASES

George E. Moore*

General Electric Company, Schenectady, N.Y.

Abstract

Three experimental programs and some of their results are described: 1) Determination of the conductivity of potassium-seeded, propane-oxygen flame gases by simple resistance measurements, using concentric graphite electrodes in a small furnace. Approximate proportionality of the conductivity to $(\text{concentration})^{\frac{1}{2}}$ was shown over the range 0.01 to almost 10% K in the gas. The temperature dependence (1900° to 2400°K) corresponds satisfactorily with that expected for ionization equilibrium. 2) Experiments with 2- x 4-in. magnetohydrodynamic channels using K-seeded $H_2$-air combustion gases at about 730 m/sec and 2100°K static temperature; the operating characteristics for both Faraday and Hall generators were measured in B-fields of 3 to 12 kgauss. Values of electron mobility were obtained in three ways and compared: from the variation of internal resistance with B for both types of generator, from the ratio of their open-circuit voltages, and from ordinary resistance measurements on the gas. Some materials problems in the operation of these generators are briefly discussed. 3) An experimental survey of the electrical conductivity of flame gases resulting from seeding with various other materials; in all, 26 elements were tried, ten of which are known to have been studied elsewhere. In addition, a number of pairs of elements were tested for enhancement effects. In general, no element or combination of elements was found which approached potassium in effectiveness as a seeding agent for application to MHD power generation.

---

Presented at the ARS Ions in Flames and Rocket Exhausts Conference, Palm Springs, Calif., October 10-12, 1962. This work was done as part of a larger program at the General Electric Research Laboratory during the past three years; most of it was taken from publications and unpublished reports by G.J. Mullaney, P.H. Kydd, W.E. Kaskan, L.P. Harris, and G.E. Moore (of the Research Laboratory), and N.R. Dibelius (of the Advanced Technology Laboratories), ably assisted by C.W. Moon, L.H. King, W.I. Foss, and J.J. Keenan.

*Manager, Combustion Research Unit, Research Laboratory.

## Introduction

The work summarized here was stimulated mainly by interest in the possibility of practical magnetohydrodynamic power generation with seeded combustion gases as the working fluid. At the time it was begun (1959) there was virtually no experimental information on the conductivity of such gases at a level near that required for acceptable power density nor much experience with, for example, electrode and insulator materials and their behavior at the required high temperatures, nor with the electrical characteristics of the generator configurations conceived. Much thought was also being given to possible methods of obtaining the required gas conductivity at lower and more reasonable temperatures, e.g., by the discovery of better seed materials. It was to these matters that the experimental work, part of which is described here, was directed; some of it has paralleled similar work in other laboratories, and much of it has been published elsewhere.

## I. Measurements of Conductivity of Seeded Combustion Gases

Compounds of K and Cs have, of course, received the most attention as additives for enhancing the conductivity of combustion gases, and potassium has been and probably still is, for economic reasons, the only practical seed element for open-cycle MHD systems. The present investigations began with determinations of the conductivity of K-seeded, propane-oxygen combustion products[1] and for the purpose of this review only work with potassium will be discussed in this section.

The apparatus is shown in Figs. 1 and 2; the burner design resulted in stable flow of the hot gas at a few liters per second and good atomization of the additive solution. Adjustment of the oxygen flow rate varied the temperature from about $1800^{o}$ to $2400^{o}K$ (equivalence ratios of 1.2 to 2.1; various losses kept the temperature considerably below the adiabatic flame temperature). By sighting through 3 x 6 mm apertures near the measuring section, the center electrode temperature was determined by optical pyrometry, and the gas temperature by sodium line reversal; they agreed within about $30^{o}$.

The electrical measurements were generally simple voltampere characteristics (Fig. 3) obtained by applying battery voltages to the rod and the three 5-cm-long rings, the upper and lower of which were guard rings. The rod was initially 1.25 cm in diameter, the rings 2.54 cm i.d.; they would last about 3 to 5 min at the highest temperatures. Although the use of such concentric electrodes avoided errors due to wall leakage, which proved to be serious with parallel plate electrodes in a square channel, their rapid erosion and gasification probably incurred some uncertainty in the effective cell dimensions used in computing $\sigma$ from the resistance measured just before terminating an experiment. This and other sources of error lead to an estimated uncertainty of the data shown in Fig. 4 of approximately 20%.

From the measured conductivity $\sigma$ one may compute the electron mobility $\mu$ ($\sigma = en_e\mu$) with the assumption that the electron concentration $n_e$ is that corresponding to thermal equilibrium and calculable from the Saha equation. Substantially all of the variation of $\sigma$ with T should be due to variation of $n_e$ with T, and the slopes of the lines for K-seeding shown in Fig. 4 yield a value of $V_i$ of about 4.39 v for potassium. $\sigma$ also varies approximately as the square root of the seed content in the range shown in Fig. 4; in fact measurements at higher seed concentrations (Fig. 5) show the same behavior, from which it was concluded that the electron-K atom collision cross section is $\sim 10^{-15}$ $cm^2$, and not much different from that of electrons with other neutrals of the gas.

For comparison with data from other experiments described in the next section, the electron mobility $\mu$ obtained from the absolute values of the conductivity is about 0.92 $m^2$/v-sec. Although the data are internally consistent and appear to give accurate indications of the temperature and composition variations, their absolute values as indicated by this value of $\mu$ may be somewhat high; this point will be further discussed in Sec. II.

Other experimental work done with this apparatus and technique included: 1) similar measurements on cesium-seeded combustion gases, in which the effects of Cl and OH in reducing the amount of ionizable Cs by CsCl and CsOH formation and by electron attachment were studied;[2,3] 2) measurement of thermionic emission limits of graphite in the presence of potassium and the effect of barium in increasing the limit;[1,4] 3) studies of the effects of alkaline earths on the conductivity of sodium-seeded combustion gases;[4] 4) the effects of large electrical power input on the behavior of K-seeded flame gases, with results similar to those independently obtained and described by Karlovitz.[5] More recently, Lawton, Payne, and Weinberg[6] have also described experiments of this type.

## II. Electrical Measurements on MHD Channels

The measurements described in this section were obtained during a continuing program of experimentation[7] intended to provide experience in the operation of combustion flows of the type that might prove useful in power generation and to yield data that can be used to check the many published predictions of performance of such flows. The apparatus, which is sketched in Fig. 6, consists mainly of a combustor 6 x 6 in. in cross section and 30 in. in length, a test section providing up to 48 in. of active length and a flow area 2 x 4 in. in cross section, and a nozzle 24 in. in length providing the transition between combustor and test section. A pressurized tank is used to inject a metered flow of 45% KOH water solution into the stream, usually to about 0.75% by weight of potassium atoms in the flow. Air preheated to 700°K and hydrogen are burned at approximately 1.8 atm pressure to yield a theoretical adiabatic flame temperature of 2550°K, a test-section Mach number of about 0.8, and a measured static temperature at the test section of 2100°K. The mass flow of 1.4 lb/sec corresponds to a total flow enthalpy of about 2 Mw.

The combustion air is supplied to the combustor by a 4-in. main coming from the air compressor, a regulating and metering apparatus, and a preheater. The hydrogen and seed solution are injected into the stream through crosses set in the air line just upstream from the combustor. Flame holding is accomplished primarily by a carborundum bar set across the flow near the combustor inlet. Both combustor and nozzle are constructed of zirconia insulating brick set into water-cooled metal housings. Bolted-flange connections at both ends of the combustor and nozzle provide ready access for inspection and maintenance operations.

The test section consists of many electrode segments set into insulating walls of commercial MgO brick that are encased in a water-cooled brass or steel jacket. Figure 7 is a photograph of one test section partially disassembled to show its construction. The metal housing is 65 in. long. Several materials have been used for the electrodes, which are intended to operate as simple thermionic emitters. (Graphite, though usable in the work of Sec. I, was unsatisfactory here.) The most successful combination, for present purposes, has consisted of a silicon-carbide block with the external electrical connection provided by a stainless steel or silicided molybdenum stud. The test section is supported between the poles of a conventional C-core electromagnet giving uniform flux densities up to 13 kgauss over the test volume. Of the 4.5-in. magnet gap, 2 in. is used by the flow and 2.5 in. is taken up by the walls of the test section. The magnet is mounted on a hydraulic lift that can be lowered to permit access all around the test section when the system is not operating.

The major variables in this study are the magnetic flux density, the electrode interconnections, the scale of electrode segmentation, the electrical loading of the test section, and the seed flow. For the experiments reported here, the magnetic field has been varied between 3 and 12 kgauss, electrical loadings from open circuit to nearly short circuit, and the seed flow from 0.4 to 1.2 wt% of potassium atoms. Both Faraday and Hall generator connections have been used in the test section. In the former, all electrode segments on the same side of the channel are shorted together, and the load resistor is connected between the resulting two terminals. In the latter, electrodes opposite each other across the channel are shorted together, and the load is connected between the terminals at inlet and outlet ends of the channel. The pitch of the electrodes has been varied from 0.5 to 2 in. or from 0.125 to 0.5 times the interelectrode span; in every case the insulating segments have occupied half of the electrode pitch. All the generator experiments reported here have been done with 12 in. of active length in the test section. Thus they refer to generators with a 3:1 aspect ratio.

Table 1 gives typical values for some characteristics of the flow. Because the power extracted from the flow in these measurements never exceeded 200 w (about 0.13 w/cm$^3$), the flow properties are independent of the measurements or the electrode configuration used. The pressure drop in the test section is essentially all caused by friction and the small temperature drop by heat transfer. There are small variations in gas properties between runs, even large changes in temperature and conductivity when the seed flow is changed substantially, and there are significant changes with time within a single run in the gas temperature and conductivity.

Figure 8 shows the temporal variations of temperature and conductivity during a typical run and indicates the reproducibility of the gas conductivities obtained in several experiments. The gas temperatures shown have been determined by optical methods. Measurements of sodium-line reversal has given temperatures of $2100^\circ \pm 40^\circ K$ at 3 to 5 min after ignition.

The conductivity variation shown in Fig. 8 is a composite of results obtained from several runs by three different methods: direct measurement of interelectrode resistance in the absence of a magnetic field; calculations based on the measured variation of the transverse currents in the Hall generator configuration when the load terminals are open-circuited; calculations based on the short-circuit currents obtained from load-line data for the Faraday generator. Beyond 5 min after ignition the conductivity remains constant near 1.8 to 1.9 mho/meter. The KOH flow indicated 0.75% by weight of potassium, yields a maximum conductivity. Small variations in seed flow from this value cause no change in conductivity.

The flow velocity has not been measured directly but can be calculated by several methods from the available data. The various results all agree that the velocity normally is in the range 730 ± 15 m/sec, provided that the assumption of frozen flow from the adiabatic flame temperature (corrected for the cooling effect of the KOH flow) is used in methods involving the energy balance for the flow. Thus it appears that the flow in the test section is roughly in ionization equilibrium at the local gas temperature, but that the state of chemical aggregation corresponds to equilibrium near the stagnation temperature. Similar conclusions were reached by Way[8] for his apparatus.

Figure 9 gives the electrode current densities obtained in a run where voltages up to 132 v were applied, with no magnetic field present, to the terminals of a guarded electrode structure set in the channel walls. The linearity of this curve up to current densities of 0.3 amp/cm$^2$ indicates that for lower currents the applied voltage is taken up entirely by the gas resistance or other ohmic elements. Also indicated in this plot are the maximum current densities encountered in normal conductivity measurements, and in Faraday and Hall generator experiments. The higher current densities in the end electrodes of the Hall generator are caused by end effects and the load current.

In runs made without the seed flow, the leakage conductances remained low, about a decade below the total conductances obtained with a seeded gas flow, and the electrodes and insulators showed little or no erosion in 4 or 5 min of operation. When the seed flow has been used the SiC electrodes have reacted to yield slight amounts of siliceous slag. These have had no material effect on the performance or life of the electrodes, but have caused rapid erosion of adjacent insulators. The end result has been a small flow of electrically conducting liquid down the channel, especially along the bottom wall, with accompanying large leakage conductances. It has been found that,to make good sense of these data, account must be taken of these conductances and their variations in time as well as the variations in gas conductivity. Experience to date has shown the desirability of keeping even small amounts of silica from the flow and suggests that the alkali seed may react with other oxides to produce large leakage conductances in sustained operation.

## Results with the Faraday Generator Configuration

Like those of previous experimenters,[8,9] the data for a single magnetic flux density, taken within a short time interval, form a linear load line (load current vs load voltage). The slopes of lines for different flux densities show internal resistances that tend to increase with flux density as expected from calculations[10] including the Hall effect. The open-circuit voltage (corrected for leakage conductances) of this configuration is proportional to flux density, as shown by Fig. 10.

The variation with time of the ratio of open-circuit voltage $V_{OC}$ to magnetic flux density B is shown in Fig. 11, which contains points obtained from 4 runs. The voltage decreases nearly linearly at a rate of 3% per min for the first 3 to 4 min, then levels out at about 85% of its initial value. The zero-time intercept of this plot indicates a flow velocity of about 720 m/sec. Since end effects should not affect this measurement, the observed trend could be caused only by increasing leakage conductances or decreasing flow velocities. Because the records of combustion pressure and mass flow rate show that the flow velocity in these runs remained constant or very slightly increased, this trend is attributed to leakage conductances, and this curve has been used in correcting the raw data for varying electrical leakage.

Figure 12 shows the effect of magnetic flux density on the internal resistance of the Faraday generator. According to simple theory[10] the internal resistance should vary as $(1 + \mu^2 B^2)$ where $\mu$ is the electron mobility in the gas flow. The plot, containing points from two runs, shows that the internal resistance does vary linearly with $B^2$, and thus determines a value of 0.42 $m^2$/v-sec for the electron mobility. The two runs shown in this figure used different electrodes with ratios of electrode pitch to channel span of 0.250 and 0.125, respectively. The close agreement of points from both runs indicates that effect of electrode segmentation on the internal resistance of this generator over the ranges of segmentation ratio and magnetic field covered in these experiments.

## Results with the Hall Generator Configuration

Figure 13 shows load characteristics plotted from the raw data obtained during a run with the Hall generator configuration. These form straight lines for each value of magnetic flux density as before, but show a stronger variation with flux density than the corresponding characteristics of the Faraday generator.

According to theory the open-circuit voltage of this configuration is proportional to the square of the magnetic flux density. This prediction is verified, for the present range of conditions, by Fig. 14 which shows the variation of $V_{oc}$ with $B^2$ for segmentation pitch-to-span ratios of 0.125, 0.250, and 0.500. As expected, the open-circuit voltage decreases steadily as the electrode pitch is increased.

The effect of changing segmentation scale is shown again in Fig. 15 where $V_{oc}/B^2$ is plotted against the ratio of electrode pitch to channel span. The three points obtained from our data are close to a straight line which can be extrapolated to zero electrode pitch to permit another calculation of electron mobility. Instead of being derived from the variation of internal resistance with flux density, this estimate depends on the ratio of open-circuit voltages of Hall and Faraday generators when both have fine electrode segmentation and operate in the same flow. The result is $\mu \simeq 0.33$ $m^2/v$-sec, somewhat lower than that found previously.

Figure 16 shows the variation of the open circuit voltage with time in the apparatus for test sections with a segmentation ratio of 0.250. The observed voltage remains fairly constant during the first 3 min of operation, then decreases rapidly until at least 6 min of operation. The differences between this behavior and that shown in Fig. 11 for the Faraday configuration probably occur because leakages along different walls are involved. The leakage conductances in the Faraday generator are those along the vertical walls constructed of relatively large slabs of MgO. In the Hall configuration the relevant leakages are along the top and bottom walls which are constructed of alternating small pieces of SiC and MgO. The different roughness characteristics of the two types of wall may account for the lack of decay in the Hall open-circuit voltage during the early part of the run. The drastic reduction in this voltage after 3 min occurs because there is then sufficient liquid slag in the channel to bridge over and around the insulating segments on top and bottom walls. To achieve good reproducibility of results for the Hall configuration, new or reclaimed MgO pieces on the top and bottom walls of the channel were used for each run, and analysis was restricted to data obtained during the first 3 min of any run, where the only corrections to the raw data needed are those for changing gas conductivity.

As with the Faraday generator, theory predicts that the internal resistance of the Hall configuration should vary as $(1 + \mu^2 B^2)$. Figure 17 shows the variation of resistance with

$B^2$ in three runs for which the segmentation ratios were 0.250 (O), 0.125 (□), and 0.500 (◆). Despite some rather wide scatter in the points shown, a straight line with a slope corresponding to an electron mobility $\mu \cong 0.41\ m^2/v\text{-sec}$ appears a reasonable approximation.

## Discussion of Results

The experimental results described here have provided two estimates of the mobility of electrons in the working fluid used, 0.33 and 0.42 $m^2/v\text{-sec}$, derived from the ratio of open-circuit voltages in the Hall and Faraday generators and from the variation of generator internal resistance with magnetic flux density. To these may be added a third, for the measured conductivity of approximately 1.85 mho/m with a gas static temperature of 2100° K and a seeding rate of 0.75 wt% potassium atoms yields, assuming equilibrium ionization at the local temperature and negligible electron attachment and KOH formation, a calculated mobility of about 0.83 $m^2/v\text{-sec}$. It may also be recalled that from the conductivity results given in Sec. I, a value of $\mu = 0.92$ was estimated. Relaxation of the assumptions to permit attachment and KOH formation, both of which must occur to some extent, would raise both of these estimates even higher. In view of the other experimental results reported here and the theoretical calculations of Frost[11] for somewhat similar gases, these values must be considered too high. Thus it appears that either the measured gas temperatures were consistently low, or that the electron density in the test section had not relaxed entirely to equilibrium with the local temperature. The latter explanation might apply to the results of this section, but not to those of Sec. I; the former would require errors of nearly 100°K in the gas temperature which seem unlikely, but are possible in both sets of experiments.

## III. A Survey of Flame Seeding for MHD

Potassium may be considered a marginal or possible seeding material for MHD in the context of present-day technology, and hence a convenient material with which to compare others in searching for additives that might provide adequate conductivity at somewhat lower temperatures. It appears that a lower equilibrium electron concentrate for a given atom concentration will be obtained for any other element (excluding Cs), unless somehow the additive should provide either a high concentration of easily ionized molecules or of low work function particles. Theoretical considerations and available experimental data suggest that molecules with ionization potentials lower than those of the constituent atoms do exist, but the prospects of finding something better than potassium are not encouraging.

In theory, one can compute the thermal (i.e., equilibrium) ionization for a given system if one knows all of the species present, including possible condensed phases, and their ionization potentials and/or work functions. However, it is not clear that the necessary information on all of the systems of possible interest are available with sufficient reliability and completeness to make a completely theoretical approach feasible. Hence, an experimental approach was taken in which various seeding materials were fed to the working fluid (flame gases) and the resulting electron concentrations were measured. Although the nature of possible compounds formed will be governed, at least in part, by the chemistry of the flame gases, they do represent an important possible working fluid for MHD power generation and they are easy to generate at high temperatures.

The general plan of the work was to test as many elements as possible, consistent with these criteria: a) The group of elements most likely to be fruitful are those in the lower left-hand part of the periodic table; b) elements which have been tried by others would be avoided. With respect to (b), Sugden and his co-workers have seeded flames with a number of elements,[12] which are indicated in Fig. 18. Although all of the results of these investigations are not known in detail, it was assumed in the absence of information to the contrary that they did not show great promise.

Criterion (a) deserves a little discussion. In the first place, the most easily ionized elements are in the lower left part of the periodic table. In crude terms, the low ionization potential of Cs can be explained by the fact that Cs has one electron outside of a closed shell and this electron is relatively far removed from the positively charged nucleus. The same argument can be made to explain the low ionization potential which has apparently been observed for BaOH,[13,4] a heavy free radical which perhaps "looks" like Cs. Criterion (a) then is simply a generalization of this argument and represents the possibility that easily ionized free radicals may be formed in large concentration from the elements under consideration.

For any element, the variables which are expected to be of significance are the temperature, concentration of seeding material, and the fuel/air ratio. The importance of temperature is obvious. The dependence of conductivity on seed concentration comes about not only through the process described by the Saha equation but also through processes which involve the formation of entities (molecules or particles) involving more than one atom of seed. Finally, the chemical properties of the flame gases, as determined by the fuel/air ratio, have an influence on what compounds are formed and at what point condensation may occur.

In this work,[19] an attempt was made to obtain data which would provide information on the effect of each of these variables, although not necessarily for each element studied. Depth of study was sacrificed to breadth of coverage. In the remainder of this section, essentially all of the data on single elements obtained will be presented, mostly in graphical form. In addition, a few experiments involving seeding by more than one element were done, in which enhancement of the conductivity was observed.

## Experimental

The aim of acquiring results quickly, conveniently, and with moderate accuracy determined the experimental method. Briefly, it involved producing a fog of an aqueous solution of the seeding material in a combustible mixture which was burned on a specially designed burner (Fig. 19), the flame gases passing through a hole in a microwave cavity. The change of the Q of the cavity was measured and related to the electron concentration; the apparatus and techniques are similar to those used by Sugden and his collaborators. By varying the distance between burner and cavity, the time from burning of the particular portion of flame gas column being observed could be varied. The system was essentially calibrated with cesium seeding. The work to be described deals with electron concentrations in the range of $10^9$ to $10^{12}$ per cc, a limitation set by the electronics. Detailed descriptions of the burner and techniques are available in other reports.[14,15]

To accomplish the desired changes in significant variables, six $H_2/O_2/N_2$ flames were used. Their properties are described in Table 2. They were selected to provide stable flames with some spread in temperature, and burned gases of reasonably typical composition.

The elements tried in one way or another are circled in Fig. 18. The particular salts employed are listed in Table 3 which is keyed to indicate the types of measurements carried out with each salt. This is in the third column, where: C = concentration dependence; T = temperature dependence; S = a single measurement only; N = no observable conductivity.

The concentration dependence data were obtained entirely with flame 1, which had the highest temperature and, hence, allowed the study of a larger concentration range than would the others. The data, obtained at a fixed cavity-burner distance of 1.6 cm, are shown in Figs. 20 a, b, and c. These are log-log plots of of the electron number density against the partial pressure the seeded metal would have had if it had evaporated completely to give a monatomic gas. These partial pressures were obtained as described elsewhere,[14] from atomizer delivery rates computed from the photometry of sodium-seeded flames. Each of these three figures contains the same potassium curve for easy comparison. The potassium curve, in effect a calibration, is in good agreement with other data for cesium seeding.[15]

At the right-hand edge of Figs. 20 a,b, and c, the symbols in boxes are placed at the approximate electron concentrations to be expected for the various elements at equilibrium assuming a metal atom partial pressure of $2 \times 10^{-5}$ atm (dotted line) using the Saha equation with known or reasonable values of the statistical weights of ions and atoms.

In many cases, it was obvious that precipitation of a condensed phase had taken place in the flame gases. This is indicated in Table 3. Presumably because of the not entirely homogeneous nature of the flame zone, in these cases condensation did not take place uniformly in the flame gases, but in a pattern related to the grid pattern of the burner.

The materials employed for seeding were very commonly contaminated with sodium and, occasionally, with potassium. In Table 3, the extent of this contamination is given. In a few cases, the observed conductivity was due entirely to the alkali present, and these data are not shown. For each of the elements shown in Figs. 20 a,b, and c, photometric measurements of the Na concentration in the flame gases were made for at least one seed concentration. In those cases where Na was observable, it was shown that the conductivity was higher than would be expected from the Na present by comparison with the conductivity curve for Na alone in Fig. 20 a. However, the addition of sodium to some materials will lead to an enhancement of conductivity, so that it is not certain that in these cases the presence of sodium is immaterial.

For a completely evaporated substance, one of two kinds of concentration dependence is expected: a "linear" region or a "square root" region, in which the conductivity is proportional to the first or one-half power of concentration, respectively. The second region is one of small fractional ionization, and the first of large, or in the limit, complete ionization. In Fig. 20 a, potassium exhibits both kinds of dependence, together with the transition region; a straight line coincident with the low concentration potassium data represents essentially the limiting electron concentration from any substance, assuming one electron per seed atom.

For the other elements considered here, one would expect square root dependence in the range of temperature and concentrations employed in this work except for the possibility of condensation, which will be discussed separately below. Reference to the figures shows that few elements obeyed this law and at least one that did, cerium, is known to have not been completely evaporated.

The occurrence of a linear dependence (Na, Li, Y, Hf) is somewhat surprising. This behavior by Na has been previously reported and investigated in detail, and has been interpreted as due to a failure to reach equilibrium near the flame.[16] Another way this sort of dependence could arise would be if the ionizing molecule contained two atoms of seed metal. Then, even though the ionization of this molecule was in the square root range, the concentration of the molecule would depend on the square of the seed concentration, thus giving a linear dependence. Finally, if the ion were formed directly in a reaction with a constituent of the flame gases, a process which could be represented by

$$M + \text{flame gases} \longrightarrow M(O,H)^+ + e$$

and if the neutral compound corresponding to the ion was essentially absent, then a linear dependence could result. In any event, the observed linear dependence cannot mean complete ionization of any stable compound present, as the electron concentrations would then have to fall on the line extrapolated from the high dilution potassium data.

## Temperature Dependence

An example of the temperature dependence data is shown in Fig. 21, and Table 4 gives the concentrations at which the measurements were made. The data were taken in groups of three points, corresponding to the three distances of 1.6, 2.1, and 2.6 cm from the burner, and are plotted at the temperatures measured by sodium line reversal at these positions (see Table 2). (The variations of temperature with distance was not the same for all the flames.)

The straight lines in the figure were drawn by visual inspection only and energies derived from such slopes are listed in Table 4. For systems in the square root range of concentration dependence (e.g., potassium here) these energies should be multiplied by two to obtain the ionization potential. As has been pointed out previously, those measurements taken for the purpose of establishing the temperature dependence were made at several times from burning, so that nonequilibrium effects might be manifested.

The behavior of potassium appears to be controlled simply by the temperature. The dependence of the conductivity on temperature within one flame is approximately the same as that from flame to flame, and the experimental ionization potential of 4.0 v is to be compared with the known value of 4.34 for the potassium atom. This agreement is considered satisfactory for the present purposes.

For most of the other elements, the low values of the temperature coefficients obviously cannot be interpreted in terms of ionization of the metal; they may suggest the presence of free radicals or molecules of moderately low ionization potential, complicated by the effects of chemical reactions involving them which may also have temperature coefficients superimposed on those of the ionization processes to make the apparent energies even smaller. In some cases, the formation of low work function particles by condensation, also with a temperature coefficient, probably was involved. Moreover, for some elements such as barium, the conductivity decreased with time in a given flame, while the temperature was still rising, suggesting a nonequilibrium process of some kind. It is probably pointless to speculate further on the nature of these processes, though possibly interesting in themselves, with the information obtained here; all that can be said is that in no case did any such effects lead to a conductivity or electron concentration high enough to be of any practical interest here.

In considering the question of electron production from low work function particles, Smith's analysis[17] was used with some simplifying assumptions to estimate work functions, number densities of particles, and number of electrons evaporated per particle for some cases in which condensation occurred in these experiments. The numbers so obtained were reasonable, but indicated that, even with the most favorable assumptions of particle size and concentration and of work function, electron concentrations comparable with those obtainable from K probably are not to be obtained in this way.

Other related experiments in the course of this work will be just briefly mentioned; the details may be found elsewhere.[19] They were concerned with ionization enhancement effects, neither of which, however, appears to occur at conductivity levels high enough to be interesting for MHD application.

Observations were made on the increase in ionization of sodium (at low concentrations) by the addition of Cl, generally confirming those of Padley, Page, and Sugden,[18] though the effect was not observed with potassium.

In some experiments in which particles of a condensed phase appeared and to which small amounts of alkali were also added, the measured electron concentrations exceeded those due to either added element (cf. Table 5); this enhancement, though interesting, tends to vanish at concentrations high enough to obtain $n_e > \sim 10^{11}$/cc. The results may be explained by reduction of the work function of the particles by adsorbed alkali.

References

[1] Mullaney, G.J., Kydd, P.H., and Dibelius, N.R., "Electrical conductivity of flame gases with large concentrations of potassium," J. Appl. Phys. 32, 668-671 (1961).

[2] Dibelius, N.R., Mullaney, G.J., and Luebke, E.A., "Electrical conductivity of flame gases seeded with alkali metals; application to MHD power plant design," Second Symposium on Engineering Aspects of MHD (Columbia University Press, 1962), pp. 307-326.

[3] Mullaney, G.J. and Dibelius, N.R., "Electrical conductivity of flame gases seeded with large concentrations of cesium," General Electric Research Lab., Schenectady, N.Y. Rept. RL-2647 C (1961).

[4] Mullaney, G.J. and Dibelius, N.R., "Electrical conductivity of flame gases seeded with large concentrations of sodium, barium, and strontium," General Electric Research Lab., Schenectady, N.Y. Rept RL-2820 C (1961).

[5] Karlovitz, B., "Flames augmented by electric power," IUPAC Symposium on Chemical and Thermodynamic Properties at High Temperatures, Montreal (1961).

[6] Lawton, J., Payne, K., and Weinberg, F., "Flame-arc combination," Nature 193, 736-738 (1962).

[7] Harris, L.P. and Moore, G.E., "Some electrical measurements on MHD channels," Third Symposium on Engineering Aspects of MHD, Rochester, New York (March 1962).

[8] Way, S., DeCorso, S.M., Hundstad, R.L., Kemeny, G.A., Stewart, W., and Young, W.E., "Experiments with MHD power generation," Trans. ASME, Ser. A (J. Eng. Power) 83, 4, 397-408 (October 1961); also, Way, S., "Comparison of theoretical and experimental results in an MHD generator," Second Symposium on Engineering Aspects of Magnetohydrodynamics " (Columbia University Press 1962), pp. 166-179.

[9] Rosa, R.J., "Physical principles of magnetohydrodynamic power generation," Phys. Fluids 4, 2, 182-194 (February 1961).

[10] Harris, L.P. and Cobine, J.D., "The significance of the Hall effect for three MHD generator configurations," Trans. ASME, Ser. A (J. Eng. Power) 83, 4, 392-396 (October 1961).

[11] Frost, L.S., "Conductivity of seeded atmospheric pressure plasmas," J. Appl. Phys. 32, 10, 2029-2036 (October 1961).

[12] Sugden,T.M.,(Cambridge Univ.)personal communication (1961).

[13] Sugden, T.M. and Wheeler, R.C., "The ions produced by traces of alkaline earths in flames," Discussions Faraday Soc. 19, 76-86 (1955).

[14] Kaskan, W.E., "Chemistry of alkali metals in flame gases," General Electric Research Lab., Schenectady, N.Y. Rept. RL-2823 C (1961).

[15] Kaskan, W.E. and Sharbaugh, A.H., "A microwave cavity technique for measuring electron concentrations in flames," General Electric Research Lab., Schenectady, N.Y. Rept. RL-2927 C (1961).

[16] Knewstubb, P.F. and Sugden, T.M., "Observations on the kinetics of the ionization of alkali metals in flame gases," Trans. Faraday Soc. 54, 372-380 (1958).

[17] Smith, F.T., "On the ionization of solid particles," J. Chem. Phys. 28, 746-747 (1958).

[18] Padley, P.J., Page, F.M., and Sugden, T.M., "Effect of halogens on the ionization in alkali-laden $H_2$ and $C_2H_2$ flames," Trans. Faraday Soc. 57, 1552-1562 (1961).

[19] Kaskan, W.E., "A survey of flame seeding," General Electric Research Lab., Schenectady, N.Y. Rept. No. RL-2952 C (1962).

Table 1 Typical channel operating conditions

| | |
|---|---|
| $\dot{m}_{air}$ | = 1.35 lb/sec at 750° to 800°F |
| $\dot{m}_H$ | = 0.044 lb/sec; Φ = 1.12 |
| $\dot{m}_{seed}$ | = 0.034 lb/sec; 45% KOH (0.7 wt% K) |
| $p_c$ | ≅ 11 psig |
| $T_0$(theor.) | = 2550°K |
| $T_{static}$ | = 2100°K (measured by line reversal) |
| $U_{channel}$ | = 720 to 740 m/sec; M ~ 0.75 to 0.8 |

Table 2 Flames and conditions used in seeding survey

| Flame no. | Burned gas composition mole fraction $H_2$ | $O_2$ | $H_2O$ | $N_2$ | z (cm) | T(°K) | Hot flow velocity cm/sec |
|---|---|---|---|---|---|---|---|
| 1 | 0.127 | ... | 0.318 | 0.556 | 1.6 | 2178 | 1796 |
| | | | | | 2.1 | 2188 | |
| | | | | | 2.6 | 2198 | |
| 2 | 0.0988 | ... | 0.2475 | 0.655 | 1.6 | 2008 | 2130 |
| | | | | | 2.1 | 2023 | |
| | | | | | 2.6 | 2035 | |
| 3 | 0.1105 | ... | 0.277 | 0.612 | 1.6 | 2088 | 1980 |
| | | | | | 2.1 | 2098 | |
| | | | | | 2.6 | 2108 | |
| 4 | 0.375 | ... | 0.232 | 0.392 | 1.6 | 1770 | 1428 |
| | | | | | 2.1 | 1770 | |
| | | | | | 2.6 | 1770 | |
| 5 | ... | 0.018 | 0.334 | 0.647 | 1.6 | 2223 | 1564 |
| | | | | | 2.1 | 2218 | |
| | | | | | 2.6 | 2208 | |
| 6 | ... | 0.0147 | 0.274 | 0.7125 | 1.6 | 2073 | 1784 |
| | | | | | 2.1 | 2073 | |
| | | | | | 2.6 | 2068 | |

Table 3 Salts used in flame seeding survey

| Element | Compound used | Measurements made | Na/seed atom ratio | Visible conden-sation |
|---|---|---|---|---|
| Li | $LiNO_3$ | C | ... | |
| Na | NaCl | C,T | ... | |
| | $Na_2CO_3$ | C | ... | |
| | $NaNO_3$ | C | ... | |
| K | $K_2CO_3$ | C,T,L | ... | |
| Mg | $MgCl_2 \cdot 6H_2O$ | N | $<10^{-4}$ | |
| Ca | $Ca(NO_3)_2$ | C | ... | |
| | $CaCl_2$ | C | ... | |
| Ba | $Ba(NO_3)_2$ | C,T,L | ... | |
| | $Ba(Cl_2) \cdot 2H_2O$ | C | ... | |
| Al | $Al(NO_3)_3 \cdot 9H_2O$ | N | ... | |
| Sc | $Sc(NO_3)_3$ | C,T,L | 0.0008 | |
| Y | $YCl_3 \cdot H_2O$ | C,T,L | 0.0014 | |
| La | $La(NO_3)_3 \cdot 6H_2O$ | C,T,L | $<10^{-4}$ | |
| Ce | $Ce(NO_3)_3 \cdot 6H_2O$ | C,T,L | 0.010 | Yes |
| | $CeCl_3$ | S | 0.0001 | Yes |
| Nd | $Nd(NO_3)_3 \cdot 6H_2O$ | C,T,L | 0.0003 | Yes |
| Ti | $TiCl_3$ | S | $<10^{-4}$ | Yes |
| Zr | $ZrOCl_2 \cdot 8H_2O$ | S | $<10^{-4}$ | |
| Hf | $HfO(NO_3)_2 \cdot 8H_2O$ | C,T,L | 0.0055 | |
| | $HfCl_4$ | S | $<10^{-4}$ | |
| Th | $Th(NO_3)_4 \cdot 4H_2O$ | S | $<10^{-4}$ | |
| V | $VOSO_4 \cdot 2H_2O$ | S | 0.0002 | Yes |
| Mo | $(NH_4)_6Mo_7O_{24} \cdot 4H_2O$ | C,T | $<10^{-4}$ | |
| W | $(NH_4)_6W_7O_{24} \cdot 6H_2O$ | S | 0.003 | Yes |
| U | $UO(NO_3)_2 \cdot 6H_2O$ | S | 0.005 | Yes |
| Fe | $FeCl_3 \cdot 6H_2O$ | S | $<10^{-4}$ | |
| Pt | $PtCl_4$ | S | $<10^{-4}$ | Yes |
| Ag | $AgNO_3$ | N | ... | |
| Cd | $Cd(NO_3)_2 \cdot 4H_2O$ | N | ... | |
| In | $InCl_3$ | C,T,L | 0.0002 | |

Table 4 Summary of temperature dependence data for some seed materials

| Element | Salt used | Conc, g/l | P, atm x $10^{-6}$ | Energy, v |
|---|---|---|---|---|
| Na | NaCl | 4.44 | 5.1 | 3.2 |
| K | $K_2CO_3$ | 2.0 | 1.95 | 2.0 |
| Ba | $BaCl_2 \cdot 2H_2O$ | 9.34 | 2.55 | 1.4 |
| Ba | $BaNO_3$ | 10.0 | 2.55 | 1.4 |
| Sc | $Sc(NO_3)_3$ | 100.0 | 29.2 | 1.6 |
| Y | $YCl_3 \cdot H_2O$ | 200.0 | 63.0 | 1.45 |
| La | $La(NO_3)_3 \cdot 6H_2O$ | 100.0 | 15.5 | 1.5 |
| Ce | $Ce(NO_3)_3 \cdot 6H_2O$ | 51.6 | 8.0 | 0.75 |
| Nd | $Nd(NO_3)_3 \cdot 6H_2O$ | 250.0 | 38.0 | 1.2 |
| Hf | $HfO(NO_3)_2 \cdot 8H_2O$ | 50.0 | 7.3 | 1.8 |
| Mo | $(NH_4)_6Mo_7O_{24} \cdot 4H_2O$ | 100.0 | 38.0 | 1.65 |
| In | $InCl_3$ | 100.0 | 30.5 | 1.3 |

Table 5 Summary of observations with pairs of seed elements

| Pairs of elements tried | | Enhancement |
|---|---|---|
| Zr + | Li | Yes |
| | Na | Yes |
| | K | No |
| | Cs | No |
| | Ba | No |
| Hf + | Na | Yes |
| | K | No |
| Ce + | Na | Yes |
| | K | Yes |
| Ba + | Na | No |
| Mg + | Na | No |
| | Ba | No |
| | La | No |
| | Th | No |
| | U | No |
| Th + | Na | No |
| | Ba | No |
| | La | No |
| | Ce | No |

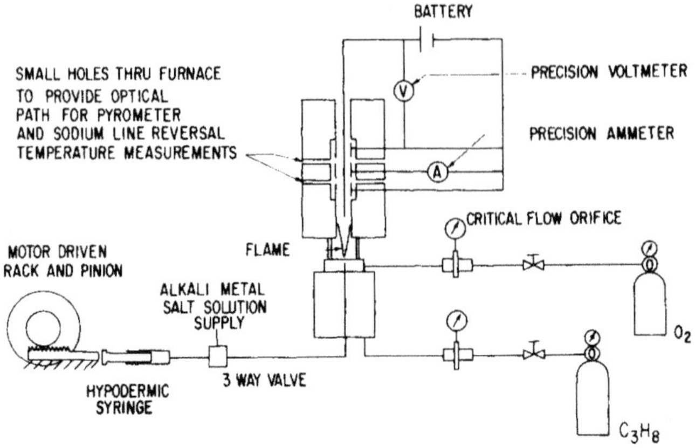


Fig. 1 Apparatus for electrical conductivity measurements in flame gases

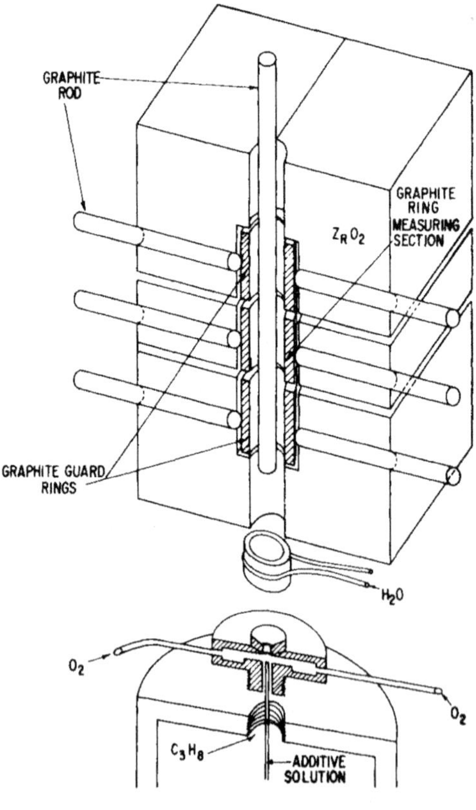


Fig. 2 Cutaway view showing the burner, electrode configuration, and zirconia insulating blocks

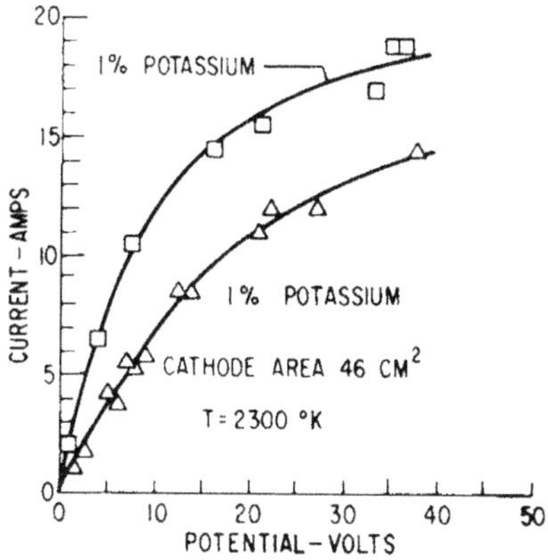


Fig. 3 Current vs voltage for a gas temperature of 2300°K with 0.1 and 1.0% by weight potassium

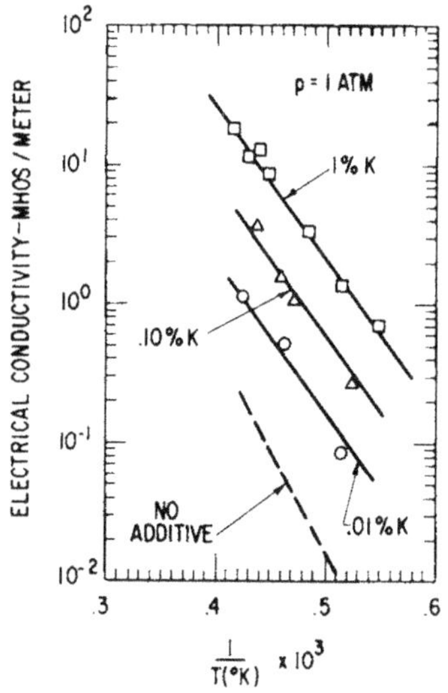


Fig. 4 Conductivity due to potassium additive in flame gases vs inverse of temperature for several additive concentrations

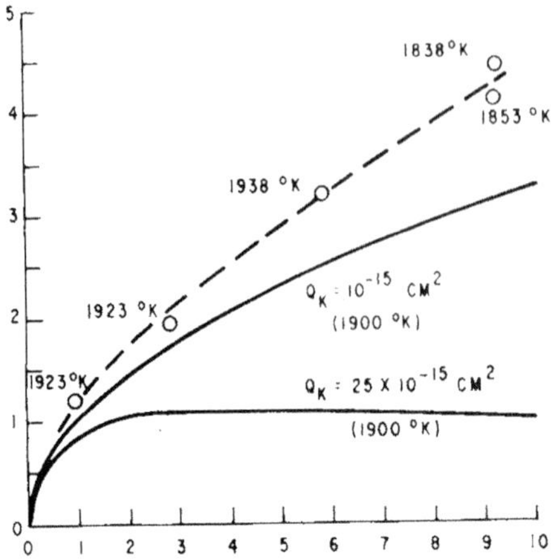


Fig. 5 Experimental conductivity as function of K concentration (--); calculated (—) using indicated cross sections

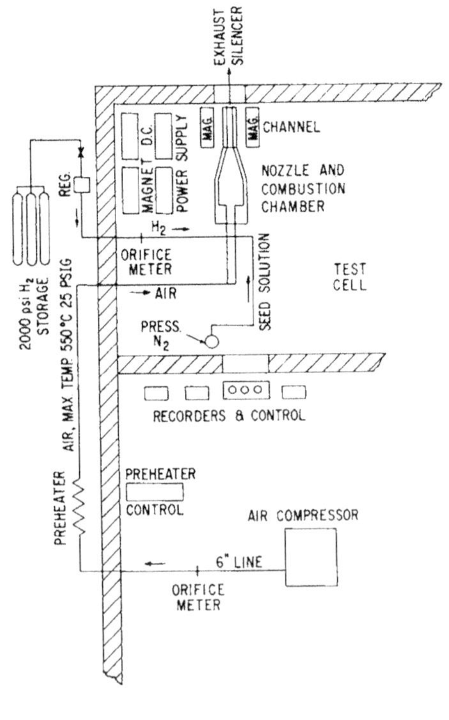


Fig. 6 Schematic of apparatus for MHD channel measurements

Fig. 7 Partially disassembled test section

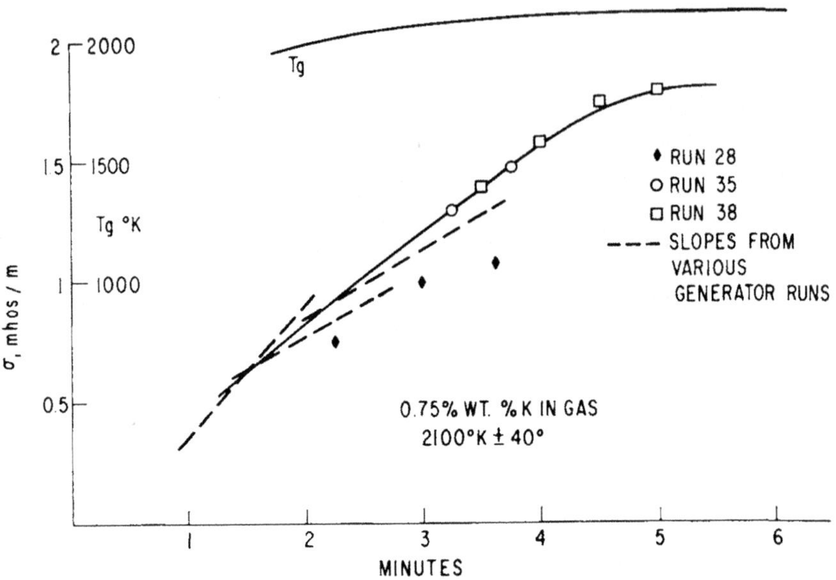


Fig. 8 Temporal variations of gas temperature and conductivity

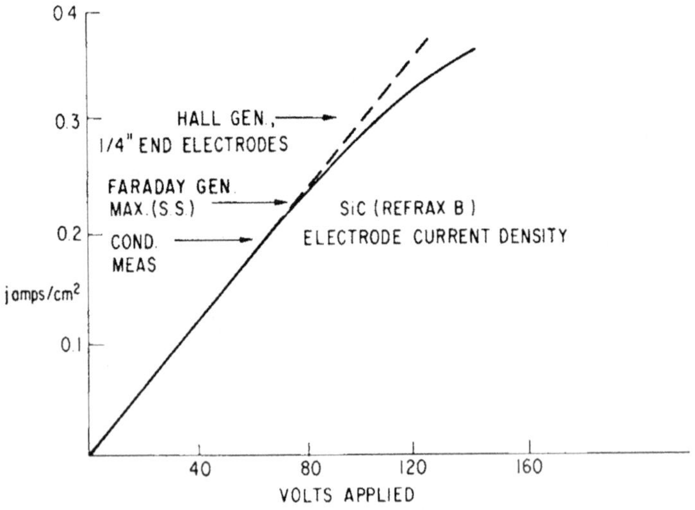


Fig. 9 Voltage-current characteristic from a conductivity test

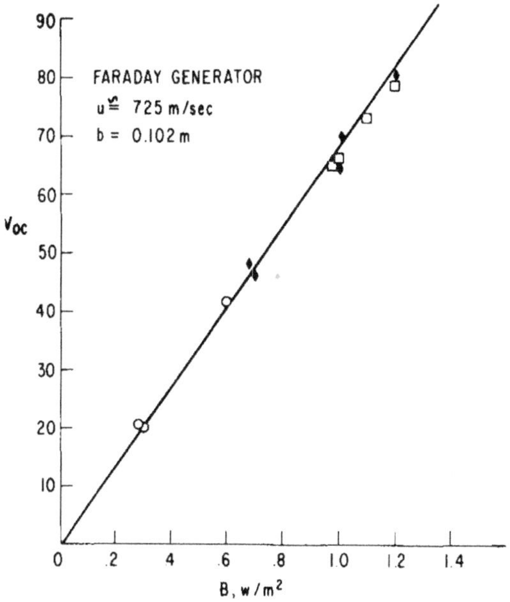


Fig. 10 Variation of open-circuit voltage with magnetic flux density for the Faraday configuration

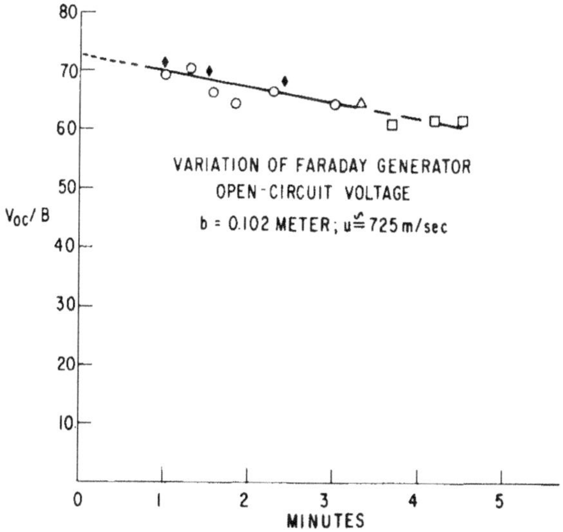


Fig. 11 Variation of open-circuit voltage with time for the Faraday configuration

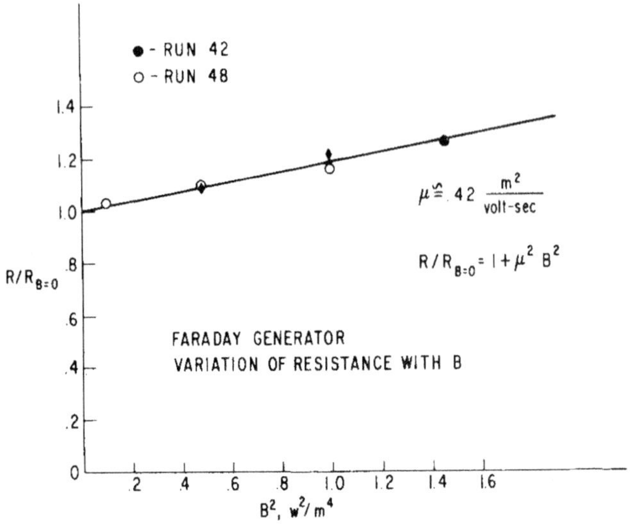


Fig. 12 Variation of internal resistance with magnetic flux density for the Faraday configuration

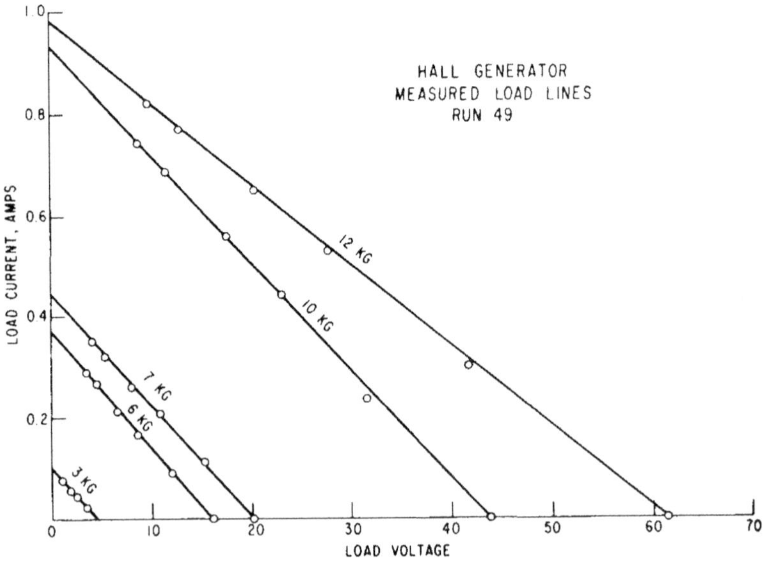


Fig. 13 Load characteristics obtained with the Hall configuration

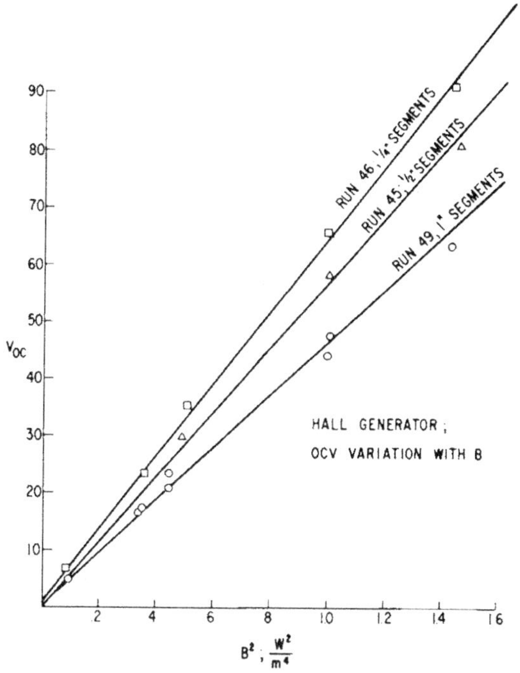


Fig. 14 Variation of open-circuit voltage with magnetic flux density for the Hall configuration

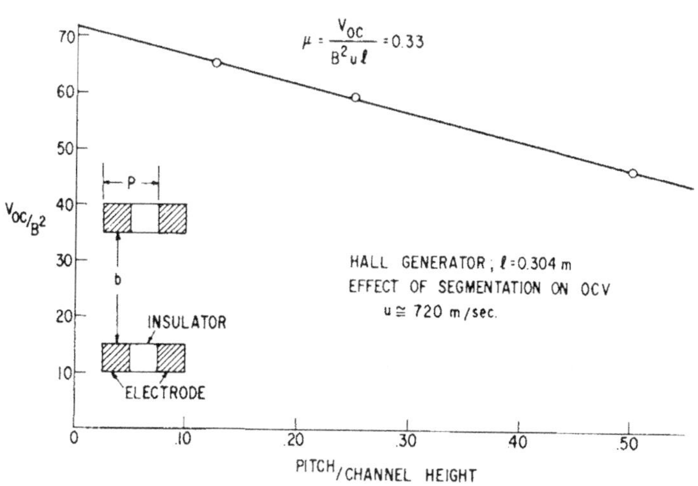


Fig. 15 Variation of open-circuit voltage with electrode segmentation for the Hall configuration

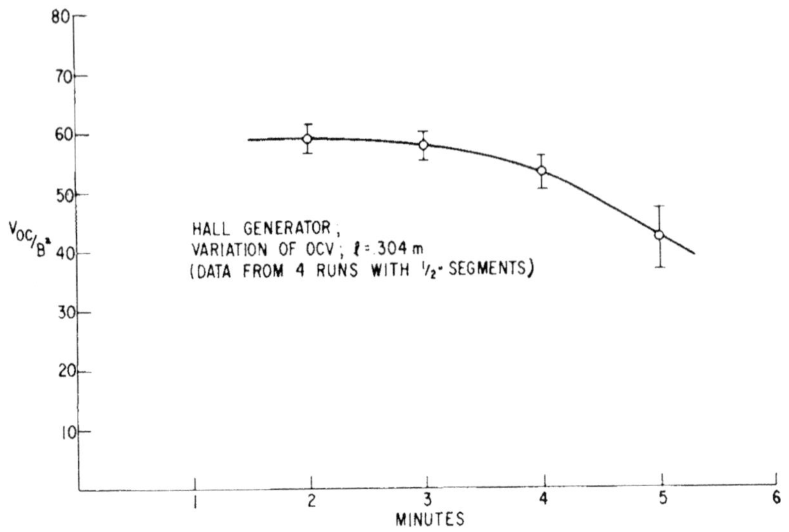


Fig. 16 Variation of open-circuit voltage with time for the Hall configuration

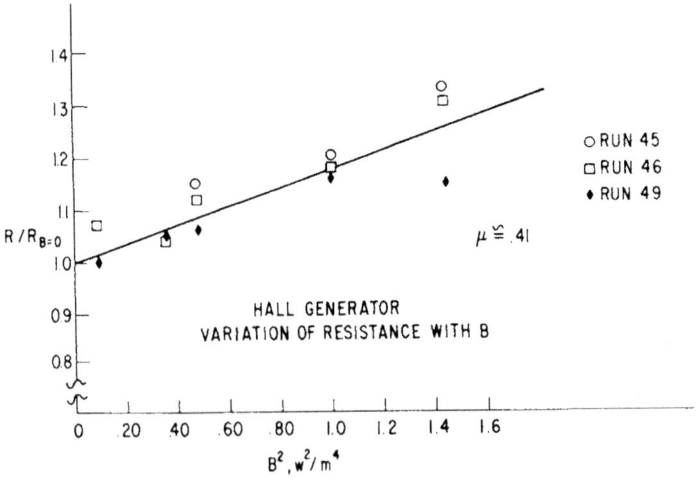


Fig. 17 Variation of internal resistance with magnetic flux density for the Hall configuration

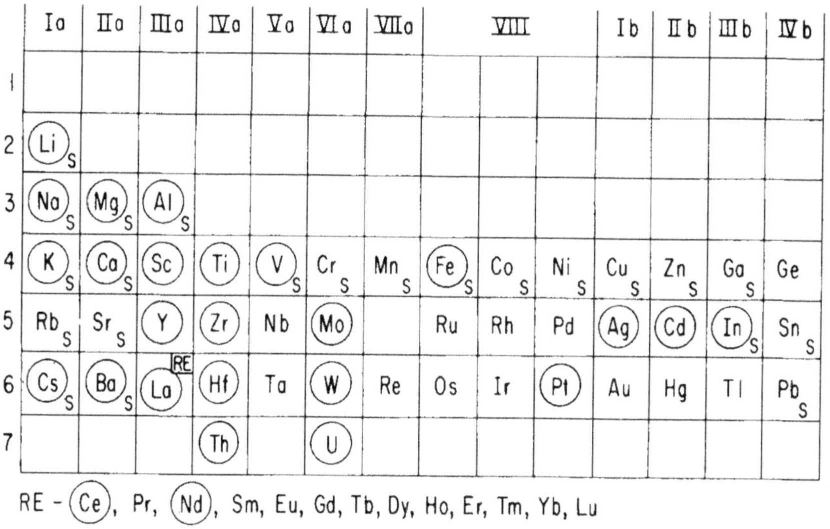


Fig. 18 Elements not eliminated by considerations of rarity, radioactivity, or electronegativity. Subscript S indicates those tried by Sugden and co-workers; circled elements indicate those tested in the present work

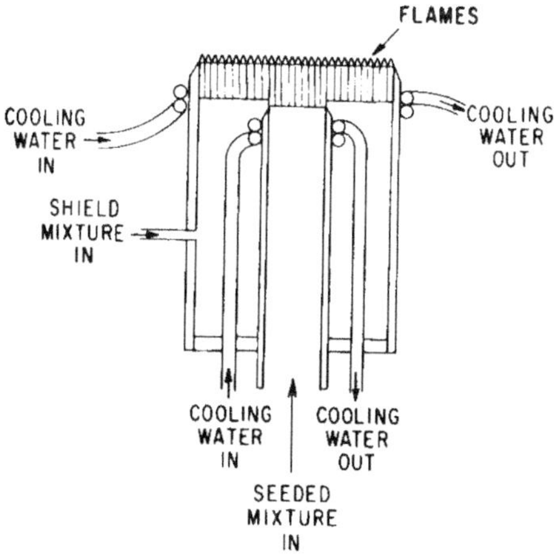


Fig. 19 Perforated copper disk burner

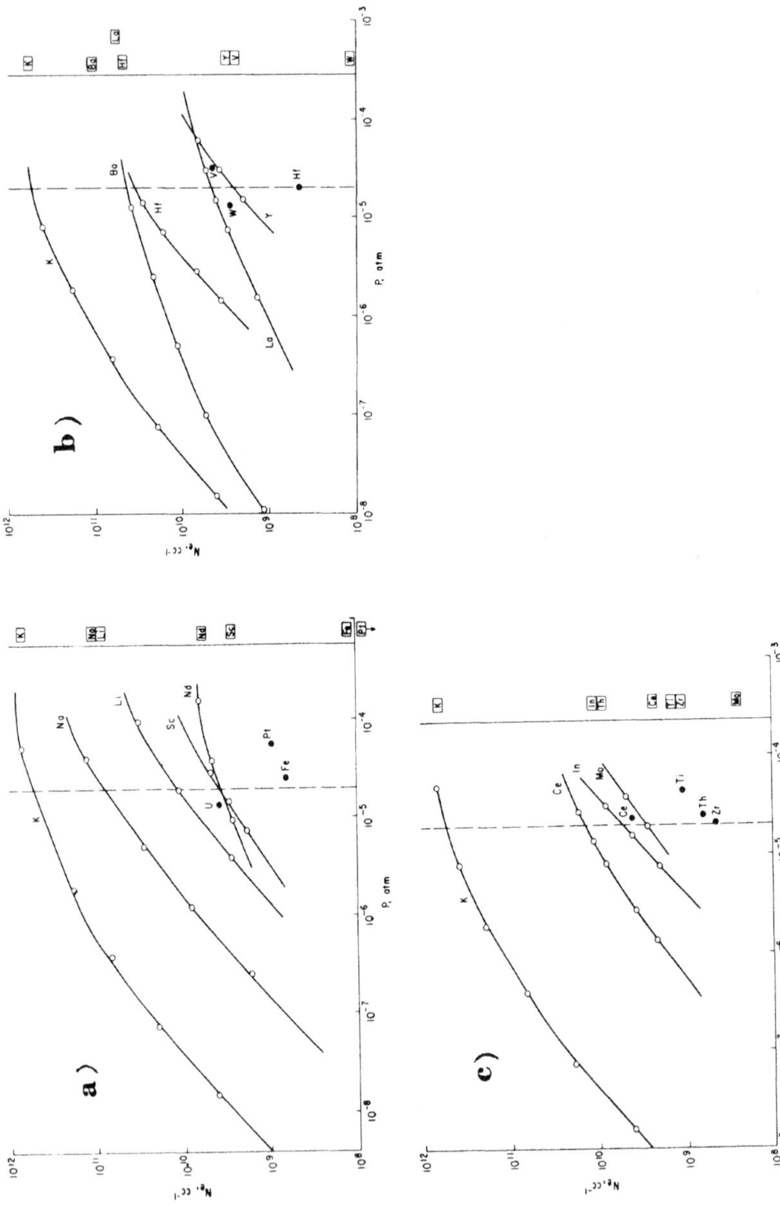


Fig. 20 a,b, and c Dependence of conductivity on seed concentration, determined 1.6 cm from the burner in flame 1

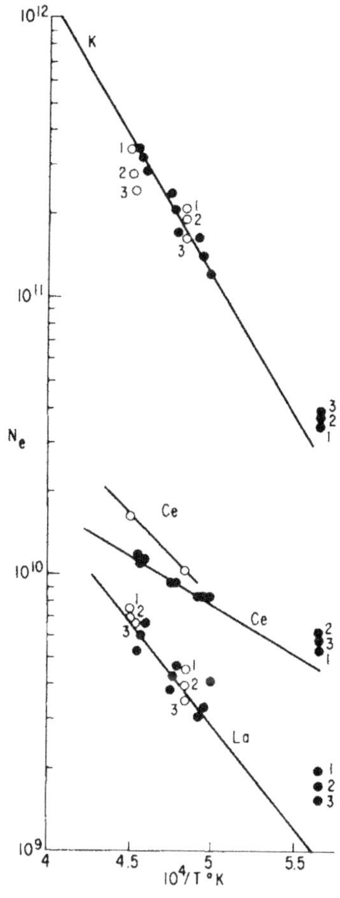


Fig. 21 Dependence of conductivity on temperature for both lean and rich flames (●:rich, o:lean)

STUDY OF ELECTRON GENERATION BY
SOLID PROPELLANT TECHNIQUE

R. Friedman,* L.W. Fagg,† T.K. Millar,‡
W.D. Charles,§ and M.C. Hughes//

Atlantic Research Corporation, Alexandria, Va.

## Abstract

Electron generators functioning by combustion of a cesium nitrate-aluminum pressed charge have been developed and test-fired in an altitude chamber. Thermodynamic analysis of the combustion and nozzle-expansion process shows that about $10^{20}$ electrons are discharged per gram of charge, the gaseous product containing about 2 mole % electrons. Spectroscopic measurements of the exhaust plume at 40-mm Hg abs confirm electron concentrations of this magnitude via Stark-broadening of cesium emission lines. Details of generator operation are given, and alternate chemical systems for electron generation are discussed.

## Introduction

The importance of the ionosphere for communication is well known. The current availability of sounding rockets raises the question of the possibility of artificial generation of local short-duration ionized regions in the upper atmosphere which will interact with radio or microwaves. The significance of such effects to upper-atmosphere research and to

---

Presented at the ARS Ions in Flames and Rocket Exhausts Conference, Palm Springs, Calif., October 10-12, 1962. This work was partially supported by Air Force Office of Scientific Research Contract AF 49(638)-651 and Air Force Cambridge Research Laboratory Contract AF 19(628)-295, in each case with support by Advanced Research Projects Agency. The voluntary cooperation of W. W. Balwanz's group at Naval Research Laboratory was greatly appreciated.

* General Manager, Research.
† Physicist, Kinetics and Combustion Division.
‡ Senior Project Engineer.
§ Senior Project Engineer.
// Project Engineer.

communications techniques is obvious. Marmo, et al.,[1] and Rosenberg and Paulson[2] have already described upper-atmosphere experiments in which explosive releases of cesium-containing chemicals have yielded artificial electron clouds.

An explosive release may not be the most efficient way to introduce electrons into the upper atmosphere because of the highly nonequilibrium nature of an explosion; i.e., some of the charge may not be heated and/or dispersed adequately. Also, for some experiments a continuous release with a duration of tens of seconds may be preferable to an instantaneous release. Therefore the authors have concerned themselves with feasibility of continuous generation of an electron-rich plasma by a solid-propellant technique. Early results have been reported previously.[3]

Free electrons rather than ions are required for strong interaction with radio or microwaves because the charge-to-mass ratio of the latter is too small. Exhaust gases from almost any rocket contain free electrons, and if the propellant is "seeded" with alkali metals (cf. Ref. 4), higher electron concentrations are obtainable. The problem to which this paper is devoted is the optimization of the electron yield per kg of charge.

Electron concentrations in the combustion chamber readily are calculated by the same well-known techniques used for analysis of any complex chemical equilibrium, if the thermodynamic parameters of the system are known and the assumption of equilibrium is valid. For the extreme temperatures and high chamber pressures of interest here, rocket experience suggests that equilibrium will be approximated closely in the chamber, although subsequent events during the nozzle expansion are in some doubt. To carry out the equilibrium calculations, an IBM-7090 program is used, all relevant known energy levels of the constituents being included. (This is the rigorous approach to the problem for which the Saha equation is the first approximation.)

By this technique, various candidate chemical systems may be screened on the basis of electron yield. In this paper the authors have limited arbitrarily their consideration to solid ingredients to conform with logistic requirements. Clearly, one wishes the highest flame temperature possible, since even cesium, the most readily ionized element, is ionized only partially in the hottest known flames at pressures reasonable for combustion (1-50 atm). In addition to achieving high temperature, it is vitally necessary to avoid species in the combustion gases such as hydroxyl, atomic chlorine, atomic fluorine, etc., which form stable negative ions. For example, in typical cesium-rich hydrogen-containing systems, the ratio $OH^-/e$ ranges from 10-100.

Although this study relates to electrons produced in a combustion chamber, it is well to remember that electrons also may be produced by photo-ionization of exhaust gases or by virtue of extreme relative velocity between exhaust plume and ambient. These effects are beyond the scope of this paper.

## Thermodynamic Survey of Candidate Systems

Cesium, being the most easily ionized element, obviously should be present as a major constituent of the system. Since cesium nitrate is an excellent oxidizer, a mixture of cesium nitrate and an oxidizable metal should yield a high flame temperature; such systems also would be free of undesirable negative ions (except for $O^-$, which is less stable than $OH^-$). The flame temperature of such a system will be slightly below the boiling point (pressure-dependent) of the product metal oxide, so metals forming stable, high-boiling oxides are desirable. Table 1 shows results of calculations for a series of such metals.

Table 1 Properties of combustion products of stoichiometric mixtures of cesium nitrate with various elements at 8.23 atm ambient pressure

| Element | Temp °K | Mole % electrons[a] | Electrons/cc | Stoichio-metric to |
|---|---|---|---|---|
| Y | 4870.7 | 4.68 | $5.80 \times 10^{17}$ | $Y_2O_3$ |
| Th | 4643.7 | 3.59 | $4.67 \times 10^{17}$ | $ThO_2$ |
| Zr | 4563.7 | 3.46 | $4.58 \times 10^{17}$ | $ZrO_2$ |
| La | 4405.6 | 2.58 | $3.54 \times 10^{17}$ | $La_2O_3$ |
| Be | 4225.3 | 2.18 | $3.12 \times 10^{17}$ | BeO |
| Al | 4162.6 | 1.90 | $2.76 \times 10^{17}$ | $Al_2O_3$ |
| Ca | 4123.4 | 1.70 | $2.48 \times 10^{17}$ | CaO |
| Mg | 3666.1 | 0.751 | $1.23 \times 10^{17}$ | MgO |
| B | 2625.8 | 0.0432 | $0.099 \times 10^{17}$ | $B_2O_3$ |

[a] Based on gaseous portion of product.

These calculations were made with an IBM-7090 computer, all known relevant equilibria being taken into account. A number of systems are seen to be capable of yielding several mole percent electrons. Of these, aluminum is particularly attractive for experimental purposes because it is available in fine powder form and has favorable handling and combustion properties. Of the metals superior to aluminum in Table 1, lanthanum is pyrophoric, beryllium is toxic, and yttrium, zirconium, and thorium are all much higher-boiling metals than aluminum and thus should not burn as well. Also the superiority of yttrium may not be real but may simply be due to inaccuracy in the vapor-pressure data for $Y_2O_3$. (Hafnium is another element that should give a very hot flame but was not included for lack of data.) The low atomic weight of aluminum gives it an advantage on a weight basis over the heavy elements, which does not show in Table 1.

The calculated properties of the aluminum-cesium nitrate system are shown in more detail in Table 2 (next page) for three values of chamber pressure. The fraction of cesium atoms ionized varies from 2.5% at 68 atm to 5.4% at 1 atm. Since the gaseous portion of the product is about half cesium vapor (the next largest constituent being nitrogen), the absolute concentration of electrons is extremely high, about two orders of magnitude higher than that achieved by seeding an oxy-hydrogen flame with cesium.[4] It must be noted, however, that a significant proportion of the product of the cesium nitrate-aluminum system is liquid aluminum oxide, in the form of μ size droplets, which constitute about 35% by weight of the product. (Because of lack of detailed knowledge, these calculations have ignored the possibility that the work function of molten aluminum oxide is low enough to contribute electrons.)

For some experiments, a system producing electron-rich products free of condensed-phase matter may be desirable, so consideration has been given to this. Clearly, metal oxides must be avoided, so a source of high energy other than metal combustion is required. A seeded cyanogen-oxygen flame would be excellent except that the ingredients are not solids. Golomb and Berger[5] have suggested that tetracyanoethylene would be a desirable solid ingredient for such systems when used with a solid oxidizer such as cesium superoxide or cesium nitrate. However, our calculations show that these systems yield much too low a flame temperature.

The authors suggest that the solid oxidizer hexanitroethane is uniquely suitable for the task in hand, since it releases

Table 2 Combustion products from 78.32% $CsNO_3$ plus 21.68% Al (All values expressed as moles).

| Species | Pressure | | |
|---|---|---|---|
| | 68 atm | 8.23 atm | 1 atm |
| O | 5.367 | 8.486 | 11.186 |
| N | 0.075 | 0.048 | 0.027 |
| e | 1.236 | 1.801 | 2.435 |
| Al | 3.385 | 4.902 | 6.265 |
| $Cs^+$ | 1.408 | 1.897 | 2.478 |
| $O^-$ | 0.171 | 0.096 | 0.043 |
| $O_2$ | 1.170 | 1.666 | 1.876 |
| $N_2$ | 26.383 | 23.907 | 22.369 |
| NO | 2.353 | 2.049 | 1.574 |
| AlN | 0.001 | -- | -- |
| AlO | 3.761 | 4.985 | 5.520 |
| $Al_2O$ | 1.434 | 1.845 | 2.017 |
| $Al_2O_2$ | 0.403 | 0.425 | 0.365 |
| Cs | 51.917 | 47.771 | 43.830 |
| $Cs_2$ | 0.935 | 0.122 | 0.016 |
| Total gas | 99.999 | 100.000 | 100.001 |
| $Al_2O_3$ (liquid) | 49.784 | 42.697 | 38.065 |
| Temp °K | 4605.3 | 4130.0 | 3715.2 |
| Mean mol. wt. of total gas | 86.59 | 80.65 | 76.50 |

its oxygen with a net liberation of energy. The cesium may be added either as cesium azide or as elemental cesium; the former has much better handling characteristics, but the latter would have less coolant and diluent effect per atom of cesium and would result in hotter flames. Calculations have been made for the system:

$$3\ C_2N_6O_{12} + 5\ C_6N_4 + 3\ CsN_3 \longrightarrow 36\ CO + 23.5\ N_2 + 3\ Cs$$

For constant-pressure combustion of 100 gm of this mixture at 1 atm, one obtains 3.041 gm-moles of gaseous products at 3498°K with composition as shown in Table 3.

Table 3 Calculated equilibrium products of tetracyano-ethylene-hexanitroethane-cesium azide system.

| Elements | Mole % |
|---|---|
| CO | 57.30 |
| $N_2$ | 37.40 |
| Cs | 4.28 |
| $Cs^+$ | 0.492 |
| e | 0.492 |
| C(g) | 0.0179 |
| N | 0.0134 |
| O | 0.000399 |
| $CO_2$ | 0.000302 |
| NO | 0.000101 |

Note that 11.5% of the cesium is ionized, and no solid carbon forms. On a weight basis, this system is nearly as good as cesium nitrate-aluminum ($0.90 \times 10^{20}$ electrons/gm vs $1.27 \times 10^{20}$ electrons/gm at 1 atm). Also, the proportions of ingredients in the above example have not been optimized, so that higher yields are presumably possible. An experimental approach to this system currently is being planned.

## Analysis of Cesium Nitrate-Aluminum System During Nozzle Expansion

Consider what will happen during isentropic nozzle expansion of the combustion products of the favored system, cesium nitrate-aluminum. The decrease in temperature will promote recombination, while the decrease in pressure will promote further ionization. The competition between these effects is seen from the following readily derivable equation which approximates the behavior of an isentropically expanding gas following the Saha relation.

$$\left(\frac{\partial \ln(n_e/n_{Cs})}{\partial \ln T}\right)_S = \frac{1}{2}\left(\frac{E}{RT} - \frac{\gamma}{\gamma-1}\right) \qquad (1)$$

Here, $n_e/n_{Cs}$ is the fraction of Cs ionized (assumed much smaller than unity), T is absolute temperature during isentropic expansion, E is ionization energy of Cs, and $\gamma$ is specific heat ratio (assumed constant).

(In addition to the just-indicated assumptions, the effects of a $T^{5/4}$ term in the Saha equation and the variation of the total number of moles during expansion were neglected.) For cesium at 4000°K, E/RT = 11.3; the right-hand-side of the equation accordingly will be negative for values of $\gamma$ smaller than 1.10, and the fractional ionization will increase during isentropic expansion. At sufficiently low temperatures, of course, E/RT will become larger than $\gamma/\gamma$-1 and the fractional ionization will pass through a maximum. For the case being considered, $\gamma$ is not, strictly speaking, a constant, but the effective $\gamma$ will be only slightly above unity because of the large liquid $Al_2O_3$ content of the stream (thermal equilibration of gas and liquid being assumed). For isentropic expansion of the $CsNO_3$:Al system from 8.23 atm to 1 atm, the temperature is calculated to decrease from 4130 to 3715°K; this corresponds to an effective $\gamma$ of 1.053, which is well below 1.10. Hence, one would expect an initial increase in ionization on expansion if the rate processes are sufficiently fast; otherwise, "frozen" flow will occur.

Some IBM-7090 calculations of the history of an isentropically expanding mixture of 78.33% $CsNO_3$ and 21.67% Al initially at 27.2 atm are presented in Table 4. (None of the forementioned assumptions were made except the equilibrium assumption.) The increase in electron fraction continues without reaching a maximum until the pressures reached are so low that thermodynamic equilibrium is surely not maintained. Thus the chamber conditions should represent a conservative estimate of electron output for units large enough so that heat loss to the walls is not excessive.

Table 4 Isentropic nozzle expansion history of $CsNO_3$:Al combustion products.

| Pressure, atm | Temperature, °K | Mole % electrons | Velocity cm/sec |
|---|---|---|---|
| 27,2 | 4428 | 1.55 | 0 |
| 1.0 | 3610 | 2.15 | 127,000 |
| 0.00052 | 2523 | 4.34 | 208,000 |
| 0.00007 | 2313 ($Al_2O_3$ solidifies) | 4.76 | 221,000 |

In practice there is clearly no need for providing a supersonic expansion cone on the nozzle for isentropic expansion to the ambient pressure; on the contrary, a simple orifice minimizes heat loss from the exhaust stream and reduces opportunities for deposit build-up. The emerging plume will undergo an irreversible expansion, involving a shock pattern, until the ambient pressure is reached. The final velocity will be lower and the final temperature higher than for the isentropic expansion to the same pressure, so of course the final equilibrium electron concentration will be higher for this reason.

The magnitude of the rate constant for dissociation during expansion must be considered. The rate constant for cesium ion recombination has been measured by various investigators at pressures from $10^{-3}$ to 1 mm Hg;[6-8] while the results for the second-order rate coefficient are not in agreement, they may all be correlated empirically by the expression,

$$\alpha = 4.7 \times 10^{-5}\ p/T \qquad (cm^3/sec) \qquad (2)$$

where $\alpha$ is the recombination rate constant for

$$Cs \underset{\alpha}{\overset{\beta}{\rightleftharpoons}} Cs^+ + e$$

and p and T are expressed in mm Hg and °K respectively. Since the equilibrium constant, equal to $\beta/\alpha$, is known from thermodynamics to be

$$\beta/\alpha = 2.36 \times 10^{15}\ T^{3/2}\ \ 10^{-19,640/T}\ \ (cm^{-3}) \quad (3)$$

the ionization rate constant $\beta$ is readily calculated from Eq. (2) and (3). (It should be remembered, however, that $\alpha$ as given by Eq. (2) may be quite unreliable.) One actually wants the half-life of the ionization reaction, given by $0.693/\beta$. Assuming expansion from 27.2 atm, one finds that at 3000°K the pressure has dropped to 11 mm Hg; at this condition the half-life for ionization as determined by the above procedure is 0.04μsec. On further expansion to 2500°K and 0.38 mm Hg, the half-life is now much longer, 24μsec. Since the flow rate is about 0.2 cm/μsec and the expansion occurs in a distance of the order of centimeters, these results imply that freezing of the ionization reaction will occur between 2500 and 3000°K.

The foregoing argument must be viewed with skepticism not only because of lack of precise ionization rate data, but also because two-phase flow is involved, and the degree of equilibration between particle and gas is unknown. If there were no energy exchange between particles and gas, the gas on expanding to vacuum could cool to below room temperature, according to the isentropic law, while the particles remained hot. This is apparently not the case, since the expanded gas at 1 mm Hg is still hot enough so that the cesium vapor is emitting a line spectrum, the intensity of which implies temperatures well above 2000°K. Also the particle size is known to be very fine, a micron or less, and the solid-to-gas ratio is high, so substantial energy transfer would be expected during expansion. However the observed luminescence may not be thermal but may be due to recombination processes. No rigorous treatment of the problem seems to be possible.

Once the pressure in the plume has equalized, further changes in electron concentration may occur because of radiant cooling of condensed particles or effects associated with mixing with ambient air, specifically cooling, dilution, and formation of $O_2^-$. Discussion of these effects is beyond the scope of this paper. At sufficient altitude the low collision rates will permit significant electron lifetime.

In summary, the cesium nitrate-aluminum electron generator should deliver a plasma to the upper atmosphere containing at least $0.86 \times 10^{20}$ electrons per gram of charge (chamber conditions at 175 psi). The electrons per unit volume in the plume obviously depends on ambient pressure and plume temperature.

## Design and Operating Characteristics of Electron Generator

To date, three sizes of cesium nitrate-aluminum electron generators have been developed and test-fired. Charge ignition is accomplished by a standard squib, and aluminized solid propellant is used as an aid. Generally, pressure peaks occur during charge ignition from simultaneous burning of the main charge and the ignition aid. This simultaneous burning has been found to be necessary for reliable charge ignition.

A typical chamber pressure vs time plot for a test firing is shown in Fig. 1. Both burning time and chamber pressure can be adjusted within limits by changing the nozzle throat diameter. Combustion performance is not as smooth as that of normal solid rocket propellants, but appears adequately steady for electron-generating purposes. Charge burning rates of 0.1 to 0.15 in./sec. at 250 psi have been observed. Burning rate is somewhat controllable with selected inorganic additives. Pressure oscillations of approximately 5-6 cps have been observed in the largest of the three units. These oscillations, which begin immediately after ignition and damp out during the first portion of the firing, have not been observed in the two smaller units.

Further development work to improve burning characteristics and to refine hardware design is presently under way.

## Measurements of Electron Yield

Two of the larger generators have been fired into an altitude chamber at the Naval Research Laboratory. Measurements of electron yield by microwave attenuation and spectrographic line-broadening techniques were made as described below.

### A. Test Firing at One Mm Hg

This generator was equipped with a graphite nozzle of 0.220-in. throat. Pulsation occurred for 3 sec. after ignition, followed by 87 sec. of steady operation at 100 psi. The exhaust jet was deflected sideways from time to time by slag, as shown by a motion picture record. The visible structure of the plume is believed to be associated with the shock-wave pattern.

The microwave absorption apparatus of W. W. Balwanz (Naval Research Laboratory) was focussed on the geometric axis of the motor. The apparatus was mounted on a track allowing traversal along this axis. The instrument registered maximum absorption along the entire traversal length throughout the firing. This indicates that the frequency of the electromagnetic radiation impinging on the plasma was less than the plasma frequency. Since the instrument was operated at 54 kmc, this means that the electron density was greater than $3.6 \times 10^{13}$/cc.

An emission spectrum obtained with an instrument of about 35Å/mm dispersion gave no observable Stark-broadening of the cesium emission lines. Even with an instrument of higher dispersion no detectable broadening would have been expected at a pressure as low as 1 mm. However calculations showed that, by maintaining a higher ambient pressure and using a higher-dispersion spectrograph, Stark-broadening should be detectable, and should lead to a value of electron concentration.

B. Test Firing at 40 Mm Hg

This generator was equipped with a tungsten nozzle with 0.187-in. throat. The initial ambient pressure was 25 mm Hg, rising to ca. 40 mm Hg soon after ignition. Combustion duration was 41 sec, with pulsation (5.8 cps) for the first eight sec, at a mean chamber pressure of 220 psi, with subsequent smoother burning at the same mean pressure until tail-off. A microwave absorption apparatus at 35 kmc located 91 cm downstream followed the 5.8- cycle pulsations, showing peak attenuations up to 7 db. This is close to the saturation of the instrument and corresponds to about $10^{13}$ electrons/cc.

A Bausch and Lomb 1.5-m grating spectrograph with 10Å/mm dispersion, focussed 15 cm downstream of the exit plane, was set to receive the spectrum from 4500 to 6500Å. Fig. 2 shows this spectrum with many broadened cesium emission lines. Widths of chosen lines were measured and compared with those calculated by Griem.[9] The reliability of these calculations recently has been confirmed experimentally by Stone and Agnew.[10]

In choosing an appropriate line for measurement it was found generally most reliable to work with isolated lines (preferably of the sharp, diffuse, or fundamental series) which were still observable above the background and which also corresponded to an upper-level quantum number as high as possible. This was done to insure optical thinness of the line; accurate line-width measurements with optically thick lines are at best very difficult to perform. Calculations and observations have shown that many of the lines corresponding to lower parent level quantum numbers are optically thick, if not reversed.

The spectrograph was aligned so that the light coming from a cross section of the plume, and in the plane of the cross section, passed through the slit. Thus an intensity profile of the plume was focussed along the length of the slit. Knowledge of the cross-sectional plume profile is necessary in order to use Abel inversion techniques to determine the intensity at the center. Actually an Abel inversion is performed at each of a series of wavelength positions covering the line under study so that an intensity profile as a function of wavelength characteristic of the plasma conditions in the center of the plume is obtained. It is the width at half maximum of this profile which is compared with the theoretical calculations used to determine the electron density.

The 5504Å ($11D_{5/2} \rightarrow 6P_{3/2}$), and 5257Å ($12D_{3/2} \rightarrow 6P_{1/2}$) lines were chosen for study and an Abel inversion performed on both. Assuming a plume temperature of 3000°K, the half-width measurements on these two lines yield values of $1.0 \times 10^{15}$ and $0.6 \times 10^{15}$ electrons/cc respectively. These values average to $0.8 \times 10^{15}$ electrons/cc which is considered to be accurate to about 50%. The method itself under the best experimental conditions is expected to be accurate to about 15%. The increase in the estimated error is given because of the transient conditions and the high particulate background in the plume.

The vertical distance along the profile photograph was calibrated against vertical distance at the position of the plume profile, so that it was determined that the plume diameter at 15 cm downstream of the orifice was about 5.3 cm, where the limits are determined by the points at which about 10% of the broadening at the center occurs.

The fact that a measurement of about $10^{15}$ electrons/cc was obtained at 15 cm downstream whereas at 91 cm the density had dropped to $10^{13}$ electrons/cc is probably indicative of considerable recombination or attachment along the length of the plume at this pressure. Less electron loss would be expected at 1 mm Hg.

The measured value of $0.8 \times 10^{15}$ electrons/cc at 40 mm Hg may be compared with the following estimate. If the temperature in the plume is taken as roughly 3000°K, then at 40 mm Hg there are $1.3 \times 10^{17}$ molecules and particles of all types per cc, according to the perfect gas law. If 2% of these are electrons (cf., thermodynamic analysis of isentropic expansion), then $2.6 \times 10^{15}$ electrons/cc would be expected. (If the plume temperature is 2000°K instead of 3000°K, then the predicted electron concentration comes out $3.9 \times 10^{15}$ electrons/cc.) The measured and calculated values are remarkably consistent when one considers that some electron loss by dilution and attachment must occur between the nozzle and a point 15 cm downstream, at this pressure.

## References

[1] Marma, F.F., Aschenbrand, L.M., and Pressman, J., "Physics of artificial electron clouds," ARS J. 30, 523-530 (1960).

[2] Rosenberg, N.W. and Paulson, J.F., Chemical Reactions in the Lower and Upper Atmosphere (Interscience Publishers, Inc., New York, 1961), pp. 331-351.

[3] Fagg, L.W. and Friedman, R., "Solid propellant source of cesium plasma," ARS J. 31, 157 (1961).

[4] Nichol, J., Siminski, V., and Wolfhard, H.G., Eighth Symposium (International) on Combustion (Williams and Wilkins, Baltimore, Md., 1962), pp. 235-241.

[5] Golomb, D. and Berger, A. W., Air Force Cambridge Research Laboratories Rept. 229, Geophysics Corporation of America (January 1961).

[6] Wada, J.Y. and Knechtli, R.C., "Generation and application of highly ionized quiescent cesium plasma in steady state," Proc. IRE, 1926-1931 (December 1961).

[7] Mohler, F.L., "Electron concentration and spectral intensity distribution in a cesium discharge," J. Res. Nat. Bur. Std. 19, 447-456 (1937).

[8] Dandurand, P. and Holt, R.B., "Electron density and light intensity decay in cesium afterglows," Phys. Rev. 82, 278-279 (1951).

[9] Griem, H.R., "Stark-broadening of isolated spectral lines from heavy elements in a plasma," Phys. Rev. 128, 515-523 (1962).

[10] Stone, P.M. and Agnew L., "Plasma-broadened cesium lines," Phys. Rev. 127, 1157-1162 (1962).

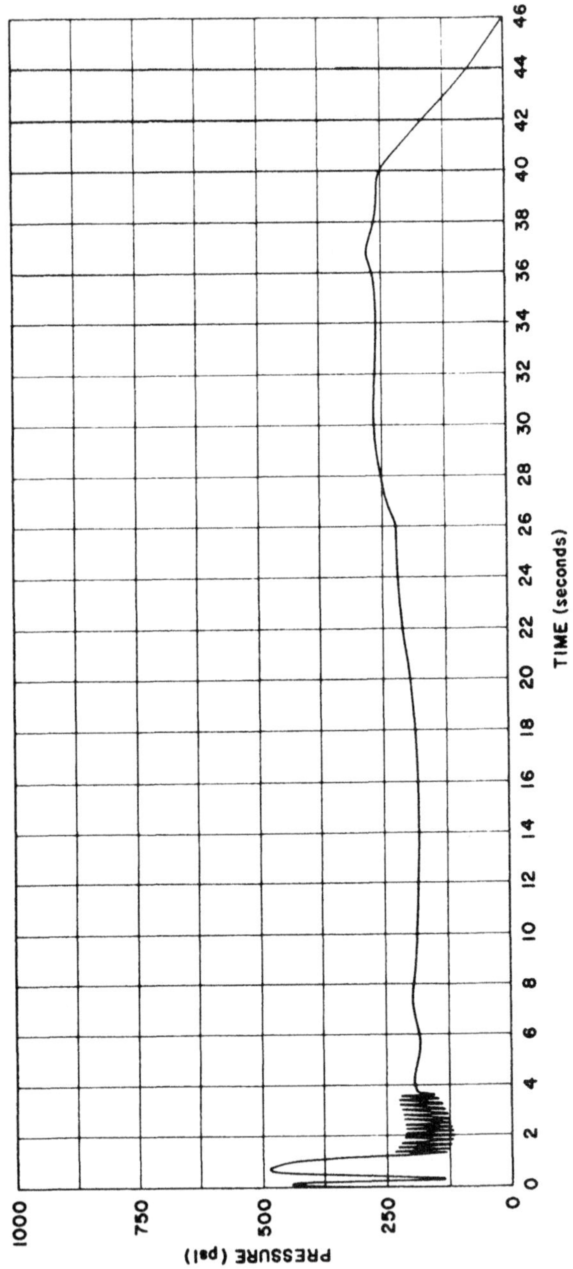


Fig. 1 Typical chamber pressure vs time curve for firing of electron generator

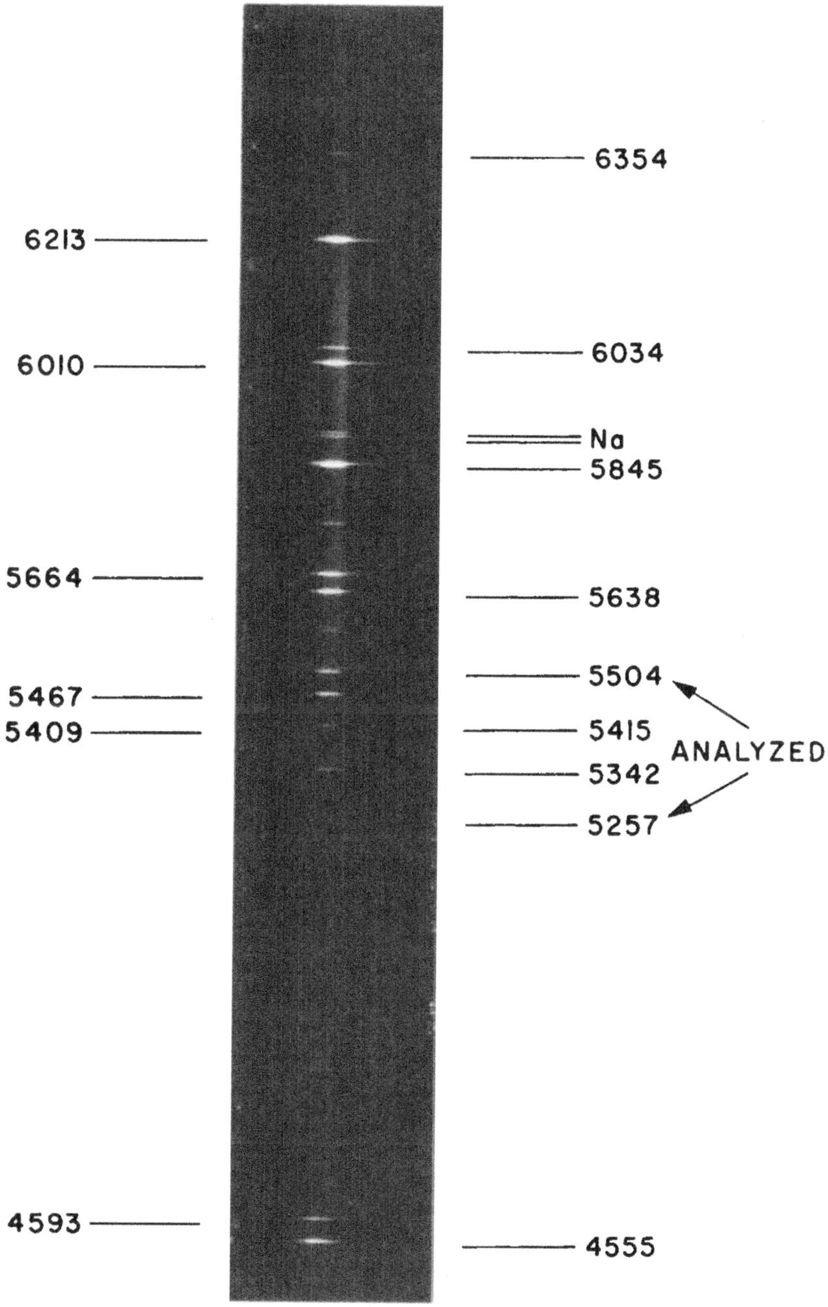


Fig. 2 Spectrum of plume at 40 mm Hg, 15 cm downstream of nozzle; cesium emission lines are labeled

# GENERATION AND PROPERTIES OF HIGH ALTITUDE CHEMICAL PLASMA CLOUDS

N. W. Rosenberg* and D. Golomb†

Air Force Cambridge Research Laboratories, Bedford, Mass.

## Abstract

Physical or chemical techniques can be used to form localized regions of high electron density in the upper atmosphere. One method is described in detail wherein a mixture of cesium nitrate, aluminum powder, and a high explosive is detonated in the 90 to 120 km altitude region of the upper atmosphere. The partially ionized reaction products expand to form a cloud from which radio waves can be reflected or scattered. In night releases the free electrons are generated in the explosion by thermal ionization; in day releases additional photoionization of cesium occurs by absorption of solar ultraviolet. The reflected radio signal duration and intensity are dependent on altitude, time of day, and wind shears. In addition to thermal ionization and photoionization, major factors in determining the usefulness of chemical plasma clouds for radio wave propagation are their size, structure, and orientation with respect to the transmission path.

## Introduction

In this paper a special application of "ions in flames" is described, i.e., the study of partially-ionized reaction products released into the upper atmosphere from vertical probe rockets. One objective of these experiments has been to create localized high electron densities (so-called electron clouds) for study of over-horizon propagation of radio-frequency signals.

Presented at the ARS Ions in Flames and Rocket Exhausts Conference, Palm Springs, Calif., October 10-12, 1962; revision submitted for publication to AIAA April 10, 1963.

*Chief, Chemical Physics Branch, Upper Atmosphere Physics Laboratory.

† Research Chemist, Chemical Physics Branch, Upper Atmosphere Physics Laboratory.

Over-horizon radio-frequency propagation can be achieved in several ways: 1) Reflection and scattering can occur from the natural E and F layers of the ionosphere, which are the normal modes of shortwave broadcasting. The transmitted electromagnetic waves are "bounced" back and forth between the ionized layers of the atmosphere and the ground until they reach the receiver. The reflection capability of these layers is dependent, among other things, on the electron density of the reflecting region which averages about $10^5$ electrons/$cm^3$ at the 100-km altitude region called the E-layer and $10^6$ electrons/$cm^3$ at the 250-km altitude region called the F-layer.

The maximum radio frequency reflected at oblique incidence from a horizontal plasma sheet is given by

$$f = 0.009\, n^{1/2} \sec \phi \qquad (1)$$

where the frequency f is given in megacycles, n is the number of electrons per $cm^3$ in the plasma sheet, and $\phi$ is the angle of incidence measured from the vertical. The maximum attainable useful incidence angles from the vertical (in the F-layer) are about 70 to 75°, permitting reflection of frequencies in the HF range, i.e., 3 to 30 Mc.

2) Reflection of higher frequencies than those reflected by the normal E and F layers is possible from naturally occurring passive reflectors of higher electron content than the ionosphere, such as meteor trails, formed by ionization of ambient air heated by meteor passage in the 90 to 110 km region. These trails attain a radar cross-section of $10^3$ to $10^4$ $m^2$ for frequencies between 30 and 100 Mc. The major disadvantage of meteor trails as reflectors is, of course, their unpredictable occurrence. Reflection is obtained for a few seconds per trail, at a rate of one trail per minute, permitting over-the-horizon propagation for several minutes per hour.

3) Reflection can be achieved by solid-surfaced passive reflectors, such as satellites, orbiting balloons, and needles. Satellites and balloons are reflectors for wavelengths smaller than their radii and their small reflecting area requires powerful transmitters and receivers. Needles are even more selective in the transmittable frequencies. Active transponders such as the Telstar and Relay satellites receive at a preselected frequency which is amplified and retransmitted.

4) Artificial gaseous plasmas are generated most effectively in a relatively narrow altitude region extending between 90 and 120 km. Below this region the high collision frequencies increase the rate of electron loss mechanisms; above this

altitude the increased diffusion rates cause a rapid dissipation of released gaseous material. Within the proper altitude region, however, the following methods are being investigated presently for plasma generation: a) striking of an electric arc from rocket-borne battery cells;[1] b) cyclotron resonance heating of a region in the ionosphere by ground-based or rocket-borne high powered transmitters of the proper frequency;[2] c) utilization of solar and chemical energy to photoionize and thermally ionize chemicals dispersed at the proper altitude. It is this last type of plasma generation which will be discussed in more detail with regard to ionization efficiency and propagation characteristics.

## Chemistry of Electron Clouds

The first successful electron cloud was formed in 1956 by release of 8 kg (275 moles) of nitric oxide at 95-km altitude in sunlight.[3] Nitric oxide, having an ionization potential of 9.25 ev, is photoionized readily by the solar Lyman -$\alpha$ , and a region of increased electron density was detected by radar echoes at the proper altitude for about 10 min. following release. In subsequent experiments[4] nitric oxide was replaced by potassium and cesium, the ionization potentials of which are 4.32 and 3.86 ev, respectively. The nitrate salt of the desired alkali metal was mixed with aluminum powder in a ratio of 2.66 moles Al to 1 mole $MNO_3$, the total weight being about 25 kg. The mixture was packed into a canister equipped with a timer-initiated igniter, set to ignite the mixture when the rocket attained an altitude of about 100 km. Upon ignition the mixture reacted, bursting the canister and expanding into a fireball composed mainly of alkali vapor, nitrogen, and condensed $Al_2O_3$. At the high explosion temperature some of the alkali vapor was ionized, thus augmenting the solar photoionization with initial thermal ionization. In a series of day and night-time releases distinct radar echoes were obtained both on backscatter and on forward scatter circuits.[5] The nighttime effect was of shorter duration than the daytime effect, due to recombination and attachment processes, without photoionization. Ground tests indicated that the alkali nitrate-aluminum reaction is deflagrative, i.e., on heating, the mixture reacts without detonation. In a closed vessel the reaction proceeds with pressure build-up until the vessel ruptures which scatters and extinguishes the unreacted portion of the mixture. Since the static burst pressure of the canister used was about 250 - 500 atm, the reaction proceeded until this pressure was attained. The void volume in the canister was such that only 3% of the above composition reacted before vessel rupture, which resulted in a low chemical yield. The estimated flame temperature of the reaction products at 500 atm is 4900°K, at which condition potassium is about 1% ionized.[6] However, substantial recombination, by as much as three orders of magnitude, can occur during the expansion process.

More recent experiments by Rosenberg and co-workers,[7,8] improved the chemical yield to near 100% in detonation of 18-kg mixtures formed by adding a high explosive, such as RDX or TNT, to the cesium nitrate and aluminum. In such mixtures the equilibrium electron yield is relatively low due to the high pressure of the reaction zone and the presence of electron attaching species such as OH. Furthermore, it is agreed generally that aluminum does not react fully in the detonation zone, that is, considerable reaction occurs during expansion of the explosion products. Therefore, it is extremely difficult to estimate the number of electrons generated and decayed in the process of detonation and expansion of the above mixture. Radar returns obtained at a number of frequencies from high altitude releases immediately after the initial expansion is at present the best way to assess the thermal ionization yield of such reactions. Such assessments place the degree of ionization of the gaseous products after expansion in the order of $10^{-4}$. Since cesium is 6% of the gaseous products, the degree of ionization of cesium is about $1.6 \times 10^{-3}$.

The major reaction products from the RDX-$CsNO_3$-Al explosion are CO, $H_2$, $N_2$, and Cs vapor, and condensed $Al_2O_3$. Altogether about 430 moles, or $2.5 \times 10^{26}$ gaseous molecules, are liberated from the 18-kg payload. These gases expand to form a cloud of 100 - 1000 m diam, depending on altitude. Optical observations show that the clouds are initially spherical, and calculations may be based on cloud models where it is assumed that the clouds have a "square-well" density distribution, or, alternately, a Gaussian density distribution, with the tails of the Gaussian being overlapped by the ambient which was pushed aside by the expanding gases (Fig. 1). After completion of the (rapid) expansion, the total number density is assumed as a first approximation to be constant throughout the region and equal to the ambient number density. The radius of a spherical "square-well" density cloud is given by

$$r = (3N/4\pi n_a)^{1/3} \qquad (2a)$$

where N is the total number of gaseous molecules released and $n_a$ is the ambient number density at the release altitude. For a Gaussian cloud, the equivalent parameter is the Gaussian half width, i.e., the radial distance from the center where the density falls to 1/e of the center point density.

The Gaussian half width, h, is given by

$$h = (N/n_a)^{1/3}/\pi^{1/2} \quad (2b)$$

Thus, r = 1.1h, i.e., these parameters differ only by 10%, and for calculation of diffusion times it is quite immaterial which model one assigns to the density distribution. In Table 1 the Gaussian half widths are calculated as a function of altitude based upon ambient density values of the CIRA standard atmosphere.[9]

Table 1

Gaussian half-width, h, for a cloud containing $2.5 \times 10^{26}$ gaseous molecules as a function of altitude

| Altitude (km) | $n_a$ (particles/cm$^3$) | h (cm) |
|---|---|---|
| 80 | $4.03 \times 10^{14}$ | $5.0 \times 10^3$ |
| 100 | $9.98 \times 10^{12}$ | $1.65 \times 10^4$ |
| 120 | $5.15 \times 10^{11}$ | $4.5 \times 10^4$ |
| 140 | $6.55 \times 10^{10}$ | $9.0 \times 10^4$ |

Based on these parameters, the optimum altitude region shall be estimated for the deposition of reasonably long-lived r.f. reflecting electron densities. The upper altitude limit is determined by rapid diffusional loss. An electron cloud becomes "invisible" to radio frequency waves of 10 Mc when the number of electrons per cubic centimeter drops below $10^6$ at its center. The diffusion time of a spherical Gaussian cloud is given by [10]

$$t_u = h^2(u^{2/3} - 1)/4D_a \quad (3)$$

where $t_u$ is the time in which the center point density reaches a value 1/u of its initial value and $D_a$ is the ambipolar diffusion coefficient at a given altitude. The time required to fall from a center point density of electrons $n_e(0) = 10^{-4} n_a$ to $n_e(t_u) = 10^6 \text{cm}^{-3}$ where $10^{-4}$ is the assumed degree of ionization, is set forth in Table 2 for various altitudes. From Table 2 it is seen that the life time of the cloud at 140 km altitude is 40 sec and this height is probably the upper limit for reasonable life. In practice it was found that this altitude is closer to 120 km.

Table 2

Initial center point electron densities and the times necessary to reduce these to $10^6/cm^3$ by ambipolar diffusion

| Altitude (km) | $n_e(0)=10^{-4}n_a$ (electrons/$cm^3$) | $D_a$ ($cm^2$/sec) | u | $t_u$ (sec) |
|---|---|---|---|---|
| 80 | $4 \times 10^{10}$ | $4.8 \times 10^3$ | $4 \times 10^4$ | $1.5 \times 10^6$ |
| 100 | $1 \times 10^9$ | $2.2 \times 10^5$ | $1 \times 10^3$ | 31000 |
| 120 | $5 \times 10^7$ | $6.9 \times 10^6$ | 50 | 920 |
| 140 | $6.5 \times 10^6$ | $1.3 \times 10^8$ | 6.5 | 40 |

Optical observations show that clouds deposited below 100 to 110 km usually grow by eddy diffusion (Fig. 2) with turbulent diffusion rates higher by as much as two orders of magnitude than molecular diffusion rates. Therefore, the diffusion times for 80 and 100 km in Table 2 are perhaps overestimated by a factor of 10 - 100.

The lowest useful altitude is determined by losses due to recombination of electrons with positive ions and attachment of electrons to neutrals. Electron densities above $10^8 cm^{-3}$ cannot be maintained for considerable time periods, because a Bates, Kingston, and McWhirter type collisional-radiative recombination[11] with an effective rate coefficient of $10^{-9} cm^3 sec^{-1}$ will take about 10 sec to decrease the electron density from $10^{10}$ to $10^8 cm^{-3}$. When the electron density is in the order of $10^8 cm^{-3}$, the predominant loss mechanism at the lower altitudes is probably a three body attachment to the explosion products and to molecular oxygen diffusing into the cloud by molecular and turbulent diffusion. The rate coefficient for this process is about $5 \times 10^{-30} cm^{-6} sec^{-1}$.[12] With a density of the attaching molecule and the third body of $\sim 4 \times 10^{14} cm^{-3}$ at 80 km, the time to decrease the electron density from $10^8$ to $10^6 cm^{-3}$ is about 6 sec; at 90 km ($n_a = 6 \times 10^{13} cm^{-3}$), the time is 250 sec. This places the lower altitude limit at about 90 km, in agreement with experience. Thus, the 90- to 120-km altitude region is where long-lived electron clouds can be generated. In defining the optimum altitude region the effect of photoionization was not considered. However, photoionization can result in the prolongation of electron cloud life time within the optimum altitude region but will not extend this region, since the arguments about diffusion and recombination losses are not altered.

It appears that improvement of the initial degree of ionization of the chemical reaction is not likely to increase the life time of electron clouds, because of rapid recombination losses during the early times when high electron densities are present. However, the size and structure of the cloud, which are dependent on the quantity of released material, the altitude of release, and atmospheric conditions (wind, turbulence, geomagnetic field, etc.) can affect the life time markedly, as discussed in the following section.

## Propagation Characteristics of Electron Clouds

The mode of r.f. signal return from ionized regions is a complex phenomenon. The return at a given frequency can be evaluated empirically in terms of its intensity, polarization, phase delay, frequency spread and shift, and angular dependence. The scattering mechanism can only be inferred from such measurements and by construction of simplified theoretical models. For the point electron clouds three major modes of signal return seem to be operative.[13,14]

1) Perfect or partial reflection from a wholly overdense volume. In this case the received signal cross-section is proportional to the physical area of the reflector, the electron density is given by Eq. 1, and no short period fading of the signal is obtained.

2) Reflective scattering from a large number of overdense irregularities. In this case the received signal cross-section is proportional to the number of irregularities and their scale and considerable short period fading is observed due to interference.

3) Booker-Gordon type scattering from underdense irregularities.[15] For this mode the scattered signal cross-section is proportional to $(\overline{\Delta n})^2$, the mean square departure of electron density from the mean, and the scale size of the (turbulent) irregularities. This mode will also result in signal fading.

Analysis of tens of electron cloud releases with regard to signal strength, duration, fading, and angular dependence indicates that for electron clouds in the 100-km altitude region all three modes are operative simultaneously, with the contribution by each mode depending on the amount of turbulence, self-induced as well as atmospheric. After tens of minutes in sunlit releases and several minutes in night releases, the overdense reflectors decay and only the third mode remains significant.

## Life Time of Electron Clouds

Figure 3 represents the life time of several electron clouds as a function of altitude and beam incidence angle for signals in the lower end of the VHF band.

Sunlit clouds observed with an incidence angle $\phi = 81.5^{\circ}$ (cf. Eq. 1) are denoted by circles. It is seen that clouds of considerable life times are obtained within the 90- to 120-km region, where signals are propagated for 100 - 3000 sec. The optimum altitude appears to be in the 100- to 110-km region. The 103-km cloud may have been the best propagation cloud because it was deposited in the optimum altitude region, and also because it was dragged out by wind shears to large horizontal dimensions, as observed by optical tracking. Therefore it seems that the size and orientation toward the transmission path are major factors in determining the life time of electron clouds as propagation links. Since wind velocities and directions may vary quite rapidly as a function of altitude, the optimum altitude for electron clouds can be expected to shift within the 100- to 110-km region, and it should not be inferred from Fig. 3 that 103 km is always the best altitude. The size of the cloud is dependent also on the amount of released gases. From recent, not yet fully analyzed experiments, it was confirmed that 4 kg charges do result in shorter lived electron clouds with smaller radio cross-section than the 18 kg charges represented in Fig. 3.

Night clouds (squares in Fig. 3) last 3 to 10 times less for a given frequency and incidence angle than sunlit clouds, depending on altitude of release. Thus, useful propagation links of several minutes duration can be maintained with night releases, where photoionization does not play a role.

The life times denoted by triangles in Fig. 3 were measured at a more obtuse incidence angle than those denoted by circles and are shorter by an order of magnitude. For the overdense reflection mode this is a consequence of Eq. (1), whereby a wave of a given frequency is reflected from lower electron densities at greater incidence angles to the normal. For instance, a 30 Mc signal is reflected from a plasma sheet containing $2 \times 10^5$ electrons per $cm^3$ at $\phi = 81.5^\circ$, whereas $1.5 \times 10^6$ electrons per $cm^3$ are necessary for $\phi = 66.5^\circ$. By the same token, electron cloud life times for given incidence angle are inversely proportional to frequency.

## Cloud Cross Sections

For the reflective cloud model, the radio cross-section at a given frequency is dependent on the effective reflection area of the overdense regions and on the scattering angle. For underdense scatter the cross-section is dependent on the mean square departure of electron density from the mean and on the scale size of turbulent eddies in the cloud. All these variables are strongly time dependent, since the cloud is in a continuous dynamic condition (expansion, diffusion, turbulence) and the electron inventory is changing by dilution, recombination, attachment and photoionization. Figure 4 gives the time function of cross-sections at about 30 Mc and $\phi = 81.5^\circ$ averaged over one minute intervals. The radar cross-sections display a similar altitude dependence as do overall signal durations: the 103 km cloud attained in general the largest cross-sections, at lower and higher altitudes the averaged cross-sections decrease. It is seen that radar cross-sections in the order of $10^6 - 10^8 m^2$ are obtained readily for tens of minutes, but with significant fading and enhancement, due to interference and reinforcement of multipath signals.

Since the signals display a considerable short term fading, no clear-cut cross-section can be assigned to the clouds which would allow the assessment of their actual physical area (or volume) at a given time. Thus, it is difficult to use radar cross-sections from turbulent electron clouds to evaluate absolute electron densities as a function of time and space and to obtain the associated diffusion, recombination and attachment coefficients. This, of course, does not affect the possible use of artificial electron clouds as propagation links with properly designed systems. The tracking by radar of the gross movement of the clouds without knowing their intricate structure also gives important information on atmospheric winds and wind shears. Any acceptable cloud model which can be used to correlate the observed behavior of artificial electron clouds will also be of significant value in interpretation of radio frequency returns by natural events such as sporadic-E and meteor trails.

## References


1 Irre, E. W., personal communication, Zimney Corporation, Monrovia, California (1962).

2 Deschamp, G., "A study of processes in the ionosphere by means of gyro-interaction experiments carried out with rockets," TR 62-1071, Air Force Cambridge Research Laboratories, Bedford, Mass. (1962).

3 Marmo, F. F., Pressman, J., Aschenbrand, L. M., Jursa,A.S. and Zelikoff, M., "Formation of an artificial ion cloud; photo ionization of NO by solar Lyman alpha at 95 km," J. Chem. Phys. 25, 187 (1956).

4 Marmo, F. F., Aschenbrand, L. M. and Pressman, J., "Physics of artificial electron clouds," ARS J. 30, 523-530 (1960).

5 Marmo, F. F., Aschenbrand, L. M. and Pressman, J., "Artificial electron clouds," Planetary Space Sc. 1, 227-237, 291-305, 306-318 (1959), 17-25, 174-186, 228-237 (1960).

6 Berger, A. W., Golomb, D. and Sullivan, J. O., "Flame temperature and composition in the aluminum-potassium nitrate reaction," J. Phys. Chem. 64, 949 (1960).

7 Rosenberg, N. W., "Artificial perturbation of the ionosphere," Space Research II (North-Holland Publishing Company, Amsterdam, 1961), pp. 1143-1158.

8 Rosenberg, N. W., et al., "Project Firefly," TR 256, Air Force Cambridge Research Laboratories, Bedford, Mass. (1961).

9 Dubin, M. et al., "U. S. Standard Atmosphere," NASA, Air Force, Weather Bureau joint publication, 1962 .

10 Brown, H. K. and Pressman, J., "Spherical Gaussian diffusion in the upper atmosphere," TR 62-3-G, Geophysics Corp. of America, Bedford, Mass. (1962).

11 Bates, D. R. and Dalgarno, A., "Electronic Recombination," Atomic and Molecular Processes (Academic Press, New York and London, 1962).

12 Chanin, L. M., Phelps, A. V. and Biondi, M. A., "Attachment of low energy electrons to oxygen molecules," Phys. Rev. 128, 219 (1962).

[13] Herman, J. R. and Noel, T. M., "Analysis of forward scatter radio data from Project Firefly," Project Firefly Vol. IV, TR 62-826, Air Force Cambridge Research Laboratories, Bedford, Mass. (1962).

[14] Camnitz, H. G., "Theoretical models of RF scatter from electron clouds," Project Firefly Vol. IV, TR 62-826, Air Force Cambridge Research Laboratories, Bedford, Mass. (1962).

[15] Booker, H. G. and Gordon, W. E., "A theory of radio scattering in the troposphere," Proc. Inst. Radio Engrs. **38**, 401-412 (1950).


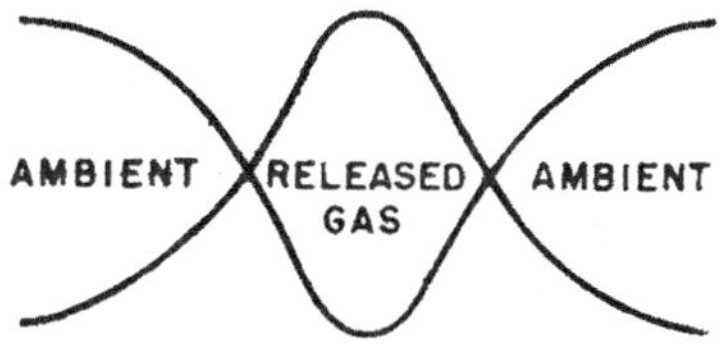


Fig. 1 Model of density profile of released gases and ambient

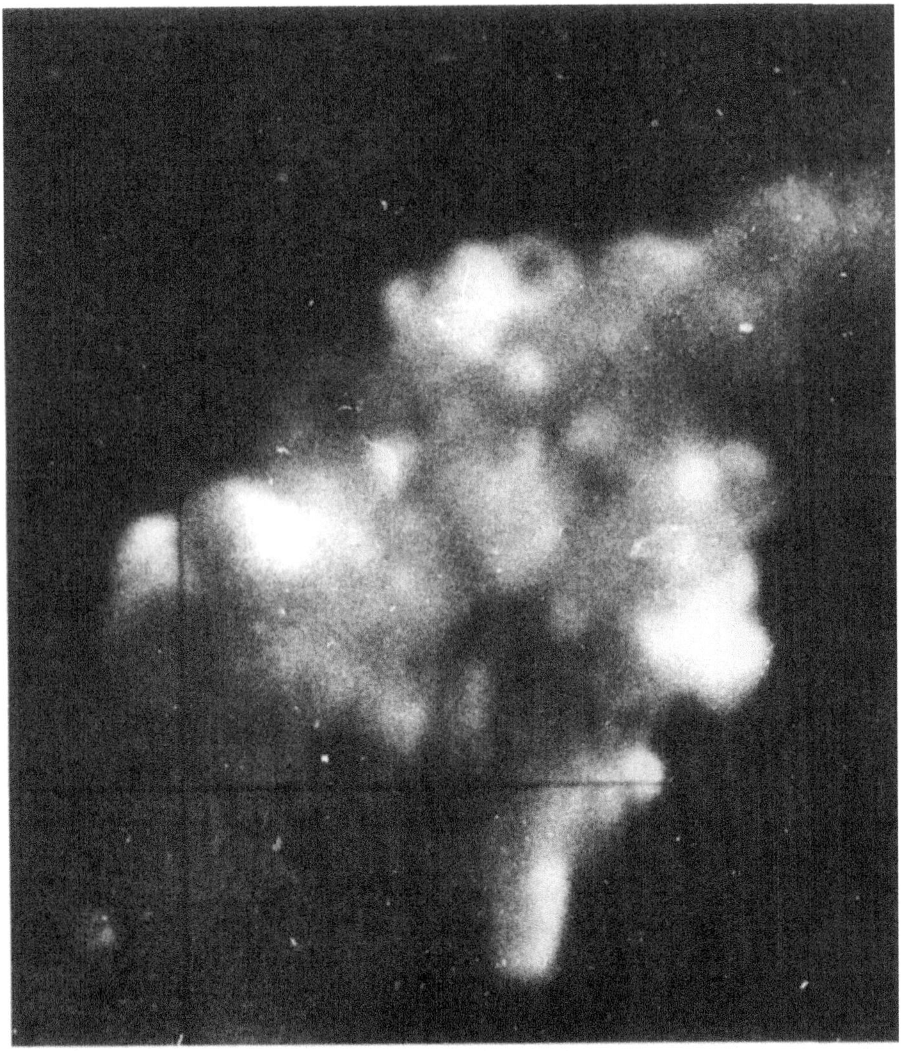

Fig. 2 Typical electron cloud in the 100-km altitude region, 250 sec after burst; sunlit against dark sky background (twilight conditions); note turbulent structure; total diameter 3 - 4 km

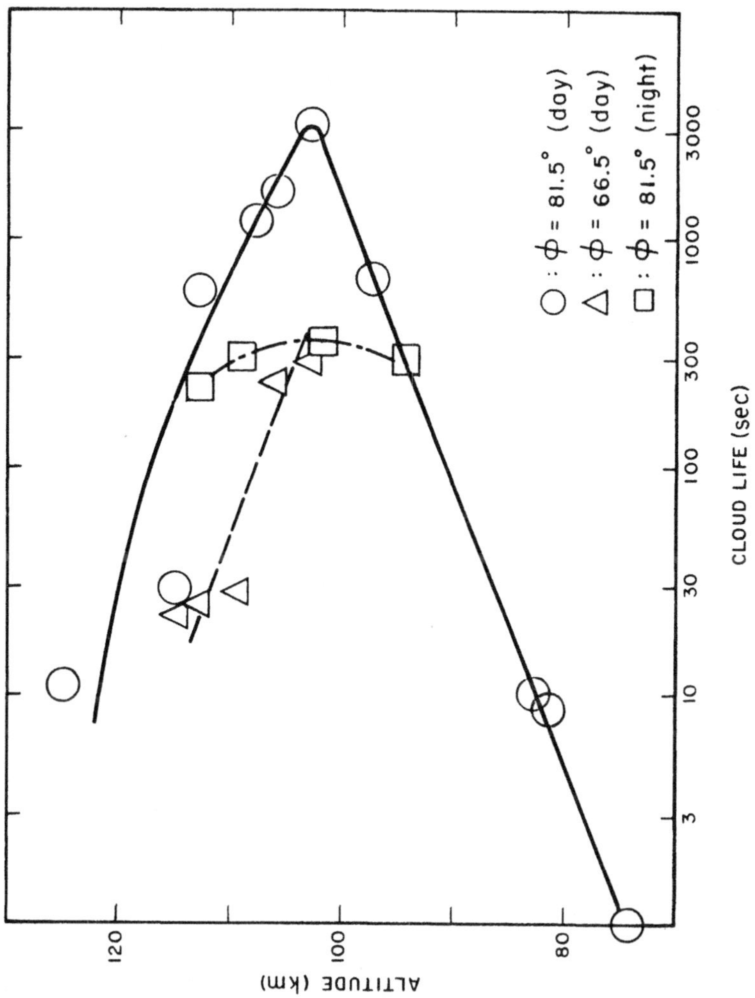


Fig. 3 Cloud life time vs altitude

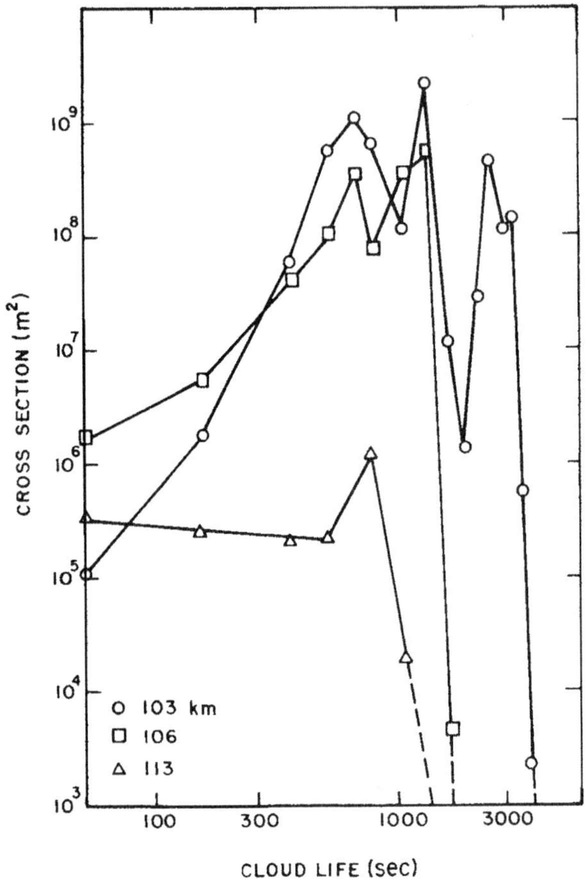


Fig. 4 Cloud radio cross-section vs time

# CONTRIBUTORS TO VOLUME 12

Printed in Dunstable, United Kingdom

91898131R00239